AF614802

Methods in Molecular Biology™

Series Editor
John M. Walker
School of Life Sciences
University of Hertfordshire
Hatfield, Hertfordshire, AL10 9AB, UK

For further volumes:
http://www.springer.com/series/7651

Neurotrophic Factors

Methods and Protocols

Edited by

Stephen D. Skaper

Department of Pharmacology and Anesthesiology, University of Padova, Padova, Italy

Editor
Stephen D. Skaper, Ph.D.
Department of Pharmacology and Anesthesiology
University of Padova
Padova, Italy
stephen.skaper@unipd.it

ISSN 1064-3745 e-ISSN 1940-6029
ISBN 978-1-61779-535-0 e-ISBN 978-1-61779-536-7
DOI 10.1007/978-1-61779-536-7
Springer New York Dordrecht Heidelberg London

Library of Congress Control Number: 2012930832

Printed on acid-free paper

Humana Press is part of Springer Science+Business Media (www.springer.com)

Preface

Neuroscience has been described by Nobel laureate and neurobiologist Eric Kandel as the "last frontier" of science. Nervous system development evolves from the well-orchestrated processes of neural induction, cell proliferation, differentiation, cell migration, survival, and synapse formation. Among these environmental cues, neurotrophic factors are secreted proteins that promote neurite outgrowth, neuronal cell differentiation and survival both in vivo and in vitro. Nerve growth factor (NGF) is the founding and best-characterized member of the neurotrophin family of neurotrophic polypeptides and was discovered more than half a century ago. Since their initial discovery, neurotrophic factors have raised expectations that their clinical application to neurodegenerative diseases might provide an effective therapy for what are now untreatable conditions.

Exploring nervous system function and dysfunction is oftentimes impractical in humans, and the availability of ex vivo and in vivo models which mimic, as closely as possible, how neural cells act and interact among themselves is of critical importance in neurobiological research. This volume of Methods in Molecular Biology aims to provide the reader, specialist and novice alike, with a selection of protocols and procedures which make use of cellular, tissue, and whole animal models which can be applied to the investigation of neurotrophic factors and other agents impacting on these systems. The book begins with a number of chapters dealing with the culture of neurons and glia from the central and peripheral nervous systems, neuron–glia coculture models, and cell-based assays for the evaluation of neuroprotective molecules, as well as assays which can be applied to the study of agents with neuroregenerative potential. Protocols describing viral- and nanoparticle-based delivery methods to neural cells are also presented, followed by chapters dealing with organotypic slice culture protocols. Lastly, several chapters are dedicated to in vivo lesion models of relevance to nervous system pathology, which can be applied to the investigation of neurotrophic factors and peptides.

I would like to gratefully acknowledge the contributors for their excellent cooperation and patience during the course of this project. While extensive, this volume is by no means intended to be all-inclusive, given the field's vastness and publication space limitations. Even so, I sincerely hope that this book will be useful to a broad spectrum of readers as they explore nervous system physiology and pathology.

Padova, Italy *Stephen D. Skaper*

Contents

Contributors

Carla Argentini • *Department of Pharmacology and Anesthesiology, University of Padova, Padova, Italy*

Lidia Bakota • *Department of Neurobiology, University of Osnabrück, Osnabrück, Germany*

Massimo Barbierato • *Department of Pharmacology and Anesthesiology, University of Padova, Padova, Italy*

Gianfranco Bazzu • *Department of Neuroscience, Medical School, University of Sassari, Sassari, Italy*

Thomas J.J. Blanck • *Department of Anesthesiology, Langone Medical Center, New York University, New York, NY, USA*

Alice Biosa • *Department of Neuroscience, Medical School, University of Sassari, Sassari, Italy*

Ricardo Borges • *Unidad de Farmacología, Facultad de Medicina, Universidad de La Laguna, La Laguna, Tenerife, Spain*

Roland Brandt • *Department of Neurobiology, University of Osnabrück, Osnabrück, Germany*

Secil Caban • *Department of Pharmaceutical Technology, Hacettepe University, Ankara, Turkey*

Giammario Calia • *Department of Neuroscience, Medical School, University of Sassari, Sassari, Italy*

Yilmaz Capan • *Department of Pharmaceutical Technology, Hacettepe University, Ankara, Turkey*

Wai Si Chan • *Department of Anesthesiology, Langone Medical Center, New York University, New York, NY, USA*

Patrick Couvreur • *Physico-Chimie, Pharmacotechnie, Biopharmacie, Facult e de Pharmacie, Universit e Paris Sud, UMR Centre National de la Recherche Scientifique 8612, Chatenay-Malabry, France*

Turgay Dalkara • *Department of Neurology, Hacettepe University, Ankara, Turkey*

Patrizia Debetto • *Department of Pharmacology and Anesthesiology, University of Padova, Padova, Italy*

Sonia Dedola • *Department of Neuroscience, Medical School, University of Sassari, Sassari, Italy*

Maria Speranza Desole • *Department of Neuroscience, Medical School, University of Sassari, Sassari, Italy*

Marc A. Dichter • *Department of Neurology, University of Pennsylvania School of Medicine, Philadelphia, PA, USA*

Natalia Domínguez • *Unidad de Farmacología, Facultad de Medicina, Universidad de La Laguna, La Laguna, Tenerife, Spain*

JULIE EGERTON • *GlaxoSmithKline R&D Ltd, Gunnels Wood Road, Stevenage, UK*
LAURA FACCI • *Department of Pharmacology and Anesthesiology, University of Padova, Padova, Italy*
TOMÁS L. FALZONE • *Instituto de Biología Celular y Neurociencias, CONICET, Facultad de Medicina, UBA, Buenos Aires, Argentina*
DONATELLA FARINA • *Department of Neuroscience, Medical School, University of Sassari, Sassari, Italy*
ELISABETTA GERACE • *Dipartimento di Farmacologia Preclinica e Clinica "Mario Aiazzi Mancini", Università di Firenze, Florence, Italy*
PIETRO GIUSTI • *Department of Pharmacology and Anesthesiology, University of Padova, Padova, Italy*
NATALIYA GOLOVYASHKINA • *Department of Neurobiology, University of Osnabrück, Osnabrück, Germany*
LLOYD A. GREENE • *Department of Pathology and Cell Biology, Columbia University, New York, NY, USA*
KIRSTEN HAASTERT-TALINI • *Hannover Medical School, Institute of Neuroanatomy, Hannover, Germany*
ELISA LANDUCCI • *Dipartimento di Farmacologia Preclinica e Clinica "Mario Aiazzi Mancini", Università di Firenze, Florence, Italy*
WILLIAM LEE • *Department of Neurobiology, Center for Glial Biology in Medicine, Atomic Force Microscopy and Nanotechnology Laboratories, Civitan International Research Center, Evelyn F. McKnight Brain Institute, University of Alabama, Birmingham, AL, USA*
J. DAVID MACHADO • *Unidad de Farmacología, Facultad de Medicina, Universidad de La Laguna, La Laguna, Tenerife, Spain*
CRISTINA MALAGELADA GRAU • *Department of Pharmacology, University of Barcelona, Barcelona, Spain*
SAAFAN Z. MALIK • *Department of Neurosurgery, University of Pennsylvania School of Medicine, Philadelphia, PA, USA*
MARGARET A. MARONSKI • *Department of Neurology, University of Pennsylvania School of Medicine, Philadelphia, PA, USA*
GIULIA MERCANTI • *Department of Pharmacology and Anesthesiology, University of Padova, Padova, Italy*
ROSSANA MIGHELI • *Department of Neuroscience, Medical School, University of Sassari, Sassari, Italy*
JOSE V. MONTOYA-GACHARNA • *Department of Anesthesiology, Langone Medical Center, New York University, New York, NY, USA*
FLAVIO MORONI • *Dipartimento di Farmacologia Preclinica e Clinica "Mario Aiazzi Mancini", Università di Firenze, Florence, Italy*
ANDREW J. MURRAY • *Department of Biochemistry and Molecular Biophysics, Columbia University, New York, NY, USA*
DAVINA E. OWEN • *Convergence Pharmaceuticals Ltd, Babraham Research Campus, Cambridge, UK*

VLADIMIR PARPURA • *Department of Neurobiology, Center for Glial Biology in Medicine, Atomic Force Microscopy and Nanotechnology Laboratories, Civitan International Research Center, Evelyn F. McKnight Brain Institute, University of Alabama, Birmingham, AL, USA*
ANDREW G. PEACE • *School of Medical Sciences, College of Life Sciences and Medicine, University of Aberdeen, Foresterhill, Aberdeen, UK*
DOMENICO E. PELLEGRINI-GIAMPIETRO • *Dipartimento di Farmacologia Preclinica e Clinica "Mario Aiazzi Mancini", Università di Firenze, Florence, Italy*
NICHOLAS M. RADIO • *Thermo Fisher Scientific, Pittsburgh, PA, USA*
ESPERANZA RECIO-PINTO • *Department of Anesthesiology, Langone Medical Center, New York University, New York, NY, USA*
ALEXANDRE CESAR SANTOS DE REZENDE • *Department of Anatomy, Cellular Biology, Physiology and Biophysics, State University of Campinas, Campinas, SP, Brazil*
GAIA ROCCHITTA • *Department of Neuroscience, Medical School, University of Sassari, Sassari, Italy*
MIRIAM RODRÍGUEZ • *Unidad de Farmacología, Facultad de Medicina, Universidad de La Laguna, La Laguna, Tenerife, Spain*
FABIO ROGERIO • *Department of Pathology, State University of Campinas, Campinas, SP, Brazil*
TANIA SCARTABELLI • *Dipartimento di Farmacologia Preclinica e Clinica "Mario Aiazzi Mancini", Università di Firenze, Florence, Italy*
PIER ANDREA SERRA • *Department of Neuroscience, Medical School, University of Sassari, Sassari, Italy*
HARI SHANKER SHARMA • *Department of Surgical Sciences, Anesthesiology and Intensive Care Medicine, Laboratory of Cerebrovascular and Pain Research, University Hospital, Uppsala University, Uppsala, Sweden*
ARUNA SHARMA • *Department of Surgical Sciences, Anesthesiology and Intensive Care Medicine, Laboratory of Cerebrovascular and Pain Research, University Hospital, Uppsala University, Uppsala, Sweden*
DERRYCK A. SHEWAN • *School of Medical Sciences, College of Life Sciences and Medicine, University of Aberdeen, Foresterhill, Aberdeen, UK*
ALEXANDRA SIDERIS • *Department of Anesthesiology, Langone Medical Center, New York University, New York, NY, USA*
STEPHEN D. SKAPER • *Department of Pharmacology and Anesthesiology, University of Padova, Padova, Italy*
YLENIA SPISSU • *Department of Neuroscience, Medical School, University of Sassari, Sassari, Italy*
GORAZD B. STOKIN • *Studenec 48 and Division of Neurology, University Psychiatric Hospital, University Medical Center, Ljubljana, Slovenia*
FREDERIK SÜNDERMANN • *Department of Neurobiology, University of Osnabrück, Osnabrück, Germany*
JHON JAIRO SUTACHAN • *Department of Anesthesiology, Langone Medical Center, New York University, New York, NY, USA*

CHRISTIAN TACKENBERG • *Department of Neurobiology, University of Osnabrück, Osnabrück, Germany*
STEVEN J. TUCKER • *School of Medical Sciences, College of Life Sciences and Medicine, University of Aberdeen, Foresterhill, Aberdeen, UK*
ANDRE SCHWAMBACH VIEIRA • *Department of Anatomy, Cellular Biology, Physiology and Biophysics, State University of Campinas, Campinas, SP, Brazil*
DEBORAH J. WATSON • *Department of Neurosurgery, University of Pennsylvania School of Medicine, Philadelphia, PA, USA*
MORENA ZUSSO • *Department of Pharmacology and Anesthesiology, University of Padova, Padova, Italy*

Chapter 1

The Neurotrophin Family of Neurotrophic Factors: An Overview

Stephen D. Skaper

Abstract

The neurotrophins are a family of closely related proteins that were first identified as survival factors for sympathetic and sensory neurons and have since been shown to control a number of aspects of survival, development, and function of neurons in both the central and peripheral nervous systems. Limiting quantities of neurotrophins during development control the numbers of surviving neurons to ensure a match between neurons and the requirement for a suitable density of target innervation. Biological effects of each of the four mammalian neurotrophins are mediated through activation of one or more of the three members of the tropomyosin-related kinase (Trk) family of receptor tyrosine kinases (TrkA, TrkB, and TrkC). In addition, all neurotrophins activate the p75 neurotrophin receptor, a member of the tumor necrosis factor receptor superfamily. Neurotrophin engagement of Trk receptors leads to activation of Ras, phosphatidylinositol 3-kinase, phospholipase C-γ1, and signaling pathways controlled through these proteins, including the mitogen-activated protein kinases. Neurotrophin availability is required into adulthood, where they control synaptic function and plasticity and sustain neuronal cell survival, morphology, and differentiation. This chapter will provide an overview of neurotrophin biology, their receptors, and signaling pathways.

Key words: Neurotrophic factors, Nerve growth factor, Brain-derived neurotrophic factor, Neurotrophin-3, Neurotrophin-4/5, Glial cell line–derived neurotrophic factor, Tropomyosin-related kinase, Receptor tyrosine kinases, Neurodegeneration, Neuroregeneration

1. Introduction

Neurotrophic factors are secreted proteins that promote neurite outgrowth, neuronal cell differentiation, and survival both in vivo and in vitro. Nerve growth factor (NGF) is the founding and best characterized member of the neurotrophin family (1) of neurotrophic polypeptides and was discovered more than half a century ago (2) during a search for survival factors that could explain the deleterious effects of target tissue ablation on the subsequent survival of motor and sensory neurons (1). NGF is also present

Stephen D. Skaper (ed.), *Neurotrophic Factors: Methods and Protocols*, Methods in Molecular Biology, vol. 846, DOI 10.1007/978-1-61779-536-7_1, © Springer Science+Business Media, LLC 2012

in the central nervous system (CNS) where it serves a trophic function in the development and maintenance of basal forebrain cholinergic neurons (3).

Neurotrophins are synthesized at a considerable distance from the cell body by peripheral tissues or neurons ("targets") that are contacted by axons of the neurotrophin-sensitive neurons. In the periphery, the tissue sources of neurotrophins are typically non-neuronal cells, whereas in the CNS, they are synthesized predominantly by neurons under physiological conditions (4). During development, a retrograde flow of a neurotrophin is established, transporting the protein from the target into the nerve terminal and up the axon to the cell body (5). Those neurons that establish this flow survive the period of neuronal cell death, while those that do not, degenerate. Once the retrograde flow of neurotrophin is established, it must continue for the lifetime of the neuron to maintain the functional differentiated state of the neuron (6). Recent studies on the expression and actions of the neurotrophin family indicate that, in addition to target-derived factor acquisition, autocrine and non-target-derived paracrine modes of factor presentation are likely to be important (7) (see Fig. 1).

Since their initial discovery, neurotrophic factors have raised expectations that their clinical application to neurodegenerative diseases might provide an effective therapy for what are now untreatable conditions. Indeed, there is an impressive volume of evidence for neuroprotective effects of neurotrophic factors in animal models of neurodegenerative diseases (see, e.g., ref. (8)). This chapter is intended to introduce the reader to the basic concepts of neurotrophic factor biology which operate to affect nervous system function.

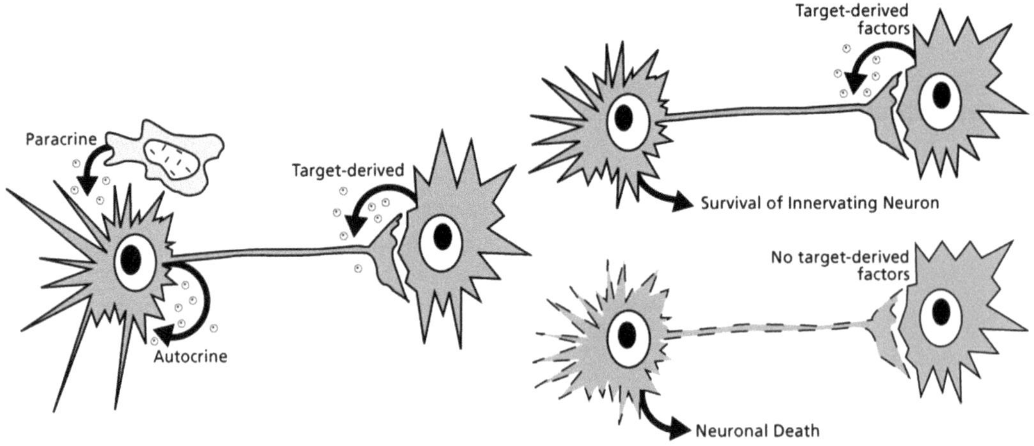

Fig. 1. Target-derived, autocrine, and paracrine modes of neurotrophin presentation. During the period of target innervation, neurotrophins support the survival of a restricted number of neurons expressing the appropriate tropomyosin-related kinase (Trk) receptors. Limiting amounts of secreted neurotrophins do not allow for the survival of all neurons, which can, however, be rescued by the administration of exogenous neurotrophins.

2. Neurotrophin Biology

The generality of the phenomenon of programmed cell death after target deprivation (axotomy) has suggested that most neurons respond to and are regulated by neurotrophic factors (9). This hypothesis was validated by the subsequent isolation from pig brain of a second neurotrophic factor, designated brain-derived neurotrophic factor (BDNF) (10). Molecular cloning of the BDNF gene (11) revealed its structural similarity to NGF, leading to the concept of the neurotrophin family. Two additional members, neurotrophin-3 (NT-3) and neurotrophin-4/5 (NT-4/5), were later identified (12). The term neurotrophin-4/5 resulted from uncertainties about whether the human neurotrophin-5 (13) was a species homolog of the NT-4 found in *Xenopus* (14). The nomenclature NT-4/5 was subsequently adopted by a number of investigators to denote the mammalian form of *Xenopus* NT-4 when it was shown that the differences between the two factors were only due to phylogenetic variation (15). Two novel neurotrophins from the platyfish and carp have been cloned and designated neurotrophin-6 (16) and neurotrophin-7 (17), respectively. These do not have orthologs in mammals or birds and appear to interact with the same receptors as the mammalian proteins. The neurotrophins exhibit actions on distinct, as well as partially overlapping, subsets of peripheral and central neurons (12, 18). Individual neurons may also be responsive to more than one neurotrophin at a given time or at subsequent times during development.

The mature neurotrophin proteins are noncovalently associated homodimers. The neurotrophins share a highly homologous structure and are members of a large superfamily of growth factors that contain a tertiary fold and cysteine "knot." These features are present in transforming growth factor-β, platelet-derived growth factor, human chorionic gonadotropin, vascular endothelial growth factor, and others. The cysteine knot consists of three disulfide bonds that form a true knot of the polypeptide chain. Two cysteines that make up the knot are missing from human neurotrophin-6. Neurotrophin residues are generally divided into two categories, conserved or variable, based on sequence alignments (12). Amino acid residues implicated in neurotrophin binding that are conserved are likely to represent a common interface to the tropomyosin-related kinase (Trk) receptors, while the unique ones may represent elements of specificity (19). The dimer interface is composed of β-strands that maintain the conformation; these hydrophobic core residues are highly conserved (20–22). In contrast to the β-strands, the β-loops are highly variable. Detailed discussions of neurotrophin structure and molecular evolution have been published (23–26).

In addition to the neurotrophins, a number of polypeptide factors have been shown to possess neurotrophic activities. These include ciliary neurotrophic factor (CNTF) (27), glial cell line–derived neurotrophic factor (GDNF) (23), insulin-like growth factor (28), and basic fibroblast growth factor (29). Transforming growth factor-β (30) and Sonic hedgehog (31) are other proteins capable of promoting survival of specific CNS neuron populations and protecting these cells from toxic insults.

3. Neurotrophin Receptors

The neurotrophins interact with two entirely distinct classes of receptors, Trks and $p75^{NTR}$, the first discovered member of the tumor necrosis factor receptor superfamily. The former was initially identified as a low-affinity receptor for NGF (32) but was subsequently shown to bind each of the neurotrophins with approximately equal nanomolar affinity (14, 33, 34). While p75 neurotrophin receptor ($p75^{NTR}$) does not contain a catalytic motif, it interacts with several proteins that relay signals important for regulating neuronal cell survival, differentiation, and synaptic plasticity. Each of the four cysteine-rich repeats of $p75^{NTR}$ participates in binding to NGF (35). $p75^{NTR}$ binds NGF along the interface between two NGF monomers, and binding results in a conformational change in NGF that alters the monomeric interface on the opposite side of the NGF dimer, eliminating the potential for binding of one NGF dimer to two $p75^{NTR}$ monomers.

In mammals, the Trk subfamily of receptor tyrosine kinases constitutes the second major class of neurotrophin receptors. The extracellular domain of each of the Trks consists of a cysteine-rich cluster followed by three leucine-rich repeats, another cysteine-rich cluster, and two Ig-like domains (see Fig. 2). Each receptor has a single transmembrane region that terminates in a cytoplasmic, tyrosine kinase-containing domain surrounded by several tyrosine residues that serve as phosphorylation-dependent docking sites for cytoplasmic adaptors and enzymes. The neurotrophins dimerize their cognate Trk receptor, resulting in activation via transphosphorylation of the cytoplasmic domain kinases. Specificity of neurotrophin action is believed to be achieved in part by the selective interaction between members of the Trk family of receptors and the different neurotrophins. Thus, NGF binds to TrkA (36, 37), TrkB binds BDNF and NT-4/5 with high affinity (38, 39), and TrkC binds NT-3 (40). NT-3 can also interact, albeit with less efficiency, with TrkA and TrkB (see Fig. 2) (15, 39).

Trk receptor function is modulated by $p75^{NTR}$ on several levels—by promoting ligand binding, by promoting accessibility to neurotrophins through promotion of axonal growth and target innervation,

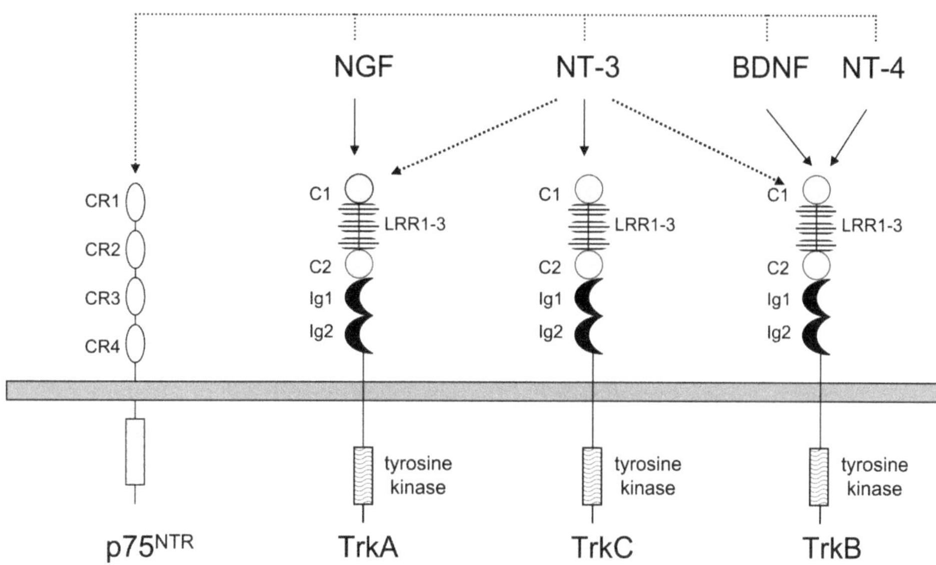

Fig. 2. Neurotrophins and their receptors. The neurotrophins display specific interactions with the three Trk receptors: NGF binds TrkA, BDNF and NT-4 bind TrkB, and NT-3 binds TrkC. In some cellular contexts, NT-3 can also activate TrkA and TrkB, albeit with less efficiency. All neurotrophins bind to and activate p75 neurotrophin receptor ($p75^{NTR}$). *CR1–CR4* cysteine-rich motifs, *C1/C2* cysteine-rich clusters, *LRR1–3* leucine-rich repeats, *Ig1/Ig2* immunoglobulin-like domains.

and by promoting endocytosis and retrograde transport to membrane compartments where internal engagement of neurotrophins with Trk receptors may promote signaling. For example, $p75^{NTR}$ inhibits activation of Trk receptors by nonpreferred neurotrophins both in vivo and in vitro (41, 42).While $p75^{NTR}$ potentiates activation of TrkA by suboptimal concentrations of NGF, it does not potentiate the activation of other Trk receptors similarly by their ligands (43). In addition, $p75^{NTR}$ collaborates with TrkA to form high-affinity (10^{-11} M) binding sites for NGF (44). Besides facilitating the binding of NGF to TrkA, $p75^{NTR}$ can promote retrograde transport of several neurotrophins (45). $p75^{NTR}$ may reduce ligand-induced Trk receptor ubiquitination, thereby delaying Trk internalization and degradation (46), or promote Trk receptor endocytosis through polyubiquitination and subsequent internalization to endosomal compartments, resulting in enhanced signaling (47). These findings each suggest a mechanism by which $p75^{NTR}$ may promote axon growth and target innervation in vivo and in vitro (48, 49). Sensory and sympathetic deficits are seen in mice lacking $p75^{NTR}$ (50–52).

The major site at which neurotrophins interact with the Trk receptors is in the membrane-proximal Ig-like domain. The three-dimensional structures of the domain in each of the Trk receptors have been solved (53). In addition, the structure of NGF bound to the TrkA Ig domain has also been determined (19). These results have made it possible to identify residues in the neurotrophins and the Trk receptors that account for the specificity observed in their interactions (54). Such information may prove useful in the design of small molecule mimetic ligands for neurotrophin receptors.

4. Neurotrophin Signaling

Many extracellular signals transduce their cellular responses by regulating tyrosine phosphorylation of their target proteins. Ligand-induced oligomerization of receptor protein tyrosine kinases and autophosphorylation have been established as a general mechanism for the activation of growth factor receptors, as well as many other families of cell surface receptors (55). The Trk receptors are typical receptor tyrosine kinases whose activation is stimulated by neurotrophin-mediated dimerization and transphosphorylation of activation loop kinases (56). Trk receptors are activated specifically by the mature and not the pro-forms of the neurotrophin gene products (57). Thus, the proteases that control processing of proneurotrophins control Trk receptor responsiveness. The cytoplasmic domains of the Trk receptors contain several additional tyrosines that are also substrates for phosphorylation by each receptor's tyrosine kinase. When phosphorylated, these residues form the cores of binding sites that serve as a scaffolding for the recruitment of a variety of adaptor proteins and enzymes that ultimately propagate the neurotrophin signal (58). Within the activated Trk molecule, the phosphotyrosines and their surrounding amino acid residues create binding sites for proteins containing phosphotyrosine-binding or Src homology 2 domains. The major pathways activated by the Trk receptors are Ras, Rac, phosphatidylinositol 3-kinase, phospholipase C-γ1, and their downstream effectors (56, 58). In addition, endocytosis and transfer of Trk receptors to different membrane compartments control the efficiency and duration of Trk-mediated signaling in part because many adaptor proteins are localized to specific membrane compartments (59). Activation of p75^{NTR} results in activation of the nuclear factor-κB and Jun kinase, as well as other signaling pathways (see Fig. 3).

Transactivation of receptor tyrosine kinases by G protein–coupled receptors (GPCRs) is now recognized as an important signaling mechanism allowing the cell to respond to a vast array of extracellular stimuli (60, 61). Activation of TrkA and TrkB receptors can also occur via a GPCR mechanism, in the absence of NGF or BDNF (62, 63). Two GPCR ligands, adenosine and pituitary adenylate cyclase–activating peptide, can activate Trk receptor activity to increase the survival of neuronal cells through stimulation of protein kinase B (Akt) activity. The effects of adenosine and pituitary adenylate cyclase–activating peptide can be blocked by K252a, an inhibitor of Trk tyrosine kinases. In contrast to

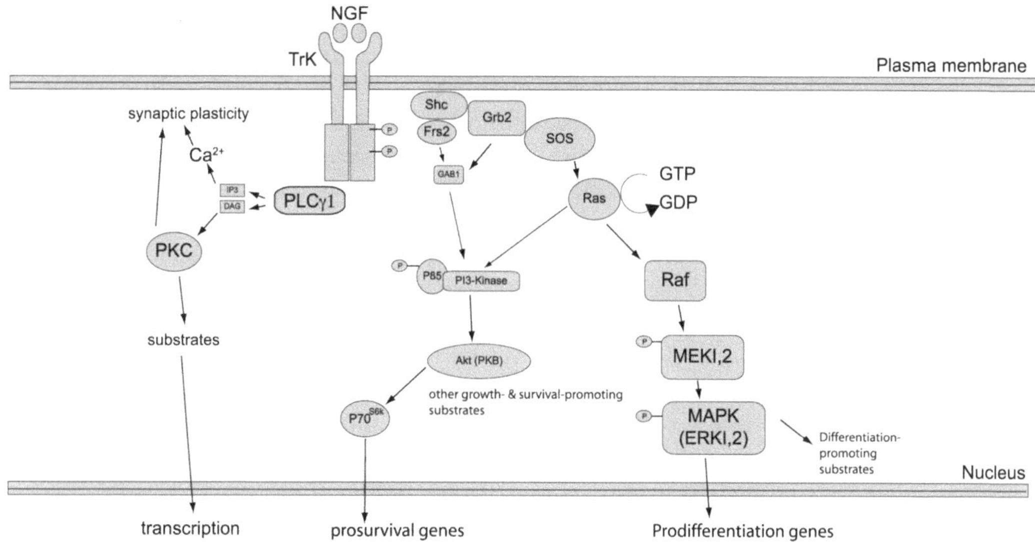

Fig. 3. Neurotrophin signaling. Depicted are interactions of NGF (exemplar neurotrophin) with Trk and major intracellular signaling pathways activated. Neurotrophin binding to Trk receptor leads to dimerization and autophosphorylation. The linker Shc binds to phospho-Y490 on Trk and to a Grb2-SOS complex. SOS is a nucleotide exchange factor that activates Ras by replacing GDP with GTP. Activated Ras interacts directly with the serine–threonine kinase Raf. The activated Raf leads to the sequential activation of MAPK kinase (MEK), the mitogen-activated protein kinase-ERK kinase (MAPK). MAPK translocates to the nucleus, where it phosphorylates transcription factors, promoting neuronal cell differentiation. Activation of phosphatidylinositol 3-kinase through Ras or Gab1 promotes survival and growth of neurons. Activation of phospholipase C-γ1 (PLC-γ1) results in activation of Ca^{2+}- and protein kinase C-regulated pathways that promote synaptic plasticity.

neurotrophin treatment, Trk receptor activation by adenosine analogs and pituitary adenylate cyclase–activating peptide is sensitive to transcriptional and translational inhibitors, with the majority of the transactivated Trk receptors found in intracellular membranes, suggesting that receptor signaling may occur and persist inside of neuronal cells (64). More recent studies indicate that the tyrosine kinase Fyn is activated by GPCR stimulation and is responsible for transactivation of Trk receptors on intracellular membranes (65).

Several excellent reviews describe recent progress in understanding the signaling pathways stimulated by the neurotrophins that affect the survival, differentiation, and function of cell within the nervous system (66–69).

5. Why Study Neurotrophins?

The power of neurotrophic factors to regulate neuronal cell survival in the developing nervous system and to promote also survival after injury or protect neurons in toxin-mediated disease models in animals has encouraged the idea that such proteins could be harnessed for the treatment of neurodegenerative disease. For example, the traditional perspective of applying neurotrophins in the context of Alzheimer's disease is based on the premise that neurotrophins are capable of upregulating cholinergic function and of rendering neurons less vulnerable to certain processes causing degeneration (70, 71). Another example is that of GDNF, which acts on dopaminergic neurons. Administered intracerebrally, GDNF has neuroprotective and neurorestorative effects in toxin-induced rodent and nonhuman primate models of Parkinson's disease (72, 73). Neurotrophic factor treatment of CNS diseases presents an especially complex problem since these polypeptides have poor pharmacokinetics and bioavailability and are not able to cross the blood–brain barrier. Neurotrophic proteins can be modified to increase blood–brain barrier transport (74), and intravenous administration of BDNF, conjugated to an anti-transferrin antibody, is able to traverse the blood–brain barrier in rats and provide neuroprotection in focal transient brain ischemia (75, 76). Viral vector or cell-based gene therapy approaches may also prove advantageous for the targeted delivery of lower doses of neurotrophic factors, and dopaminergic neuroprotection by viral-mediated delivery of GDNF has been demonstrated in animals (72, 77). In the case of spinal cord injury, the transplantation of genetically modified cells (e.g., viral expression of neurotrophins either in vivo or ex vivo in stem cells) to lesion sites has been investigated as a means to augment axonal regeneration (78, 79). A cell-based approach for NGF delivery has shown promise in a phase I trial in Alzheimer's disease (80). In the case of CNTF, a fusion protein consisting of mature human CNTF and an 11-amino acid protein transduction domain rescued the learning and memory impairments induced by amyloid β-peptide in mice (81), as did infusion of recombinant cells secreting CNTF encapsulated in alginate polymers (82).

Peripheral nervous system disorders should also be amenable to a neurotrophic protein therapeutic. NGF, which is selectively trophic for small fiber sensory and sympathetic neurons, was selected as a potential therapy for diabetic polyneuropathy because of the serious consequences associated with degeneration of those neuronal populations in this condition. In addition, a reduced NGF availability may contribute to the pathogenesis of diabetic neuropathy (83), and animal models of neuropathy respond to the exogenous administration of NGF (84). Two sets of phase II clinical trials suggested that recombinant human NGF administration was effective in ameliorating the symptoms associated with both diabetic polyneuropathy (85) and human immunodeficiency virus–related neuropathy (86). Clinical trials are continuing in this direction.

Clearly, methodologies which are directed to understanding neurotrophic protein–receptor interactions, biological activities, and neuroprotective/neurorestorative effects are of great importance in advancing this critical area of neuroscience. The protocols detailed in this volume have been written with this goal in mind.

References

1. Levi-Montalcini R (1987) The nerve growth factor 35 years later. Science 237, 1154–1162
2. Hamburger V, and Levi-Montalcini R (1949) Proliferation, differentiation and degeneration in the spinal ganglia of the chick embryo under normal and experimental conditions. J Exp Zool 111, 457–501
3. Dreyfus CF (1989) Effects of nerve growth factor on cholinergic brain neurons. Trends Pharmacol Sci 10, 145–149
4. Thoenen H (1995) Neurotrophins and neuronal plasticity. Science 270, 593–598
5. Ginty DD, and Segal RA (2002) Retrograde neurotrophin signaling: Trk-ing along the axon. Curr Opin Neurobiol 12, 268–274
6. Barde Y-A (1989) Trophic factors and neuronal survival. Neuron 2, 1525–1534
7. Bothwell M (1995) Functional interactions of neurotrophins and neurotrophin receptors. Annu Rev Neurosci 18, 223–253
8. Skaper SD (2008) The biology of neurotrophins, signalling pathways, and functional peptide mimetics of neurotrophins and their receptors. CNS & Neurol Disorders – Drug Targets 7, 46–62
9. Oppenheim RW (1991) Cell death during development of the nervous system. Annu Rev Neurosci 14, 453–501
10. Barde Y-A, Edgar D, and Thoenen H (1982) Purification of a new neurotrophic factor from mammalian brain. EMBO J 1, 549–553
11. Leibrock J, Lottspeich F, Hohn A, Hofer M, Hengerer B, Masiakowski P, et al (1994) Molecular cloning and expression of brain-derived neurotrophic factor. Nature 341, 149–152
12. Ibáñez CF (1995) Neurotrophic factors: from structure-function studies to designing effective therapeutics. Trends Biotech 13, 217–227
13. Berkemeier LR, Winslow JW, Kaplan DR, Nicolics K, Goeddel DV, and Rosenthal A (1991) Neurotrophin-5: a novel neurotrophic factor that activates trk and trkB. Neuron 7, 857–866
14. Hallböök F, Ibáñez CF, and Persson H (1991) Evolutionary studies of the nerve growth factor family reveal a novel member abundantly expressed in *Xenopus* ovary. Neuron 6, 845–858
15. Ip NY, Stitt TN, Tapley P, Klein R, Glass DJ, Fandl J, et al. (1993) Similarities and differences in the way neurotrophins interact with the Trk receptors in neuronal and nonneuronal cells. Neuron 10, 137–149
16. Götz R, Köster R, Winkler C, Raulf F, Lottspeich F, Schartl M, et al. (1994) Neurotrophin-6 is a new member of the nerve growth factor family. Nature 372, 266–269
17. Lai KO, Fu WY, Ip FCF, and Ip NY (1998) Cloning and expression of a novel neurotrophin, NT-7, from carp. Mol Cell Neurosci 11, 64–76
18. Eide FF, Lowenstein DH, and Reichardt LF (1993) Neurotrophins and their receptors-current concepts and implications for neurologic disease. Exp Neurol 121, 200–214
19. Wiesmann C, Ultsch MH, Bass SH, and de Vos AM (1999) Crystal structure of nerve growth factor in complex with the ligand-binding domain of the TrkA receptor. Nature 401, 184–188
20. McDonald NQ, Lapatto R, Murray-Rust J, Gunning J, Wlodawer A, and Blundell TL (1991) New protein fold revealed by a 2.3-A resolution crystal structure of nerve growth factor. Nature 354, 411–414
21. Robinson RC, Radziejewski C, Stuart DI, and Jones EY (1995) Structure of the brain-derived neurotrophic factor/neurotrophin 3 heterodimer. Biochemistry 34, 4139–4146
22. Robinson RC, Radziejewski C, Spraggon G, Greenwald J, Kostura MR, Burtnick LD, et al (1999) The structures of the neurotrophin 4 homodimer and the brain-derived neurotrophic factor/neurotrophin 4 heterodimer reveal a common Trk-binding site. Protein Sci 8, 2589–2597
23. Lindsay RM, and Yancopoulos GD (1996) GDNF in a bind with known orphan: accessory implicated in new twist. Neuron 17, 571–574
24. Barde Y-A (1994) Neurotrophic factors: an evolutionary perspective. J Neurobiol 25, 1329–1333
25. Kullander K, Carlson B, and Hallböök FJ (1997) Molecular phylogeny and evolution of

the neurotrophins from monotremes and marsupials. Mol Evol 45, 311–321
26. Butte MJ (2001) Neurotrophic factor structures reveal clues to evolution, binding, specificity, and receptor activation. Cell Mol Life Sci 58, 1003–1013
27. Ip NY, and Yancopoulos GD (1996) The neurotrophins and CNTF: two families of collaborative neurotrophic factors. Annu Rev Neurosci 19, 491–415
28. Doré S, Kar S, and Quirion R (1997) Rediscovering an old friend, IGF-I: potential use in the treatment of neurodegenerative diseases. Trends Neurosci 20, 326–331
29. Walicke P, Cowan WM, Ueno N, Baird A, and Guillemin R (1986) Fibroblast growth factor promotes survival of dissociated hippocampal neurons and enhances neurite extension. Proc Natl Acad Sci USA 83, 3012–3016
30. Farkas LM, Dünker N, Roussa E, Unsicker K, and Krieglstein K (2003) Transforming growth factor-βs are essential for the development of midbrain dopaminergic neurons in vitro and in vivo. J Neurosci 23, 5178–5186
31. Miao N, Wang M, Ott JA, D'Alessandro JS, Woolf TM, Bumcrot DA, et al (1997) Sonic hedgehog promotes the survival of specific CNS neuron populations and protects these cells from toxic insult in vitro. J Neurosci 17, 5891–5899
32. Radeke MJ, Misko TP, Hsu C, Herzenberg LA, and Shooter E (1987) Gene transfer and molecular cloning of the rat nerve growth factor receptor. Nature 325, 593–597
33. Rodriguez-Tébar A, Dechant G, and Barde Y-A (1990) Binding of brain-derived neurotrophic factor to the nerve growth factor receptor. Neuron 4, 487–492
34. Rodríguez-Tébar A, Dechant G, Götz R, and Barde Y-A (1992) Binding of neurotrophin-3 to its neuronal receptors and interactions with nerve growth factor and brain-derived neurotrophic factor. EMBO J 11, 917–922
35. He XE, and Garcia KC (2004) Structure of nerve growth factor complexed with the shared neurotrophin receptor p75. Science 304, 870–875
36. Kaplan DR, Hempstead BL, Martin-Zanca D, Chao MV, and Parada LF (1991) The trk proto-oncogene product: a signal transducing receptor for nerve growth factor. Science 252, 554–558
37. Klein R, Jing SQ, Nanduri V, O'Rourke E, and Barbacid M (1991) The trk proto-oncogene encodes a receptor for nerve growth factor. Cell 65, 189–197
38. Klein R, Nanduri V, Jing SA, Lambelle F, Tapley P, Bryant S, et al (1991) The trkB tyrosine protein kinase is a receptor for brain-derived neurotrophic factor and neurotrophin-3. Cell 66, 395–403
39. Squinto SP, Stitt TN, Aldrich TH, Davis S, Bianco SM, Radziejewski C, et al (1991) trkB encodes a functional receptor for brain-derived neurotrophic factor and neurotrophin-3 but not nerve growth factor. Cell 65, 885–893
40. Lamballe F, Klein R, and Barbacid M (1991) trkC, a new member of the trk family of tyrosine protein kinases, is a receptor for neurotrophin-3. Cell 66, 967–979
41. Benedetti M, Levi A, and Chao MV (1993) Differential expression of nerve growth factor receptors leads to altered binding affinity and neurotrophin responsiveness. Proc Natl Acad Sci USA 90, 7859–7863
42. Bibel M, Hoppe E, and Barde Y-A (1999) Biochemical and functional interactions between the neurotrophin receptors trk and p75NTR. EMBO J 18, 616–622
43. Mahadeo D, Kaplan L, Chao MV, and Hempstead BL (1994) High affinity nerve growth factor binding displays a faster rate of association than $p140^{trk}$ binding. J Biol Chem 269, 6884–6891
44. Esposito D, Patel P, Stephens RM, Perez P, Chao MV, Kaplan DR, et al (2001) The cytoplasmic and transmembrane domains of the p75 and Trk A receptors regulate high affinity binding to nerve growth factor. J Biol Chem 276, 32687–32695
45. Curtis R, Adryan KM, Stark JL, Park JS, Compton DL, Weskamp G, et al (1995) Differential role of the low affinity neurotrophin receptor (p75) in retrograde axonal transport of the neurotrophins. Neuron 14, 1201–1211
46. Makkerh JP, Ceni C, Auld DS, Vaillancourt F, Dorval G, and Barker PA (2005) p75 neurotrophin receptor reduces ligand-induced Trk receptor ubiquitination and delays Trk receptor internalization and degradation. EMBO J 6, 936–941
47. Geetha T, Jiang J, and Wooten MW (2005) Lysine 63 polyubiquitination of the nerve growth factor receptor TrkA directs internalization and signaling. Mol Cell 20, 301–312
48. Bentley CA, and Lee KF (2000) p75 is important for axon growth and schwann cell migration during development. J Neurosci 20, 7706–7715
49. Harrison SMW, Jones ME, Uecker S, Albers KM, Kudrycki KE, and Davis BM (2000) Levels of nerve growth factor and neurotrophin-3 are affected differentially by the presence of p75 in sympathetic neurons in vivo. J Comp Neurol 424, 99–110

50. Lee K-F, Li E, Huber LJ, Landis SC, Sharpe AH, Chao MV, et al (1992) Targeted mutation of the gene encoding the low affinity NGF receptor p75 leads to deficits in the peripheral sensory nervous system. Cell 69, 737–749
51. Lee KF, Bachman K, Landis S, and Jaenisch R (1994) Dependence on p75 for innervation of some sympathetic targets. Science 263, 1447–1449
52. Stucky CL, and Koltzenburg M (1997) The low-affinity neurotrophin receptor p75 regulates the function but not the selective survival of specific subpopulations of sensory neurons. J Neurosci 17, 4398–4405
53. Ultsch MH, Wiesmann C, Simmons LC, Henrich J, Yang M, Reilly D, et al (1999) Crystal structures of the neurotrophin-binding domain of TrkA, TrkB and TrkC. J Mol Biol 290, 149–159
54. Urfer R, Tsoulfas P, O'Connell L, Hongo JA, Zhao W, and Presta LG (1998) High resolution mapping of the binding site of TrkA for nerve growth factor and TrkC for neurotrophin-3 on the second immunoglobulin-like domain of the Trk receptors. J Biol Chem 273, 5829–5840
55. Weiss A, and Schlessinger J (1998) Switching signals on or off by receptor dimerization. Cell 94, 277–280
56. Huang EJ, and Reichardt LF (2003) Trk receptors: roles in neuronal signal transduction. Annu Rev Biochem 72, 609–642
57. Lee R, Kermani P, Teng KK, and Hempstead BL (2001) Regulation of cell survival by secreted proneurotrophins. Science 294, 1945–1948
58. Segal RA, and Greenberg ME (1996) Intracellular signaling pathways activated by neurotrophic factors. Annu Rev Neurosci 19, 463–489
59. York RD, Molliver DC, Grewal SS, Stenberg PE, McCleskey EW, Stork PJS (2000) Role of phosphoinositide 3-kinase and endocytosis in nerve growth factor-induced extracellular signal-regulated kinase activation via Ras and Rap1. Mol Cell Biol 20, 8069–8083
60. Daub H, Weiss FU, Wallasch C, and Ullrich A (1996) Role of transactivation of the EGF receptor in signalling by G-protein-coupled receptors. Nature 379, 557–560
61. Luttrell LM, Daaka Y, and Lefkowitz RJ (1999) Regulation of tyrosine kinase cascades by G-protein-coupled receptors. Curr Opin Cell Biol 11, 177–180
62. Lee FS, and Chao MV (2001) Activation of Trk neurotrophin receptors in the absence of neurotrophins. Proc Natl Acad Sci USA 98, 3555–3560
63. Lee FS, Rajagopal R, Kim AH, Chang PC, and Chao MV (2002) Activation of Trk neurotrophin receptor signaling by pituitary adenylate cyclase-activating polypeptides. J Biol Chem 277, 9096–9102
64. Rajagopal R, Chen ZY, Lee FS, and Chao MV (2004) Transactivation of Trk neurotrophin receptors by G-protein-coupled receptor ligands occurs on intracellular membranes. J Neurosci 24, 6650–6658
65. Rajagopal R, and Chao MV (2006) A role for Fyn in Trk receptor transactivation by G-protein-coupled receptor signaling. Mol Cell Neurosci 33, 36–46
66. Sofroniew MV, Howe CL, and Mobley WC (2001) Nerve growth factor signaling, neuroprotection, and neural repair. Annu Rev Neurosci 24, 1217–1281
67. Chao MV (2003) Neurotrophins and their receptors: a convergence point for many signalling pathways. Nat Rev Neurosci 4, 299–309
68. Lu B, Pang PT, and Woo NH (2005) The yin and yang of neurotrophin action. Nat Rev Neurosci 6, 603–614
69. Reichardt LF (2006) Neurotrophin-regulated signalling pathways. Phil Trans Royal Soc B 361, 1545–1564
70. Hefti F (1986) Nerve growth factor promotes survival of septal cholinergic neurons after fimbrial transactions. J Neurosci 6, 2155–2162
71. Williams LR, Varon S, Peterson GM, Wictorin K, Fischer W, Bjorklund A, et al (1986) Continuous infusion of nerve growth factor prevents basal forebrain neuronal death after fimbria fornix transaction. Proc Natl Acad Sci USA 83, 9231–9235
72. Kordower JH, Emborg ME, Bloch J, Ma SY, Chu Y, Leventhal L, et al (2000) Neurodegeneration prevented by lentiviral vec tor delivery of GDNF in primate models of Parkinson's disease. Science 290, 767–773
73. Yasuda T, and Mochizuki H (2010) Use of growth factors for the treatment of Parkinson's disease. Expert Rev Neurother 10, 915–924
74. Podulso JF, Curran GL, and Gill JS (1998) Putrescine-modified nerve growth factor: bioactivity, plasma pharmacokinetics, blood-brain/nerve barrier permeability, and nervous system biodistribution. J Neurochem 71, 1651–1660
75. Wu D, and Pardridge WM (1999) Neuroprotection with noninvasive neurotrophin delivery to the brain. Proc Natl Acad Sci USA 96, 254–259
76. Zhang Y, and Pardridge WM (2001) Neuroprotection in transient focal brain ischemia after delayed intravenous administration of brain-derived neurotrophic factor conjugated

to a blood-brain barrier drug targeting system. Stroke 32, 1378–1384

77. Choi-Lundberg DL, Lin Q, Chang YN, Chiang YL, Hay CM, Mohajeri H, et al (1997) Dopaminergic neurons protected from degeneration by GDNF gene therapy. Science 275, 838–841
78. Zhou L, Baumgartner BJ, Hill-Felberg SJ, McGowen LR, and Shine HD (2003) Neurotrophin-3 expressed in situ induces axonal plasticity in the adult injured spinal cord. J Neurosci 23, 1424–1431
79. Blesch A, Yang H, Weidner N, Hoang A, and Otero D (2004) Axonal responses to cellularly delivered NT-4/5 after spinal cord injury. Mol Cell Neurosci 27, 190–201
80. Tuszynski MH, Thal L, Pay M, Salmon DP, U HS, Bakay R, et al (2005) A phase 1 clinical trial of nerve growth factor gene therapy for Alzheimer disease. Nat Med 11, 551–555
81. Qu HY, Zhang T, Li XL, Zhou JP, Zhao BQ, Li Q, et al (2008) Transducible P11-CNTF rescues the learning and memory impairments induced by amyloid-beta peptide in mice. Eur J Pharmacol 594, 93–100
82. Garcia P, Youssef I, Utvik JK, Florent-Béchard S, Barthélémy V, Malaplate-Armand C, et al (2010) Ciliary neurotrophic factor cell-based delivery prevents synaptic impairment and improves memory in mouse models of Alzheimer's disease. J Neurosci 30, 7516–7527
83. Brewster WJ, Fernyhough P, Diemel LT, Mohiuddin L, and Tomlinson RR (1994) Diabetic neuropathy, nerve growth factor and other neurotrophic factors. Trends Neurosci 17, 321–325
84. Apfel SC (2002) Nerve growth factor for the treatment of diabetic neuropathy: what went wrong, what went right, and what does the future hold? Internatl Rev Neurobiol 50, 393–413
85. Apfel SC, Kessler JA, Adornato BT, Litchy WJ, Sanders C, and Rask CA (1998) Recombinant human nerve growth factor in the treatment of diabetic polyneuropathy. NGF Study Group. Neurology 51, 695–702
86. McArthur JC, Yiannoutsos C, Simpson DM, Adornato BT, Singer EJ, Hollander H, et al (2000) A phase II trial of nerve growth factor for sensory neuropathy associated with HIV infection. Neurology 54, 1080–1088

Chapter 2

Neuronal Growth-Promoting and Inhibitory Cues in Neuroprotection and Neuroregeneration

Stephen D. Skaper

Abstract

During development of the nervous system, neurons extend axons over considerable distances in a highly stereospecific fashion in order to innervate their targets in an appropriate manner. This involves the recognition, by the axonal growth cone, of guidance cues that determine the pathway taken by the axons. These guidance cues can act to promote and/or repel growth cone advance. The directed growth of axons is partly governed by cell adhesion molecules (CAMs) on the neuronal growth cone that bind to CAMs on the surface of other axons or nonneuronal cells. In vitro assays have established the importance of the CAMs ((neural cell adhesion molecule NCAM), N-cadherin, and L1) in promoting axonal growth over cells. Compelling evidence implicates the fibroblast growth factor receptor tyrosine kinase as the primary signal transduction molecule in the CAM pathway. CAMs are important constituents of synapses, and they appear to play important and diverse roles in regulating synaptic plasticity associated with learning and memory. Synthetic NCAM peptide mimetics corresponding to the binding site of NCAM for the fibroblast growth factor receptor promote synaptogenesis, enhance presynaptic function, and facilitate memory consolidation. Dimeric versions of functional binding motifs of N-cadherin behave as N-cadherin agonists, promoting both neuritogenesis and neuronal cell survival. Negative extracellular signals that physically direct neurite growth have also been described. The latter include the myelin inhibitory proteins, Nogo, myelin-associated glycoprotein, and oligodendrocyte-myelin glycoprotein. Potentiation of outgrowth-promoting signals, together with antagonism of myelin proteins or their convergent receptor, NgR, and its second messenger pathways, may provide new opportunities in the rational design of treatments for acute brain injury and neurodegenerative disorders.

Key words: Cell adhesion molecules, Fibroblast growth factor receptor, Neuroprotection, Neuroregeneration, Myelin, Nogo, Rho kinase

1. Introduction

Nervous system development evolves from the well-orchestrated processes of neural induction, cell proliferation, differentiation, cell migration, survival, and synapse formation (1, 2). The directed growth of axons is fundamental to synapse formation. Receptors for guidance molecules are present on growth cones and their

Stephen D. Skaper (ed.), *Neurotrophic Factors: Methods and Protocols*, Methods in Molecular Biology, vol. 846, DOI 10.1007/978-1-61779-536-7_2, © Springer Science+Business Media, LLC 2012

filopodia, and interactions with cognate ligands could result in promotion, inhibition, attraction, or repulsion of the growth cone. A broad spectrum of putative growth-promoting and/or guidance molecules has been identified to date. These include neurotrophic factors (3, 4), chemoattractants such as netrin 1 and 2 (5), chemorepellents from the collapsin/semaphorin family (6–8), and cell adhesion molecules (CAMs) (2, 9). The importance of the last class of molecules is evidenced by the severe neurological deficits found in humans with mutations in the *L1* gene, which can include a complete absence of corticospinal tracts (10).

Growth inhibitory molecules present in the central nervous system (CNS) myelin contribute, at least in part, to the inability of mammalian CNS axons to regenerate upon injury (11). In recent years, some molecular determinants of axonal regeneration and plasticity in the adult brain have been defined (12). Three proteins, Nogo-A (13–15), myelin-associated glycoprotein (MAG) (16, 17), and oligodendrocyte-myelin glycoprotein (OMgp) (18), appear to be responsible for this inhibition of axonal growth. This myelin inhibitory biology leads to the hypothesis that modulation of the interactions of these myelin proteins with their axonal receptor(s) would overcome the inhibitory effects of CNS myelin and promote axon regeneration, leading to improved functional recovery after CNS injury (see Fig. 1). Much recent work has focused upon the identification and characterization of signal transduction pathways for both CAMs and myelin-associated inhibitors.

2. CAMs, Neuronal Plasticity, and Neuroprotection

Cell–cell interactions mediated by CAMs are fundamental to numerous developmental processes. In the nervous system, the ability of neurons to extend axons and innervate their targets in an appropriate manner is governed to a large extent by the binding of CAMs on the surface of other axons or nonneuronal cells (19, 20). The neuronal receptors important for general cell contact–dependent axonal growth are the β1-integrins, which recognize extracellular matrix molecules; and in mammals, three CAMs, namely, the neural cell adhesion molecule (NCAM), N-cadherin, and L1, promote axonal growth during development (2).

N-Cadherin is a member of the classical cadherin family of transmembrane glycoproteins that mediate cellular recognition via homophilic (binding with other N-cadherin molecules on neighboring cells) interaction (21). In the nervous system, N-cadherin function has been implicated in cell migration (22), axonal growth and guidance (23), and synapse formation and synaptic plasticity (24, 25). In addition to homophilic binding, cadherins have been shown to interact with many adaptor or signaling molecules,

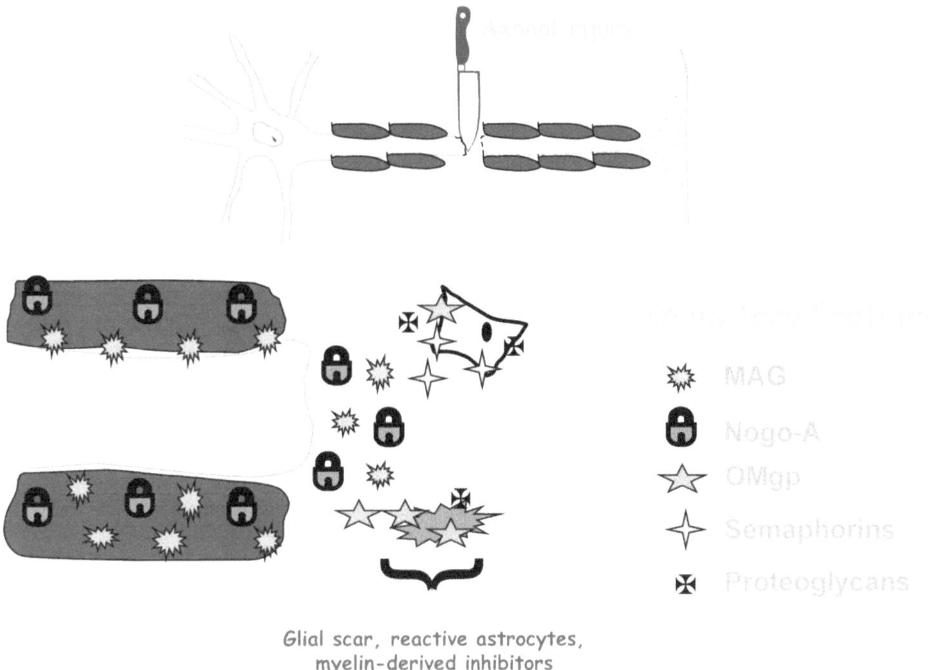

Fig. 1. Strategies for central nervous system regeneration. Mechanical injury or traumatic injury, or chronic neurodegenerative disease, can result in neurons being "disconnected" from their innervation fields, leading to a loss of critical tissue-derived trophic support. Regeneration may be promoted by replacement of neurotrophic factors, by stimulation of axonal outgrowth and/or cell survival with adhesion molecules, and by agents capable of overriding the outgrowth inhibitory environment. *MAG* myelin-associated glycoprotein, *OMgp* oligodendrocyte-myelin glycoprotein.

including the fibroblast growth factor receptor (FGFR). Neurite outgrowth stimulated by N-cadherin is inhibited by a variety of agents that block FGFR function in neurons (26–28). Soluble forms of some adhesion molecules (29), including N-cadherin (30), are effective also in promoting axonal growth. The homophilic binding site resides in extracellular domain 1 (ECD1) (31), and peptide mimetics of two linear sequences from ECD1 (HAVDI and INPISGQ) function as specific N-cadherin antagonists (32, 33). This information was then used to design cyclic peptides containing a tandem repeat of the individual motifs, which function as N-cadherin agonists and stimulate neurite outgrowth (34). A recent study has shown that a dimeric version (SW4 peptide) of a short N-cadherin binding motif (HAVDI) is capable of promoting the survival of several populations of CNS neurons, under distinct paradigms of injury, in an FGFR-dependent manner (35). An example of this is illustrated in Fig. 2 for hippocampal neurons subjected to glutamate-induced excitotoxicity. The neuroprotective

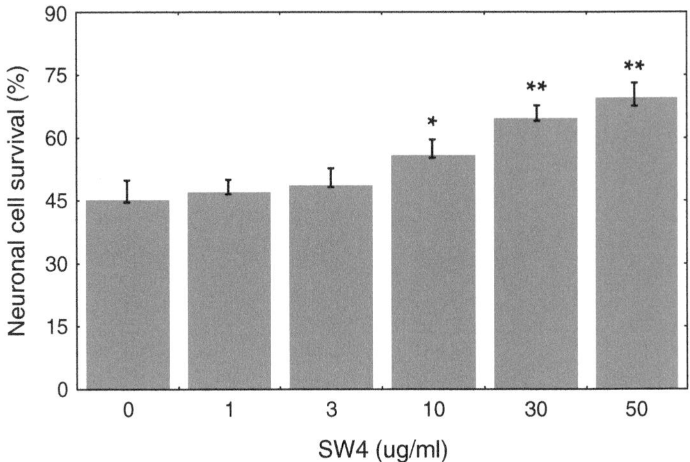

Fig. 2. The dimeric HAVDI peptide SW4 attenuates glutamate-induced toxicity in culture hippocampal neurons. Cultures (days in vitro, 7–9) were pretreated with the indicated concentrations of SW4 peptide for 48 h and then exposed to 50 μM glutamate, and neuronal cell survival was quantified 24 h later by a colorimetric reaction (MTT). Data are means ± SD (three experiments) expressed relative to no glutamate (100%). ** $P < 0.05$ or ** $P < 0.01$ vs. glutamate only. Reproduced with permission from Elsevier (75).

effects of SW4 displayed a concentration-dependence similar to those inducing neuritogenesis (34). The responses to the dimeric agonist peptide were inhibited by a monomeric version of the same motif (itself a highly specific N-cadherin antagonist) (32) and by a specific FGFR antagonist. These data suggest that the dimeric agonist peptide functions by binding to and clustering N-cadherin in neurons, and thereby activating an N-cadherin/FGFR signaling cascade (35). The NCAMs, L1 and CHL1 (36), and the synthetic NCAM peptide ligand, C3d (37), have been reported to display survival-promoting effects (albeit modest) on cultured CNS neurons. The neurotropic and neurotrophic effects of CAMs strengthen the emerging role for adhesion molecules in synaptic plasticity (38, 39).

CAMs are important constituents of synapses, with well-recognized roles in building and maintaining synaptic structure during brain development. Growing evidence indicates that CAMs play important and diverse roles in regulating synaptic plasticity associated with learning and memory (40, 41). The role of NCAM in cognitive processes is demonstrated by studies in which interference with NCAM function through the administration of antibodies or gene inactivation has resulted in impaired long-term potentiation (42, 43) and learning and memory deficits (44, 45). As with N-cadherin, synthetic NCAM peptide mimetics have now become available (46). After homophilic binding, NCAM promotes neurite outgrowth through mechanisms involving interaction with the FGFR (47, 48). The recently identified FG loop

(FGL) peptide, a 15-amino acid sequence corresponding to the binding site of NCAM for the FGFR1, has been shown to bind to and activate FGFR1 and to stimulate neurite outgrowth (48). Intracerebroventricular administration of the NCAM mimetic FGL peptide increases memory strength in rats and enhances presynaptic function in primary hippocampal neurons (49). These results provide the first evidence for a memory-facilitating effect resulting from a treatment that mimics NCAM function.

3. Growth Inhibitory Molecules

In contrast to fish, amphibia, and the mammalian peripheral nerves and developing central nerves, adult central mammalian neurons do not regrow functional axons after damage. It is not the absence of growth-promoting molecules in the CNS, but rather the presence of axonal outgrowth inhibitors in CNS astroglial "scars" (50) and CNS myelin (51, 52) that suppresses the regrowth of damaged axons. In particular, two components present in CNS myelin have been characterized as potent inhibitors of axonal growth: MAG (16, 17) and Nogo-A, the largest transcript of the recently identified *nogo* gene (formerly called NI-220) (13–15). Other inhibitors include chondroitin sulfate proteoglycans (53) and OMgp (18). Nogo is a member of the Reticulon family and occurs in three forms, Nogo-A, -B, and -C, which are generated from alternate splicing (13–15). Nogo-A, -B, and -C all contain a 66-amino acid ECD (Nogo-66) that alone can inhibit neurite outgrowth and induce growth cone collapse (13–15). MAG is a type I membrane protein composed of five extracellular immunoglobulin (Ig)–like domains (16, 17), whereas OMgp is a glycosylphosphatidylinositol-anchored protein (18). Remarkably, a single neuronal protein, the Nogo-66 receptor (NgR), binds Nogo, MAG, and OMgp (54–57). The p75 low-affinity neurotrophin receptor protein has been implicated in transducing a myelin/NgR signal to the axonal interior (58, 59) (see Fig. 3).

The presence of myelin-derived inhibitors suggests that blocking their action might allow the intrinsic growth potential of CNS axons to be unmasked (60). Indeed, neutralizing Nogo-A with IN-1 antibody induced CNS axon regeneration and improved recovery after various lesions (61, 62). Targeting the axonal NgR with a competitive antagonist compound has the potential to block the action of the three known myelin inhibitors. Both intrathecal (63) and delayed systemic application (64) of NEP1–40 (Nogo extracellular peptide, residues 1–40) produced significant axonal regrowth after spinal cord hemitransection injury, as well as enhanced locomotor recovery (64). However, these reagents only target a single myelin protein, Nogo, which may not be sufficient to

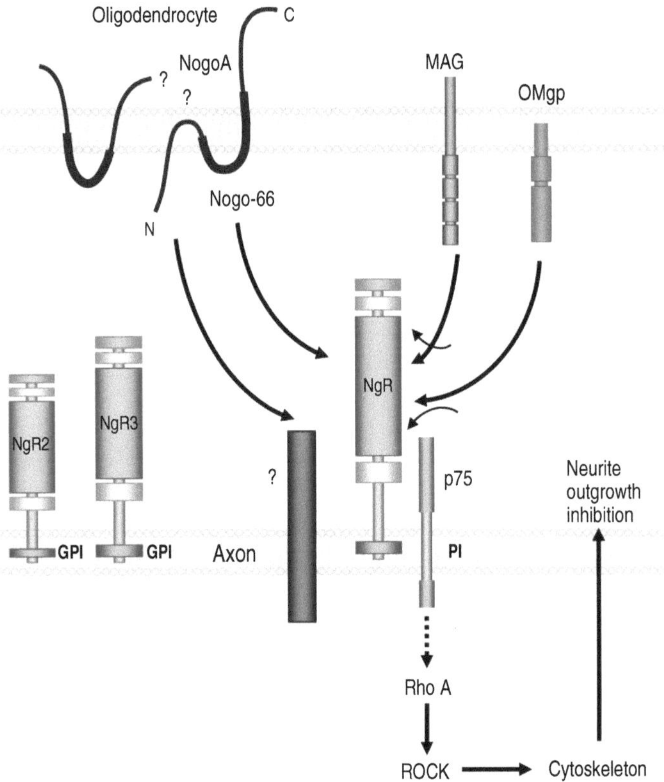

Fig. 3. The Nogo receptor and inhibition of axonal regeneration. The leucine-rich repeat domains of the Nogo receptor (NgR) are necessary for interaction with Nogo-66, MAG, and OMgp. NgR does not transduce signals directly but utilizes coreceptor molecules, for example, p75 or others. Coreceptor activation, in turn, activates the Rho and ROCK pathway to modulate the cytoskeleton and neurite growth. While structurally similar to NgR1, NgR2 and NgR3 display essentially no binding to known myelin-derived NgR ligands. *GPI* glycosylphosphatidylinositol, *ROCK* Rho-associated kinase.

facilitate maximal CNS axonal regeneration because other inhibitors such as MAG and OMgp are present in the CNS myelin environment. In rats with middle cerebral artery occlusion, both the recovery of motor skills and corticofugal axonal plasticity are promoted by intracerebroventricular administration of a function-blocking NgR fragment, sNgR310-Fc (an Ig-fusion protein containing a soluble fragment of NgR encompassing the ligand binding domain (sNgR310)) (65). Stroke lesion size was not significantly reduced in the sNgR310-Fc protein–treated group (65), suggesting that reduction of NgR function after stroke allows increased anatomical plasticity and improved motor performance that are not attributable to neuroprotection. Comparable results were obtained in *ngr*$^{-/-}$ mice with photothrombotic cortical lesion (65). A newly described neutralizing anti-Nogo-66 receptor monoclonal antibody 7E11 was reported to be a more potent inhibitor than sNgR310-Fc in in vitro competition binding assays and in a neurite outgrowth assay against CNS myelin (66). Further experiments in relevant animal models may prove the relevance of anti-NgR antibodies in promoting CNS axonal regeneration (67).

The signaling mechanisms responsible for the transduction of the inhibitory properties of Nogo-A and MAG domains are not well understood. Recent evidence supports the notion that cytoskeletal components required for proper axonal pathfinding and the formation of axons and dendrites are differentially regulated by members of the Rho family, including RhoA, Rac1, and Cdc42 (68). Rho proteins serve as a molecular switch by cycling between an inactive GDP-bound state and an active GTP-bound state. The most important effector of RhoA in the growth cone is probably the serine–threonine kinase Rho-kinase ROCK (69). Data indicate that MAG activates RhoA by increasing the proportion of the protein bound to GTP (70); Nogo-66 and myelin utilize this same signaling pathway (see Fig. 3) (71, 72). Pharmacological studies in vitro and in vivo indicate that ROCK plays a prime role in mediating myelin-induced inhibition via NgR (72, 73) and that Rho pathway inactivation can promote spinal cord repair (73, 74) and enhance axonal regeneration after corticospinal tract lesions in the adult rat (71).

4. Nerve Regeneration and Neurotrophic Factors

Information is encoded in the CNS through networks of neurons that are functionally connected by synapses. Upon injury, factors in CNS myelin inhibit neurite outgrowth. Understanding of how signals from extracellular factors associated with myelin and the injury site are integrated with outgrowth-promoting signals and neurotrophic factors to regulate axonal elongation should further facilitate the development of interventions to improve the outcome of both acute CNS injury and chronic neurodegenerative disorders. Methodologies which can be applied to understanding how neurite stimulatory signals and neurotrophic factors interact will be of considerable importance in advancing nerve regeneration efforts. This volume describes as well protocols which have this goal in mind.

References

1. Tessier-Lavigne M, and Goodman CS (1996) The molecular biology of axonal guidance. Science 274, 1123–1133
2. Walsh FS, and Doherty P (1997) Neural cell adhesion molecules of the immunoglobulin superfamily: role in axonal growth and guidance. Annu Rev Cell Dev Biol 13, 425–456
3. Ip NY, and Yancopoulos GD (1996) The neurotrophins and CNTF: two families of collaborative neurotrophic factors. Annu Rev Neurosci 19, 491–515
4. Skaper SD, and Walsh FS (1998) Neurotrophic molecules: strategies for designing effective therapeutic molecules in neurodegeneration. Mol Cell Neurosci 12, 179–193
5. Serefini T, Kennedy TE, Galko MJ, Mirzayan C, Jessell TM, and Tessier-Lavigne M (1994) The netrins define a family of axon outgrowth-promoting proteins homologous to *C. elegans* UNC-6. Cell 78, 409–424
6. Mueller BK (1999) Growth cone guidance: first steps toward a deeper understanding. Annu Rev Neurosci 22, 351–388
7. Fournier AE, and Strittmatter SM (2001) Repulsive factors and axon regeneration in the CNS. Curr Opin Neurobiol 11, 89–94
8. Kolodkin AL (1996) Semaphorins: mediators of repulsive growth cone guidance. Trends Cell Biol 6: 15–22
9. Goodman CS (1996) Mechanisms and molecules that control growth cone guidance. Annu Rev Neurosci 19, 341–377
10. Kenwrick S, Watkins A, and De Angelis E (2000) Neural cell recognition molecule L1: relating biological complexity to human disease mutations. Hum Mol Genet 9, 879–886
11. Cajal RS (1928) Degeneration and Regeneration of the Nervous System. Hafner. New York.
12. McGee AW, and Strittmatter SM (2003) The Nogo-66 receptor: focusing myelin inhibition of axon regeneration. Trends Neurosci 26, 193–198
13. Chen MS, Huber AB, Van Der Haar ME, Frank M, Schnell L, Spillmann AA, et al (2000) Nogo-A is a myelin-associated neurite outgrowth inhibitor and an antigen for monoclonal antibody IN-1. Nature 403, 434–439
14. GrandPré T, Nakamura F, Vartanian T, and Strittmatter SM (2000) Identification of the Nogo inhibitor of axon regeneration as a Reticulon protein. Nature 403, 439–444
15. Prinjha R, Moore SE, Vinson M, Blake S, Morrow R, Christie G, et al (2000) Inhibitor of neurite outgrowth in humans. Nature 403, 383–384
16. McKerracher L, David S, Jackson DL, Kottis V, Dunn RJ, and Braun PE (1994) Identification of myelin-associated glycoprotein as a major myelin-derived inhibitor of neurite outgrowth. Neuron 13, 805–811
17. Mukhopadhyay G, Doherty P, Walsh FS, Crocker PR, and Filbin MT (1994) A novel role for myelin-associated glycoprotein as an inhibitor of axonal regeneration. Neuron 13, 757–767
18. Wang KC, Koprivica V, Kim JA, Sivasankaran R, Guo Y, Neve RL, et al (2002) Oligodendrocyte-myelin glycoprotein is a Nogo receptor ligand that inhibits neurite outgrowth. Nature 417, 941–944
19. Dodd J, and Jessell JM (1988) Axon guidance and the patterning of neuronal projections in vertebrates. Science 242, 692–699
20. Goodman CS, and Shatz CJ (1993) Developmental mechanisms that generate precise patterns of neuronal connectivity. Cell/Neuron 72(suppl. 10), 77–98
21. Takeichi M (1995) Morphogenetic role of classic cadherins. Curr Opin Cell Biol 7, 619–627
22. Barami K, Kirschenbaum B, Lemmon V, and Goldman SA (1994) N-Cadherin and Ng-CAM/8D9 are involved serially in the migration of newly generated neurons into the adult songbird brain. Neuron 13, 567–582
23. Matsunaga M, Hatta K, Nagafuchi A, and Takeichi M (1988) Guidance of optic nerve fibers by N-cadherin adhesion molecules. Nature 334, 62–64
24. Inoue A, and Sanes JR (1997) Lamina-specific connectivity in the brain: regulation by N-cadherin, neurotrophins, and glycoproteins. Science 276, 1428–1431
25. Bozdagi O, Shan W, Tanaka H, Benson DL, and Huntley GW (2000) Increasing numbers of synaptic puncta during late-phase LTP: N-cadherin is synthesized, recruited to synaptic sites, and required for potentiation. Neuron 28, 245–259
26. Saffell JL, Williams EJ, Mason IJ, Walsh FS, and Doherty P (1997) Expression of a dominant negative FGF receptor inhibits axonal growth and FGF receptor phosphorylation stimulated by CAMs. Neuron 18, 231–242
27. Williams E-J, Furness J, Walsh FS, and Doherty P (1994) Activation of the FGF receptor underlies neurite outgrowth stimulated by L1, N-CAM, and N-cadherin. Neuron 13, 583–594
28. Williams E-J, Williams G, Howell FV, Skaper SD, Walsh FS, and Doherty P (2001)

Identification of an N-cadherin motif that can interact with the fibroblast growth factor receptor and is required for axonal growth. J Biol Chem 276, 43879–43886
29. Doherty P, Williams E, and Walsh FS (1995) A soluble form of the L1 glycoprotein stimulated neurite outgrowth. Neuron 14, 57–66
30. Utton MA, Eickholt B, Howell FV, Wallis J, and Doherty P (2001) Soluble N-cadherin stimulates fibroblast growth factor receptor dependent neurite outgrowth and N-cadherin and the fibroblast growth factor receptor co-cluster in cells. J Neurochem 76, 1421–1430
31. Koch AW, Bozic D, Pertz O, and Engel J (1999) Homophilic adhesion by cadherins. Curr Opin Struct Biol 9, 275–281
32. Williams E, Williams G, Gour BJ, Blaschuk OW, and Doherty P (2000) A novel family of cyclic peptide antagonists suggests that N-cadherin specificity is determined by amino acids that flank the HAV motif. J Biol Chem 275, 4007–4012
33. Williams EJ, Williams G, Gour B, Blaschuk O, and Doherty P (2000) INP, a novel N-cadherin antagonist targeted to the amino acids that flank the HAV motif. Mol Cell Neurosci 15, 456–464
34. Williams G, Williams E-J, and Doherty P (2002) Dimeric versions of two short N-cadherin binding motifs (HAVDI and INPISG) function as N-cadherin agonists. J Biol Chem 277, 4361–4367
35. Skaper SD, Facci, L, Williams G, Williams EJ, Walsh FS, and Doherty P (2004) A dimeric version of the short N-cadherin binding motif HAVDI promotes neuronal cell survival by activating an N-cadherin/fibroblast growth factor receptor signalling cascade. Mol Cell Neurosci 26, 17–23
36. Chen S, Mantei N, Dong L, and Schachner M (1999) Prevention of neuronal cell death by neural cell adhesion molecules L1 and CHL1. J Neurobiol 38, 428–439
37. Ditlevsen DK, Køhler LB, Pedersen MV, Risell M, Kolkova K, Meyer M, et al (2003) The role of phosphatidylinositol 3-kinase in neural cell adhesion molecule–mediated neuronal differentiation and survival. J Neurochem 84, 546–556
38. Goda Y (2002) Cadherins communicate structural plasticity of presynaptic and postsynaptic terminals. Neuron 35, 1–3
39. Togashi H, Abe K, Mizoguchi A, Takaoka K, Chisaka O, and Takeichi M (2002) Cadherin regulates dendritic spine morphogenesis. Neuron 35, 77–89
40. Schachner M (1997) Neural recognition molecules and synaptic plasticity. Curr Opin Cell Biol 9, 627–634
41. Benson DL, Schnapp LM, Shapiro L, and Huntley GW (2000) Making memories stick: cell adhesion molecules in synaptic plasticity. Trends Cell Biol 10, 473–482
42. Lüthi A, Laurent JP, Figurov A, Muller D, and Schachner M (1994) Hippocampal long-term potentiation and neural cell adhesion molecules L1 and NCAM. Nature 372, 777–779
43. Rønn LCB, Bock E, Linnemann D, and Jahnsen H (1995) NCAM-antibodies modulate induction of long-term potentiation in rat hippocampal CA1. Brain Res 677, 145–151
44. Doyle E, Nolan PM, Bell R, and Regan CM (1992) Intraventricular infusions of anti-neural cell adhesion molecules in discrete posttraining period impair consolidation of a passive avoidance response in the rat. J Neurochem 59, 1570–1573
45. Cremer H, Lange R, Christoph A, Plomann M, Vopper G, Roes J, et al (1994) Inactivation of the NCAM gene in mice results in size reduction of the olfactory bulb and deficits in spatial learning. Nature 367, 455–459
46. Rønn LCB, Olsen M, Ostergaard S, Kiselyov V, Berezin V, Mortensen MT, et al (1999) Identification of a neuritogenic ligand of the neural cell adhesion molecule using a combinatorial library of synthetic peptides. Nat Biotech 17, 1000–1005
47. Saffell JL, Walsh FS, and Doherty P (1994) Expression of NCAM containing VASE in neurons can account for a developmental loss in their neuritic outgrowth response to NCAM in a cellular substratum. J Cell Biol 125, 427–436
48. Kiselyov VV, Skladchikova G, Hinsby AM, Jensen PH, Kulahin N, Soroka V, et al (2003) Structural basis for a direct interaction between FGFR1 and NCAM and evidence for a regulatory role of ATP. Structure 11, 691–701
49. Cambon K, Hansen SM, Venero C, Herrero AI, Skibo G, Berezin V, et al (2004) A synthetic neural cell adhesion molecule mimetic peptide promotes synaptogenesis, enhances presynaptic function, and facilitates memory consolidation. J Neurosci 24, 4197–4204
50. Fawcett JW, and Asher RA (1999) The glial scar and central nervous system repair. Brain Res Bull 49, 377–391
51. Schwab ME, and Thoenen H (1985) Dissociated neurons regenerate into sciatic, but not optic nerve explants irrespective of neurotrophic factors. J Neurosci 5, 2415–2423
52. Schwab ME, and Caroni P (1988) Oligodendrocytes and CNS myelin are nonpermissive substrates for neurite growth and fibroblast spreading in vitro. J Neurosci 8, 2381–2393

53. Niederöst BP, Zimmermann DR, Schwab ME, and Bandtlow CE (1999) Bovine CNS myelin contains neurite growth-inhibitory activity associated with chondroitin sulphate proteoglycans. J Neurosci 19, 8979–8989
54. Fournier AE, GrandPre T, and Strittmatter SM (2001) Identification of a receptor mediating Nogo-66 inhibition of axonal regeneration. Nature 409, 341–346
55. Domeniconi M, Cao Z, Spencer T, Sivasankaran R, Wang K, Nikulina E, et al (2002) Myelin-associated glycoprotein interacts with the nogo66 receptor to inhibit neurite outgrowth. Neuron 35, 283–290
56. Liu BP, Fournier A, GrandPré T, and Strittmatter SM (2002) Myelin-associated glycoprotein as a functional ligand for the Nogo-66 receptor. Science 297, 1190–1193
57. Wang KC, Koprivica V, Kim JA, Sivasankaran R, Guo Y, Neve RL, et al (2002) Oligodendrocyte-myelin glycoprotein is a Nogo receptor ligand that inhibits neurite outgrowth. Nature 417, 941–944
58. Wang KC, Kim JA, Sivasankaran R, Segal R, and He Z (2002) p75 interacts with the Nogo receptor as a co-receptor for Nogo, MAG, and OMgp. Nature 420, 74–78
59. Wong ST, Henley JR, Kanning KC, Huang KH, Bothwell M, and Poo MM (2002) A p75 (NTR) and Nogo receptor complex mediates repulsive signaling by myelin-associated glycoprotein. Nat Neurosci 5, 1302–1308
60. Lee DHS, Strittmatter SM, and Sah DWY (2003) Targeting the Nogo receptor to treat central nervous system injuries. Nat Rev Drug Discovery 2, 872–879
61. Bregman BS, Kunkel-Bagden E, Schnell L, Dai HN, Gao D, and Schwab ME (1995) Recovery from spinal cord injury mediated by antibodies to neurite growth inhibitors. Nature 378, 498–501
62. Thallmair M, Metz GA, Z'Graggen WJ, Raineteau O, Kartje GL, and Schwab ME (1998) Neurite growth inhibitors restrict plasticity and functional recovery following corticospinal tract lesions. Nat Neurosci 1, 124–131
63. GrandPre T, Li S, and Strittmatter SM (2002) Nogo-66 receptor antagonist peptide promotes axonal regeneration. Nature 417, 547–551
64. Li S, and Strittmatter SM (2003) Delayed systemic Nogo-66 receptor antagonist promotes recovery from spinal cord injury. J Neurosci 23, 4219–4227
65. Lee J-K, Kim J-E, Sivula M, and Strittmatter SM (2004) Nogo receptor antagonism promotes stroke recovery by enhancing axonal plasticity. J Neurosci 24, 6209–6217
66. Li W, Walus L, Rabacchi SA, Jirik A, Chang E, Schauer J, (2004) A neutralizing anti-Nogo66 receptor monoclonal antibody reverses inhibition of neurite outgrowth by central nervous system myelin. J Biol Chem 279, 43780–43788
67. Yu P, Huang L, Zou J, Yu Z, Wang Y, Wang Z, et al (2008) Immunization with recombinant Nogo-66 receptor (NgR) promotes axonal regeneration and recovery of function after spinal cord injury in rats. Neurobiol Dis 32, 535–542
68. Luo L (2000) Rho GTPases in neuronal morphogenesis. Nat Rev Neurosci 1, 173–180
69. Bito H, Furuyashiki T, Ishihara H, Shibasaki Y, Ohashi K, Mizuno K (2000) A critical role for Rho-associated kinase, p160ROCK, in determining axonal outgrowth in mammalian CNS neurons. Neuron 26, 431–441
70. Yamashita T, Higuchi H, and Tohyama M (2002) The p75 receptor transduces the signal from myelin-associated glycoprotein to Rho. J Cell Biol 157, 565–570
71. Fournier AE, Takizawa BT, and Strittmatter SM (2003) Rho kinase inhibition enhances axonal regeneration in the injured CNS. J Neurosci 23, 1416–1423
72. Niederöst B, Oertle T, Fritsche J, McKinney RA, and Bandtlow CE (2002) Nogo-A and myelin-associated glycoprotein mediate neurite growth inhibition by antagonistic regulation of RhoA and Rac1. J Neurosci 22, 10368–10376
73. Dergham P, Ellezam B, Essagian C, Avedissian H, Lubell WD, and McKerracher L (2002) Rho signalling pathway targeted to promote spinal cord repair. J Neurosci 22, 6570–6577
74. Kubo T, Hata, K, Yamaguchi A, and Yamashita T (2007) Rho-ROCK inhibitors as emerging strategies to promote nerve regeneration. Curr Pharm Des 13, 2493–2499
75. Skaper, S.D., Facci, L., Gilliams, G., Williams, E.-J., Walsh, F.S. and P. Doherty (2004) A dimeric version of the short N-cadherin binding motif HAVDI promotes neuronal cell survival by activating an N-cadherin/fibroblast growth factor receptor signalling cascade. Mol and Cell Neurosc 26, 17–23

Chapter 3

Culture of Rat Cerebellar Granule Neurons and Application to Identify Neuroprotective Agents

Laura Facci and Stephen D. Skaper

Abstract

In primary culture of the early postnatal cerebellum, glutamatergic granule cells are highly enriched and recapitulate many properties characteristic of developing granule neurons in vivo. For example, withdrawal of K^+ from differentiated rat primary cerebellar granule neurons results in the apoptotic death of the majority of cells after 48 h. Removal of cerebellar granule neurons from depolarizing culture conditions with high K^+ is thought to reflect the regulation of trophic action of neuronal activity and has found widespread application as a model for studying the mechanisms of survival factor withdrawal-induced neuronal cell apoptosis and the neuroprotective action of trophic agents. This chapter presents a protocol for the culture of postnatal rat cerebellar granule neurons and results in a preparation containing 95% glutamatergic granule cells and its application to the evaluation of corticotropin receptor agonists as neuroprotective agents.

Key words: Cerebellum, Granule neurons, Cell culture, Rat, Depolarization, Development, Phosphatidylinositol 3-kinase, Glycogen synthase kinase-3, Apoptosis, Neuroprotection, Neurotrophic factors

1. Introduction

Like many other cell types, neurons require survival factors to suppress apoptotic machinery and prevent death (1). Apoptosis is observed not only in neurodegenerative disorders but also during nervous system development. In the developing cerebellum, granule cells proliferate at the outer part of the external granular layer and become postmitotically differentiated at the inner part of this layer (2). In primary culture of the early postnatal cerebellum, granule cells are highly enriched and recapitulate many properties characteristic of developing granule neurons in vivo (3, 4). For example, withdrawal of K^+ from differentiated rat primary cerebellar granule neurons (CGNs) results in the apoptotic death of the majority of

Stephen D. Skaper (ed.), *Neurotrophic Factors: Methods and Protocols*, Methods in Molecular Biology, vol. 846,
DOI 10.1007/978-1-61779-536-7_3,

cells after 48 h (3, 5, 6). Removal of CGN from depolarizing culture conditions with high K^+ is thought to reflect the regulation of trophic action of neuronal activity and has found widespread application as a model for studying the mechanisms of survival factor withdrawal-induced neuronal cell apoptosis (7–11). Moreover, the in vivo correlate of CGN apoptosis has been described (12, 13).

Apoptosis is a morphologically distinct form of cell death characterized by cellular fragmentation, chromatin condensation and margination, and internucleosomal cleavage of genomic DNA (14). The apoptosis of mature CGNs deprived of depolarizing levels of extracellular K^+ is prevented by insulin-like growth factor I, brain-derived neurotrophic factor, basic fibroblast growth factor, or cyclic AMP (11, 15–17). Neuronal survival in this model is abolished by the phosphatidylinositol 3-kinase (PI 3-kinase) inhibitors LY294002 or wortmannin (7, 8), indicating that PI 3-kinase is a central component of the survival machinery of these neurons. Recombinant gene overexpression studies have indicated that protein kinase B (PKB; also known as Akt), a downstream effector of PI 3-kinase, can play a critical role in regulating the survival of CGNs (18). Among several known targets of PKB, glycogen synthase kinase-3 is inhibited by PKB-mediated phosphorylation (19). Recombinant overexpression studies using active or dominant-negative proteins (10, 20) or small molecule inhibitors (8) have led to the proposal that glycogen synthase kinase-3 may be a relevant effector of PI 3-kinase-mediated cell survival (see Fig. 1).

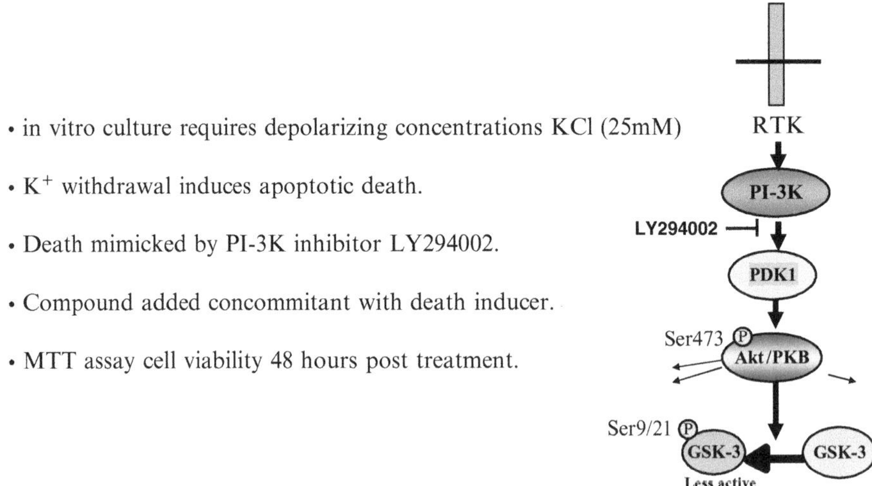

Fig. 1. Schematic representation of the phosphatidylinositol 3-kinase (PI3K)/glycogen synthase kinase (GSK-3) signaling pathway involved in cerebellar granule neuron survival promoted by a high-potassium environment. Phosphorylation of protein kinase B (PKB or Akt) on Ser473 leads to its activation, with subsequent phosphorylation (and inactivation) of GSK-3 on Ser9 and Ser21. The active state of GSK-3 promotes apoptosis, while GSK-3 inactivation is antiapoptotic. *RTK* receptor tyrosine kinase; *PDK1* phosphoinositide-dependent kinase-1.

This chapter describes a protocol for culturing CGNs from postnatal rat brain, followed by a procedure for inducing their apoptotic death based on pharmacological inhibition of PI 3-kinase, combined with the MTT (3-(4,5-dimethylthiazol-2-yl)-2,5-diphenyltetrazolium bromide) assay for cell viability. An example of neuroprotection is also presented, using corticotropin-releasing factor.

2. Materials

2.1. Equipment and Labware

1. Stereo dissecting microscope (backlighting of stage is preferred) with fiber optic light source
2. Laminar flow cabinet for dissections
3. Laminar flow biological safety cabinet (CL2)
4. Humidified, water-jacketed culture incubator at 37°C and 5% CO_2/95% air
5. Water bath set at 37°C
6. Dissecting tools (Fine Science Tools)
7. Bench centrifuge to accommodate 15- and 50-mL tubes
8. 15- and 50-mL polypropylene plastic centrifuge tubes (sterile)
9. 0.5-mL microfuge tubes
10. Vacuum-driven filter units (0.45 μm pore size, 250 and 500 mL capacity) from Millipore (Stericup®)
11. 10 and 6 cm ∅ sterile Petri plastic dishes
12. 10 cm ∅ sterile tissue culture dishes
13. Microtest 96 (96-well plate), tissue culture plastic
14. 0.22 and 0.45 μm filters (Millipore)
15. Neubauer hemocytometer (Fisher Scientific)
16. 96-well poly-D-lysine-treated tissue culture plates (BD Biosciences)

2.2. Reagents

1. Water, sterile, tissue culture grade (Invitrogen)
2. Phosphate-buffered saline, pH 7.4 (Invitrogen)
3. Basal medium Eagle (BME) (Invitrogen)
4. Neurobasal-A medium (Invitrogen)
5. Neurobasal-A medium, phenol red-free (Invitrogen)
6. B27 supplement, 50×, with antioxidants (Invitrogen)
7. Fetal calf serum (FCS) (Invitrogen) (see Note 1)
8. L-Glutamine (200 mM stock), sterile, for cell culture (Invitrogen)

9. Sodium pyruvate (100 mM stock), sterile, for cell culture (Sigma)
10. Penicillin/streptomycin, 10,000 U/mL penicillin + 10,000 μg/mL streptomycin (100× stock), sterile, for cell culture (Invitrogen)
11. Gentamicin, 50 mg/mL, sterile, for cell culture (Sigma)
12. HEPES, 1 M, sterile (Sigma)
13. Cytosine β-D-arabinofuranoside (Ara-C) (Sigma)
14. Papain digestion kit (Worthington Biochemicals)
15. Trypsin, hog pancreas, for cell culture (Sigma)
16. Trypsin inhibitor, soybean, for cell culture (Sigma)
17. DNase from bovine pancreas, type II (Sigma)
18. Bovine serum albumin (BSA) (Sigma)
19. Poly-L-lysine, sterile, for cell culture (MW 30,000–70,000) (Sigma)
20. Phenol red (0.5% stock) (Sigma)
21. Trypan blue stain 0.4% (Invitrogen)
22. Erythrosin B (Sigma)
23. LY294002 (Merck4Biosciences)
24. MTT formazan (Sigma)

2.3. Culture Media and Other Solutions

1. *Krebs Buffer* (10×). Add the following components and amounts: NaCl (17.67 g), KCl (0.900 g), KH_2PO_4 (0.415 g), D-glucose (6.425 g), $NaHCO_3$ (5.350 g), 0.5% phenol red (5 mL), dissolve in water and bring to 250 mL, and filter-sterilize. The solution can be stored at 4°C for 3 months.
2. *Solution K.* Dilute the 10× Krebs buffer stock 1:10 in double-distilled water, add $MgCl_2$ to 0.5 mM and 0.3% (w/v) BSA. Filter-sterilize. Can be stored for 1 month at 4°C.
3. *0.15 M Borate buffer, pH 8.4.* Dissolve 28.6 g of sodium borate ($Na_2B_4O_7 \cdot 10\ H_2O$) in 500 mL water (pH will be ~9.2). Adjust pH to 8.4 with 5 N HCl. Filter-sterilize and store at 4°C (up to 6 months).
4. *Ara-C stock solution (10 mM).* Prepare Ara-C as 10-mM stock solution in PBS, filter-sterilize, and store as aliquots at −20°C for up to 6 months. Once thawed, the aliquot tube may be kept at 4°C for 4 weeks (see Note 2).
5. *Serum-free culture medium.* This consists of Neurobasal-A medium (see Note 3) with the addition of B27 supplements (1:50 of a 50× stock solution), 1 mM sodium pyruvate (from a 100-mM stock solution), 2 mM L-glutamine (from a 200-mM stock solution), 20 mM KCl (1:50 from a 1-M stock solution), 100 U/mL penicillin-100 μg/mL streptomycin

(1:100 from a stock solution containing 10,000 U/mL penicillin and 10,000 μg/mL streptomycin). Medium should be stored at 4°C and used within 2 months.

6. *Serum-containing culture medium*. Culture medium consists of BME (with Earle's salts) and glutamine, and supplemented by adding 20 mM KCl (1:50 from a 1-M stock solution), $NaHCO_3$ (2.2 g/L), gentamicin (1 mL/L of 50 mg/mL stock), and 10% (v/v) heat-inactivated FCS (see Note 1). Medium should be stored at 4°C and used within 2 months.

7. *MTT stock solution*. Prepare a 3-mg/mL stock solution of MTT in PBS and filter through a 0.22-μm filter attached to a syringe (syringe size depends on volume being filtered). Aliquot and store at −20°C.

2.4. Poly-L-Lysine Substratum

1. Dissolve a 5-mg bottle of poly-L-lysine in 5 mL of 0.15 M borate buffer, pH 8.4 (see Note 4). Dilute this solution to 50 μg/mL with sterile, tissue culture-grade water (see Note 5). Coat desired dishes or multiwall plates with 150–200 μL/cm^2 of surface area. Place the dishes/plates in a 37°C incubator overnight (see Note 6). Aspirate poly-L-lysine solution and leave plates or dishes uncovered in a vertical laminar flow hood to dry. Coated and dried plates/dishes can be stored at 4°C for a month.

2. For serum-free cultures, the day prior to use, incubate poly-L-lysine-coated dishes/plates with culture medium (choice of medium is not important) containing 10% FCS. Keep in 37°C incubator until used (aspirate FCS-containing medium immediately before seeding with cells—do not allow to dry).

2.5. Papain Dissociation System: Solutions

1. The kit consists of the following components: vial (1)—100 mL EBSS (Earle's balanced salt solution); vial (2)—papain containing L-cysteine and EDTA; vial (3)—DNase; vial (4)—ovomucoid protease inhibitor with BSA.

2. Add 2 mL of 1 M HEPES buffer to the entire contents of vial (1)—this will still be referred to as "EBSS."

3. *Solution A*: Add 5 mL of EBSS to vial (2) (20 U of papain/mL in 1 mM L-cysteine with 0.5 mM EDTA).

4. *Solution B*: Add to vial (3) 0.5 mL of EBSS (2,000 U/mL DNase).

5. *Solution C*: Add 32 mL of EBSS to vial (4) (10 mg ovomucoid inhibitor and 10 mg BSA/mL). This solution is stable when stored at 2–8°C.

6. *Solution D*: Add 250 μL of solution B to 5 mL of solution A.

2.6. Stock Solutions for Trypsin Dissociation (Prepare on Day of Use)

1. Solution B: Dissolve 20 mg of hog pancreas trypsin in 25 mL solution K and filter-sterilize.
2. Solution C: Add 2.4 mg DNase, 15.6 mg trypsin inhibitor, and 0.3 mL of 0.15 M $MgSO_4 \cdot 7\ H_2O$ to solution K, and bring volume to 30 mL with solution K. Filter-sterilize.
3. Solution D: Combine 4 mL of solution C and 21 mL of solution K.
4. Solution E: To 25 mL of solution K, add 0.2 mL of 0.15 $MgSO_4 \cdot 7\ H_2O$ and 0.2 mL of 0.012 M $CaCl_2\ 2\ H_2O$.

3. Methods

3.1. Tissue Dissection and Dissociation (Papain Dissociation Procedure)

1. Dissect out the cerebellum from 7- to 8-day-old rat pups of both sexes (CD strain, Sprague Dawley) under aseptic conditions following decapitation. Pups are sacrificed in accordance with appropriate institutional and national guidelines for the care and use of laboratory animals. Standard techniques and anatomical landmarks are followed.
2. Make an incision under the skin along the midline of the dorsal surface and peel back the skin. Make a second incision close to the midline along the exposed dorsal surface (the cut should be along the complete anteroposterior axis of the lateral ventricles). Splay open the hemispheres along the cut surface and continue cut along the dorsal surface of the hemisphere.
3. Remove the cerebellum to a 10-cm Ø dish containing solution K. Free the tissue of meninges with the use of a stereo dissecting microscope.
4. Remove solution K and mince the tissue with a flamed razor blade. Collect the tissue pieces with solution K and transfer to a 15-mL centrifuge tube.
5. Allow the tissue pieces to settle by gravity, remove the medium, and add 5 mL of solution D. Incubate for 15 min at 37°C with occasional agitation.
6. Vigorously titrate with a cotton-plugged, sterile Pasteur pipette (attached to an automatic pipettor) until devoid of clumps (see Notes 7 and 8).
7. Centrifuge at $200 \times g$ for 5 min.
8. During step 5, prepare two tubes with 2.7 mL EBSS + 0.3 mL solution C + 0.15 mL solution B (3.15 mL total = EBSS Mix); 5 mL solution C.
9. After centrifugation, discard the supernatant and resuspend the pellet in 3 mL EBSS Mix. Using a Pasteur pipette, carefully layer the cell suspension on top of the 5 mL ovomucoid.
10. Centrifuge at $150 \times g$ for 5 min.

11. Remove and discard the supernatant and resuspend the pellet in culture medium (either serum-containing or serum-free).
12. It will be necessary to dilute the cell suspension before counting. A density of 1.5–2 million cells/mL for hemacytometer counting is convenient. See Subheading 3.3 for details on counting with the hemacytometer. Typical yields are 20–23 million cells/cerebellum for 7–8-day-old animals.
13. Dilute the cell suspension with complete culture medium to give the desired plated density. Generally, it is advisable to plate $2.5–3.0 \times 10^5$ cells/cm^2 of culture surface area. Survival of CGN is density-dependent, and plating the cells below 1×10^5 cells/cm^2 will result in reduced cell viability (see Notes 9–11).
14. Add Ara-C 18–22 h after cell plating to a final concentration of 10 μM (1:1,000 from a 10-mM stock solution in PBS). Ara-C is an antimitotic and is added to inhibit the growth of nonneurons.

3.2. Preparation of CGNs Using Trypsin Dissociation

1. Collect cerebellar tissue from 7- to 8-day-old rat pups into solution K, as described in Subheading 3.1. Mince tissue with a razor blade and collect in a 15-mL tube, then add solution B (1 mL/cerebellum). If more than 10 cerebella are processed, it is advisable to perform this step using a 50-mL tube.
2. Incubate for 15 min (37°C) with gentle swirling.
3. Add solution D (1 mL/cerebellum) to the tube and centrifuge at $300 \times g$ for 3 min.
4. Remove as much as possible the supernatant; the pellet may not be very compact. It is possible to transfer the pellet to a new tube, thereby eliminating most of the supernatant.
5. Add 2 mL of solution C. Pipette C down 5 times with a 5-mL graduated pipette to break up the largest tissue pieces. Dissociate the tissue using a long (9-in.) glass Pasteur pipette (cotton plugged, sterile) with the tip slightly flame-constricted—25 up-and-down strokes should be sufficient (see Note 12).
6. Add 3 mL of solution C per tube and apply a further 15 strokes with the Pasteur pipette.
7. Add 6 mL of solution E per tube.
8. Centrifuge at $200 \times g$ for 7 min.
9. Resuspend the cell pellet in 2–3 mL complete serum-containing culture medium, using a Pasteur pipette as above. The remaining steps are the same as steps 10–12 in Subheading 3.1.

3.3. Cell Counting with a Hemocytometer

1. Hemocytometers are modified glass slides engraved with grids of fixed area at each side of a central trough, which are covered with a coverslip prior to use. The sample of cells to be counted is harvested and applied to the slide surface with a Pasteur pipette or pipetman, and allowed to move under the coverslip by capillary action.

2. The most commonly used counting chamber is a neubauer hemocytometer. When the coverslip is in the correct position, the area within the grid corresponds to 1 mm^2. If all the cells on the central 5×5 grid are counted, the number of cells in 1 cm^3 is equal to the cell count $\times 10^4$. Since there will always be cells located on the lines of the square, the convention is to count all cells on the top and left-hand lines, and ignore all those on the bottom and right-hand lines. A minimum of 200 cells should be counted where possible to minimize inaccuracies of the technique.
3. Cell viability is easily checked by dye exclusion, which relies on the premise that living cells prevent certain agents crossing the membrane, while dead cells are permeable to a selection of stains. Erythrosin B is the dye of choice here, although the most commonly used is trypan blue due to ease (although it may give a high background in serum-containing media) (see Notes 13 and 14). We have not noticed this when dealing with primary neurons, and routinely utilize trypan blue. A 0.4% solution of erythrosin B is prepared by dissolving 0.4 g erythrosin B in 95 mL PBS (heated to boiling point). When dissolved, cool and bring volume to 100 mL. Filter-sterilize and aliquot.
4. Thoroughly clean surface of the hemocytometer with 70% ethanol. Secure a clean coverslip onto the top of the slide centrally by moistening the edges of the coverslip and pressing down firmly (see Note 15).
5. Using a 0.4-mL Eppendorf tube, mix 10 μL of a 0.4% (v/v) trypan blue solution and 90 μL of cell suspension using a 200-μL pipetman.
6. Carefully pipette a small volume (10 μL seems to work fine) onto the surface of the slide to draw the cells into the counting chamber. Repeat on the opposite edge of the coverslip.
7. Use a light microscope at low magnification to focus on the counting chamber (see Note 16).
8. Count both the number of cells in the 5×5 grid that exclude trypan blue (viable) and the number of cells that do not exclude trypan blue (they are blue and not viable). Count both grids.
9. Calculate viability as a percentage by dividing the number of live cells counted by the number of dead cells, and multiplying the answer by 100.
10. Calculate the cell count by multiplying the live cells counted by 10^4 to give a cell number/mL and multiply by 1.1 to allow for the dilution factor of adding the dye.
11. Clean the hemocytometer immediately, by gently wiping off with a Kimwipe moistened with 70% ethanol.

3.4. Setting Up the CGN Cultures for Neuroprotection Assay

1. Prepare one or more 96-well tissue culture plastic plates coated with poly-L-lysine.
2. Dilute the CGN cell dissociate to 400,000 cells/mL, using BME supplemented with 10% heat-inactivated FCS, 2 mM L-glutamine, 50 μg/mL gentamicin, and 25 mM KCl (see Note 17).
3. Add to each culture well 0.1 mL of the above cell suspension. This will give a density of 40,000 CGN/well.
4. Incubate cultures at 37°C in a humidified atmosphere with 5% CO_2.
5. Eighteen to twenty-four hours after cell plating, add cytosine β-D-arabinofuranoside (Ara-C) to a final concentration of 10 μM. To do this, remove 10 μL of medium from each well and add 10 μL of fresh medium containing 100 μM Ara-C (diluted 1:100 from a 10-mM stock solution).

3.5. Induction of Apoptosis

1. At 8 days in vitro, completely exchange the CGN culture medium for fresh medium lacking Ara-C (100 μL/well) and containing 5 mM KCl or 25 mM KCl.
2. Alternatively, exchange the medium for fresh serum-free medium containing 25 mM KCl ± 75 μM LY294002 (added from a 37.5-mM stock solution in dimethyl sulfoxide). Under these conditions, maximal CGN cell death is achieved with 50–75 μM LY294002 (7) (see Note 18).
3. Include the test compound with the medium change at this time (see Note 19).
4. Continue incubation for 48 h.
5. Assess cell viability by the MTT assay.

3.6. MTT Analysis of Cell Viability of LY294002-Treated CGN: Protection by Corticotropin-Releasing Factor

Ideally, a colorimetric assay for living cells should utilize a colorless substrate that is modified to a colored product by any living cell, but not by dead cells, their lytic debris, or tissue culture medium. Tetrazolium salts are attractive candidates for this purpose since they measure the activity of various dehydrogenase enzymes (21). The tetrazolium ring is cleaved in active mitochondria, and so the reaction occurs only in living cells. A rapid colorimetric assay is based on the tetrazolium salt MTT that measures only living cells and can be read on a scanning multiwall spectrophotometer (ELISA reader) (22, 23).

1. At 8 days in culture, exchange the culture medium for serum-free BME (or Neurobasal-A) containing 25 mM KCl ± 75 μM LY294002 ± 30 nM corticotropin-releasing factor. After 6, 24, and 48 h, cellular vitality is evaluated by MTT assay.
2. Remove the culture medium (aspirate under vacuum, using a Pasteur pipette).

3. Add to each well 100 μL of serum-free BME or Neurobasal-A (with 25 mM KCl) containing 0.15 mg/mL MTT (which represents a 1:20 dilution of the 3 mg/mL stock) (see Note 20).
4. Incubate the MTT-containing cultures for 1 h at 37°C (see Note 21).
5. Examination of the cultures under a light microscope will show that viable cells contain blue crystals. This is the formazan reaction product.
6. Remove the culture medium by aspiration and add 100 μL of dimethyl sulfoxide/well. This is to dissolve the reaction product. Gently tap the microwell plate from underneath to mix the blue reaction product uniformly with the solvent (see Note 22).
7. Read the plate on a microplate (ELISA) reader. We use a SpectraMax M2 microplate reader from Molecular Devices, although there are a number of quality plate readers on the market. For optimal results, the plate should be read at a wavelength of 570 nm (test wavelength), followed by 630 nm (reference wavelength). The difference (A_{570}–A_{630}) is used as the final absorbance value generated by the sample. Correcting for optical imperfections in the microplates by subtracting A_{630} is recommended, but is not an essential procedure.
8. Normalize the absolute MTT values by scaling to the mean of cultures in serum-free medium with 25 mM KCl (defined as 100%).
9. An example of the outcome of this illustrative experiment is shown in Table 1. Note the time-dependent loss of cell viability in the LY294002-treated cultures.
10. Cell vitality should be routinely confirmed by morphological observation using a light microscope with phase contrast optics.

3.7. Cell Culture Safety Issues

1. A horizontal laminar flow hood is recommended for dissection of tissues. Once dissection is complete, all work should be moved to a Class 2 vertical laminar flow safety cabinet.
2. The antibiotics and enzymes used should be considered as noxious agents. Suitable precautions should be taken in the use and disposal of these products.
3. All liquids containing live cells should be aspirated into a waste vessel containing MicroSol +3 (a broad spectrum disinfectant) (Anachem), and pipettes placed in containers with MicroSol. Plasticware is then disposed of via autoclaving.
4. Accumulated culture liquid waste should then be transferred to a 20-L carboy containing ~100 g of Virkon powder (VK734, Medisave, UK). Virkon is a multipurpose disinfectant and contains oxone (potassium peroxymonosulfate), sodium dodecylbenzenesulfonate, sulfamic acid, and inorganic buffers. Virkon

Table 1
Corticotropin-releasing factor protects cerebellar granule neurons from LY294002-induced death

Treatment	MTT
6 h	
LY	77 ± 6✠
LY + CRF	92 ± 3*
24 h	
LY	64 ± 5†
LY + CRF	87 ± 6**
48 h	
LY	28 ± 1
LY + CRF	84 ± 5***

Cerebellar granule neurons were cultured for 8 days, then shifted to serum-free BME containing 25 mM KCl and 75 μM LY294002 ("LY") ± 30 nM corticotropin-releasing factor (CRF). After the times indicated, cellular vitality was evaluated by MTT assay. Data are means ± S.D. ($n = 3$)
*$p < 0.05$; **$p < 0.01$; ***$p < 0.0001$ (LY + CRF vs. LY); ✠$p < 0.05$ or $p < 0.001$ vs. LY at 24 and 48 h, respectively; †$p < 0.001$ vs. LY at 48 h
Adapted (with substantial modification) from *Neuropharmacology* 45(5), L. Facci, D.A. Stevens, M. Pangallo, D. Franceschini, S.D. Skaper, P.J.L.M. Strijbos, Corticotropin-releasing factor (CRF) and related peptides confer neuroprotection via type 1 CRF receptors, 623–636 (Table 3), Copyright© (2003), with permission from Elsevier

has a wide spectrum of activity against viruses, some fungi, and bacteria. The pink color is useful in that it helps gage the concentration when preparing the Virkon, and importantly, as the Virkon ages, it discolors, making it obvious when it needs to be replaced. The solution is generally stable for 5–7 days. Virkon is relatively safe in terms of skin contact but can cause eye damage and should not be used as a hand-washing liquid. Wear appropriate personal protective clothing.

4. Notes

1. Heat inactivation of fetal calf serum is recommended to destroy heat-labile complement. Thaw the bottle of serum in advance, using a 37°C water bath. Next, immerse the serum bottle in the water bath after reequilibrating to 56°C and leave for 30 min. Swirl the bottle occasionally to ensure proper mixing. Allow the serum to cool to room temperature, aliquot into 50-mL tubes, and store at –20°C.

2. Cytosine arabinoside is an antimitotic and potential carcinogen. Usage should be restricted to fume cupboards and biological safety cabinets. Avoid weighing out powder, or if not possible, then weigh out in a suitable closed environment. Culture media can be disposed of by normal routes. Stock/concentrated solutions should be enclosed within a leakproof container and put into a burn bin for disposal.
3. There are two Neurobasal medium formulations—without and with the suffix "A." The principal difference is in their osmolarity: Neurobasal "A" is intended for postnatal and adult neuron cultures, which the other formulation is preferable for culturing embryonic neurons.
4. Poly-L-lysine (and poly-L-ornithine) is very hygroscopic and difficult to weigh. It is advisable to purchase the material as a 5–10-mg quantity, and then dissolve the entire amount in borate buffer. The solutions can be stored at 4°C for up to 6 months.
5. Working solutions of poly amino acids should be prepared in sterile tissue culture plastic polypropylene tubes. TAKE CARE: The poly amino acid will adhere to glass surfaces, and also to polystyrene plastic.
6. In a pinch, the coating time with poly-L-lysine can be limited to 2 h—the poly amino acid sticks rapidly to tissue culture plastic-treated surfaces.
7. Avoid pulling air through the Pasteur pipette and creating bubbles/foam—this can cause cell damage.
8. If necessary, and to clear the suspension of any remaining small clumps, this may be passed through a cell strainer (70 μm, BD Falcon #352350) placed over the mouth of a 50-mL centrifuge tube.
9. To maintain long-term CGN cultures, it is necessary to add glucose after the first 10 days and weekly thereafter. Glucose is prepared as a 0.5-M stock in water, filter-sterilized, and added at 1:100 to the cultures. Alternatively, a 0.5-M glucose solution for cell culture can be purchased from Sigma.
10. It may be a good idea to check periodically the volume of the cultures. Should evaporation be noted, the lost volume can be replaced by adding sterile water. This is more likely to occur if the cultures are carried for more than 2 weeks. Never change the culture medium on CGN, especially when culturing in serum-containing medium—the CGNs condition their medium, and serum contains glutamate which is toxic to mature CGN.
11. On the subject of evaporation, when culturing in 96-well plates, the empty wells should be filled with water—this will help to reduce volume loss on the edge wells.

12. Depending on the dissociation volume, it may be advisable to divide the suspension into 2–3 parts prior to dissociation. The dissociation efficiency of the Pasteur pipette is reduced when the volume exceeds 4 mL.
13. Erythrosin B is cytotoxic, and cells should not be left in the dye for prolonged times prior to counting.
14. Be sure to add the trypan blue to the cells just before counting—after several minutes, it is toxic to the cells.
15. As an alternative, you can lightly moisten the edges of the hemocytometer before placing the coverslip—this will prevent the cover from sliding off prior to loading with cells.
16. The hemocytometer chamber will accept about 10 μL—never overfill, as this can cause erroneous cell counts.
17. Alternatively, you can use the Neurobasal-A-based serum-free culture medium. In this case, remember to pretreat overnight the polylysine-coated plates with medium containing 10% FCS, prior to cell seeding.
18. If at all possible, we recommend not exceeding a dimethyl sulfoxide concentration of 0.4%. When treating cells with LY294002 and test compound, each is added from a 500× stock solution. When only LY294002 or test compound is used, the dimethyl sulfoxide concentration should be brought to 0.4% final by addition of solvent.
19. If the compound being tested is made up in nonaqueous medium (e.g., in dimethyl sulfoxide or ethanol), then control cultures should be included (medium with 5 mM and 25 mM KCl) which contain the same amounts of organic solvent.
20. The stock solution (3 mg/mL) of MTT, once thawed, can be refrozen and used at least three additional times without compromising its efficacy.
21. The incubation time for obtaining optimal MTT color yield will vary as a function of cell type and plating density, as well as MTT concentration and time of incubation (14). We have established the 1-h time used here for CGN under the culture conditions specified. The experimenter is advised to conduct trials with the cell type being used to establish optimal conditions for color yield. Note that excessive incubation times and/or concentrations of MTT can be cytotoxic.
22. The initial MTT procedure for cultured primary neurons used a solution of isopropanol-0.08 N HCl to dissolve the MTT reaction product. We find that this procedure is very slow to dissolve the blue crystals and requires rather vigorous agitation of the microwell plate. In contrast, the action of dimethyl sulfoxide is quite rapid, taking only a matter of seconds.

Acknowledgments

L. Facci was supported by Fondazione CARIPARO Progetto Dottorati di Ricerca Anno 2009.

References

1. Raff MC, Barres BA, Burne J, Coles HS, Ishizaki Y, and Jacobson, MD (1993) Programmed cell death and the control of cell survival – Lessons from the nervous system. Science 262, 695–700
2. Hatten ME, and Heintz N (1995) Mechanisms of neural patterning and specification in the developing cerebellum. Annu Rev Neurosci 18, 385–408
3. Gallo V, Kingsbury A, Balázs R, and Jørgensen OS (1987) The role of depolarisation in the survival and differentiation of cerebellar granule cells in culture. J Neurosci 7, 2203–2213
4. Lasher RS, and Zagon IS (1972) The effect of potassium on neuronal differentiation in cultures of dissociated newborn rat cerebellum. Brain Res 41, 482–488
5. Thangnipon W, Kingsbury A, Webb M, and Balázs R (1983) Observations on rat cerebellar granule cells in vitro: influence of substratum, potassium concentration and relationship between neurones and astrocytes. Brain Res 313, 177–189
6. Ginham R, Harrison DC, Facci L, Skaper SD, and Philpott K (2001) Upregulation of death pathways in rat cerebellar granule neurons undergoing apoptosis. Neurosci Lett 302, 113–116
7. Miller TM, Tansey MG, Johnson EM Jr, and Creedon DR (1997) Inhibition of phosphatidylinositol 3-kinase activity blocks depolarisation- and insulin-like growth factor 1-mediated survival of cerebellar granule cells. J Biol Chem 272, 9847–9853
8. Cross DAE, Culbert AA, Chalmers KA, Facci L, Skaper SD, and Reith AD (2001) Selective small-molecule inhibitors of glycogen synthase kinase-3 activity protect primary neurones from death. J Neurochem 77, 94–102
9. Dudek H, Datta SR, Franke TF, Birnbaum MJ, Yao R, Cooper GM, et al (1997) Regulation of neuronal survival by the serine-threonine protein kinase Akt. Science 275, 661–665
10. Pap M, and Cooper GM (1998) Role of glycogen synthase kinase-3 in the phosphatidylinositol 3-kinase/Akt cell survival pathway. J Biol Chem 273, 19929–19932
11. D'Mello SR, Galli C, Ciotti C, and Calissano P (1993) Induction of apoptosis in cerebellar granule neurons by low potassium: inhibition of cell death by insulin-like growth factor I and cAMP. Proc Natl Acad Sci USA 90, 10989–10993
12. Wood KA, Dipasquale B, and Youle RJ (1993) In situ labelling of granule cells for apoptosis-associated DNA fragmentation reveals different mechanisms of cell loss in developing cerebellum. Neuron 11, 621–632
13. Krueger BK, Burne JF, and Raff MC (1995) Evidence for large-scale astrocyte death in the developing cerebellum. J Neurosci 15, 3366–3374
14. Bredesen DE (1995) Neuronal apoptosis. Ann Neurol 38, 839–851
15. Fernandez-Sanchez MT, and Novelli A (1993) Basic fibroblast growth factor protects cerebellar neurons in primary culture from NMDA and non-NMDA receptor mediated neurotoxicity. FEBS Lett 335, 124–131
16. Kubo T, Nonomura T, Enokido Y, and Hatanaka H (1995) Brain-derived neurotrophic factor (BDNF) can prevent apoptosis of rat cerebellar granule neurons in culture. Dev Brain Res 16, 249–258
17. Shimoke K, Yamagishi S, Yamada M, Ikeuchi T, and Hatanaka H (1999) Inhibition of phosphatidylinositol 3-kinase activity elevates c-Jun N-terminal kinase activity in apoptosis of cultured cerebellar granule neurons. Dev Brain Res 112, 245–253
18. Crowder RJ, and Freeman RS (2000) Glycogen synthase kinase-3b activity is critical for neuronal death caused by inhibiting phosphatidylinositol 3-kinase or Akt but not for death caused by nerve growth factor withdrawal. J Biol Chem 275, 34266–34271
19. Cross DAE, Alessi DR, Cohen P, Andjelkovich M, and Hemmings BA (1995) Inhibition of glycogen synthase kinase-3 by insulin mediated by protein kinase B. Nature 378, 785–789

20. Hetman M, Cavanaugh JE, Kimelman D, and Xia Z (2000) Role of glycogen synthase kinase-3β in neuronal apoptosis induced by trophic withdrawal. J Neurosci 20, 2567–2574
21. Slater TF, Sawyer B, and Straeuli UD (1963) Studies on succinate-tetrazolium reductase systems. III. Points of coupling of four different tetrazolium salts. Biochim Biophys Acta 77, 383–393
22. Mosmann T (1983) Rapid colorimetric assay for cellular growth and survival: application to proliferation and cytotoxicity assays. J Immunol Methods 65, 55–63
23. Manthorpe M, Fagnani R, Skaper SD, and Varon S (1986) An automated colorimetric microassay for neuronotrophic factors. Dev Brain Res 25, 191–198

Chapter 4

Isolation and Culture of Neural Progenitor Cells from Rat Postnatal Cerebellum

Morena Zusso and Patrizia Debetto

Abstract

The mammalian brain contains undifferentiated, mitotically active, and multipotent neural stem/progenitor cells that in vivo contribute new neurons and glia to specific areas of the mature brain. When isolated under the appropriate conditions, these cells maintain in vitro the ability to proliferate and differentiate into cells that express neuronal and glial markers. Neural stem/progenitor cells have been identified and isolated from many regions of the embryonic, postnatal, and adult central nervous system, including cerebellum. This chapter details techniques to isolate and culture neural progenitor cells from rat postnatal cerebellum, which can be used as an in vitro model to study the molecular mechanisms underlying proliferation and differentiation into mature neural cells induced by various stimuli including pharmacological agents.

Key words: Neurogenesis, Neural stem/progenitor cells, Cerebellum, Postnatal, Rat

1. Introduction

During development of the mammalian central nervous system (CNS), neural cells are generated by the proliferation of multipotent stem or progenitor cells that migrate, find their site of final destination, and ultimately differentiate. Neural stem/progenitor cells are most abundant at early developmental stages; however, they persist throughout development, and it is now known that the in vivo genesis of new neurons from endogenous pools of mitotically active neural stem/progenitors (i.e., neurogenesis) continues in the adult CNS of all mammalian organisms, including humans (1). Adult neurogenesis mainly occurs in two constitutive neurogenic brain regions: the subventricular zone (SVZ) of the lateral ventricles, which generates neurons destined for the olfactory bulb, and the subgranular zone of the dentate gyrus of the hippocampus, which generates neurons that functionally integrate into the existing circuitry (2–5). However, neural stem/progenitor cells have been

Stephen D. Skaper (ed.), *Neurotrophic Factors: Methods and Protocols*, Methods in Molecular Biology, vol. 846,
DOI 10.1007/978-1-61779-536-7_4, © Springer Science+Business Media, LLC 2012

identified also in several mammalian embryonic and adult brain regions that are normally nonneurogenic, including cerebral cortex, substantia nigra, spinal cord, and cerebellum (6–9). Neural stem/progenitor cells can be isolated, propagated, and differentiated in vitro both from constitutive and latent areas. In particular, the mammalian cerebellum is one of the brain regions exhibiting protracted postnatal neurogenesis. Cerebellar neurons are generated from different sources and at different developmental stages. Granule cells, the most abundant interneurons, are generated almost entirely during the first 2 postnatal weeks from proliferating progenitors located in the external granule layer (for review, see ref. (10)). This germinal layer is active during early postnatal periods whose duration is strictly species dependent (11–14). Recently, it has been shown that stem cells can be isolated from both postnatal (15, 16) and adult (17) cerebellum, suggesting that latent neurogenic potential may be present in this brain structure.

In this chapter, we describe a method that we developed for the isolation, culturing, and differentiation of neural progenitor cells from rat postnatal cerebellum. This method typically yields 1×10^6 viable progenitor cells from a single postnatal cerebellum. Once isolated and cultured, cerebellar progenitors provide a suitable in vitro model to identify the critical molecules that regulate cell proliferation and subsequent differentiation induced by pharmacological stimuli (e.g., serotonergic antidepressant drugs) (16).

2. Materials

2.1. Dissection and Cell Cultures

1. Animals: Sprague Dawley rat pups; age of pups = postnatal day 7. We define postnatal day 0 as the day of birth.
2. HBSS/BSA: Add 3 mg/mL bovine serum albumin (BSA) to calcium- and magnesium-free Hank's balanced salt solution (HBSS; Gibco, Life Technologies, Rockville, MD) and filter sterilize through 0.22 μm filter. Store at 4°C for up to 1 month.
3. Trypsin solution: Add 0.8 mg/mL trypsin to HBSS/BSA and filter sterilize (see Note 1).
4. Trypsin inhibitor solution: Add 0.5 mg/mL trypsin inhibitor and 0.1 mg/mL deoxyribonuclease I (DNase) to HBSS/BSA. Filter sterilize (see Note 1).
5. Growth factor stock solutions: Reconstitute epidermal growth factor (EGF) (human recombinant) and basic fibroblast growth factor (bFGF) (Sigma-Aldrich) in order to have a 500 μg/mL

stock of each. Aliquot into sterile tubes and store at −20°C. The shelf life of the stock solution is 1 year at −20°C.

6. Culture medium: Serum-free *growth medium* (500 mL). To 480 mL of Neurobasal medium (Gibco), add 10 mL 50× B-27 supplement (Gibco), 5 mL 200-mM L-glutamine, 5 mL 100× Pen-Strep (10,000 IU/mL penicillin and 10,000 μg/mL streptomycin) (Gibco), and 20 μL EGF and bFGF stock solutions (final concentrations 20 ng/mL). Store at 4°C for <1 month; prewarm to 37°C before use.

2.2. Differentiation of Cerebellar Neural Progenitor Cells

1. 100× Poly-D-lysine stock solution: Dissolve 20 mg/mL poly-D-lysine hydrobromide (MW > 40,000; Sigma-Aldrich) in sterile tissue culture water. Sterilize by 0.22-μm filter and store at 4°C for up to 1 month.
2. Poly-D-lysine-coated chamber slides: Dilute the poly-D-lysine stock solution to a final concentration of 200 μg/mL in sterile tissue culture water. Add enough diluted solution to entirely cover the surface of chamber slides (Permanox slide, Lab Tek, Nalge Nunc International) and incubate at room temperature for 1 h. Remove the poly-D-lysine solution and allow to dry under the laminar flow hood.
3. Culture medium: *Differentiation medium* (500 mL). To 480-mL Basal Medium Eagle (Sigma-Aldrich), add 10 mL 10% heat-inactivated fetal bovine serum (see Note 2), 5 mL 200-mM L-glutamine, 5 mL 2.5-M KCl, and 5 mL 100× Pen-Strep (10,000 IU/mL penicillin, 10,000 μg/mL streptomycin). Store at 4°C for <1 month; prewarm to 37°C before use.

2.3. Immunostaining of Cerebellar Neural Progenitor Cells

1. Phosphate-buffered saline (PBS): To 800-mL double distilled water, add 8 g NaCl, 0.2 g KCl, 1.44 g Na_2HPO_4, and 0.24 g KH_2PO_4. Dissolve, adjust the pH to 7.4 with HCl, and bring to 1,000 mL with water. Store at room temperature.
2. 4% Paraformaldehyde solution: Heat 50 mL distilled water to 70°C (do not boil), add 4 g paraformaldehyde, and then add 1-N NaOH drop by drop, until the solution clears. When solution is cool, adjust the pH to 7.4 with HCl and bring to 100 mL with PBS (see Note 3).
3. Antibodies for characterization of cell phenotype: Primary antibodies: mouse antineurofilament 200 (Sigma-Aldrich), mouse antimicrotubule-associated protein-2 (Millipore Corporation), mouse anti-β-tubulin isotype III (Millipore Corporation), rabbit antiglial fibrillary acidic protein (Millipore Corporation), and mouse anti-O4 (Millipore Corporation). Fluorescent secondary antibodies: Alexa Fluor 555–conjugated goat anti-mouse and anti-rabbit and Alexa Fluor 488–conjugated goat anti-mouse and anti-rabbit (Molecular Probes) (see Note 4).

4. 4′,6-Diamidino-2-phenylindole (DAPI) stock solution: Dissolve 1 mg/mL DAPI (Sigma-Aldrich) in ultrapure water. Aliquot and store at −20°C. The stock solution is stable for several months and can be used repeatedly if stored protected from light.

3. Methods

3.1. Dissection of Cerebellum from Postnatal Rats and Establishment of Primary Cultures

1. Rapidly decapitate 7-day-old rat pups (see Note 5) by cutting the neck with sterile scissors large enough to ensure the head is removed in one cut. Discard the body into a plastic bag for disposal. Transfer the heads to a sterile 100-mm ∅ Petri dish containing cold HBSS/BSA and cut the skin on top of the head (see Note 6).
2. Under the dissection hood, cut the skull along the midline, starting from the back of the head to the eyes. Make a superficial cut and be careful not to damage the underlying brain tissue. Using forceps, lift up the skull to expose the brain.
3. Using a pair of curved forceps, remove the cerebellum and immediately transfer it to a sterile Petri dish containing cold HBSS/BSA (see Note 7). Repeat step 2 until all cerebella have been collected.
4. Under the dissecting microscope, carefully remove the meninges from each cerebellum using two forceps. The meninges will appear as a delicate membrane containing blood vessels. Place the cleaned cerebella into a new dish containing cold HBSS/BSA and keep on ice.
5. Once all the cerebella have been processed, under the tissue culture hood, remove the HBSS/BSA from the dish and mince the tissue into small pieces (0.5–1 mm^3) with a flamed razor blade (see Note 8). Transfer dissected tissues into a 50-mL conical tube containing 1-mL HBSS/BSA/cerebellum and centrifuge at $1{,}000\times g$ for 3 min.
6. Carefully aspirate the HBSS/BSA (see Note 9), resuspend the tissue pieces in 1-mL trypsin solution/cerebellum, and incubate for 15 min in a 37°C water bath under gentle shaking to keep the tissue pieces suspended.
7. Stop the enzymatic reaction by adding the trypsin inhibitor solution (1 mL/cerebellum). Pellet the tissue by centrifugation at $1{,}000\times g$ for 7 min and remove the supernatant by gentle aspiration (see Note 9).
8. Add 5 mL HBSS/BSA and dissociate the tissue by triturating with a sterile, fire-polished, cotton-plugged glass Pasteur pipette

until the suspension looks cloudy and only small pieces of tissue are left (see Note 10).

9. Pellet the cells by centrifugation at 1,000 × *g* for 7 min. Discard the supernatant and resuspend the cells in 10 mL of prewarmed (37°C) culture medium. Make an appropriate dilution of an aliquot of the cell suspension in trypan blue solution and count the number of viable cells using a hematocytometer (see Note 11).
10. Plate the cells at a density of 2×10^5 viable cells/cm^2 in uncoated tissue culture plates, in culture medium containing 20 ng/mL of EGF and bFGF.
11. Incubate at 37°C, 5% CO_2, in a humidified incubator.
12. After 5–6 days, replace half of the medium with fresh medium. Add a double amount of EGF and bFGF to give a final concentration of 20 ng/mL.

3.2. Culture Propagation

Cerebellar progenitors proliferate to form floating aggregates, which are usually ready for subculturing 10–12 days after plating:

1. Dislodge aggregates by taping sides of plates; transfer content of the plates to 15-mL sterile, conical tubes, using a sterile plastic pipette. Use fresh medium to rinse plates and add rinse to the tube.
2. Pellet cell suspension by centrifugation at 1,000 × *g* for 7 min. Remove the supernatant without disturbing the cell pellet.
3. Resuspend cells in 1 mL culture medium. Using a P200 (200 μL) Pipetman, gently dissociate the pellet (30–40×) to produce a single cell suspension (see Note 12).
4. Count viable cells by trypan blue exclusion and plate cells at the density of 5×10^3 cells/cm^2 onto uncoated tissue culture plates, in culture medium containing appropriate growth factors (EGF and bFGF, 20 ng/mL) (see Note 13).
5. Subculture when the aggregates start to lift off and float in suspension. This will require 10–12 days (see Note 14).

3.3. Differentiation of Cerebellar Neural Progenitors

When maintained in *growth medium*, cerebellar neural progenitors display a round shape (Fig. 1a) and an undifferentiated state, as evidenced by their immunoreactivity for several immature neuronal cell markers (Fig. 1b). Nevertheless, plating cerebellar progenitors onto an adhesive substrate and growing in serum-containing medium without growth factors (*differentiation medium*) is sufficient to promote a differentiation process into the three CNS cell types (Fig. 1c, d):

1. Dislodge aggregates by taping sides of plates and transfer content of the plates to 15-mL sterile, plastic, conical tubes, using

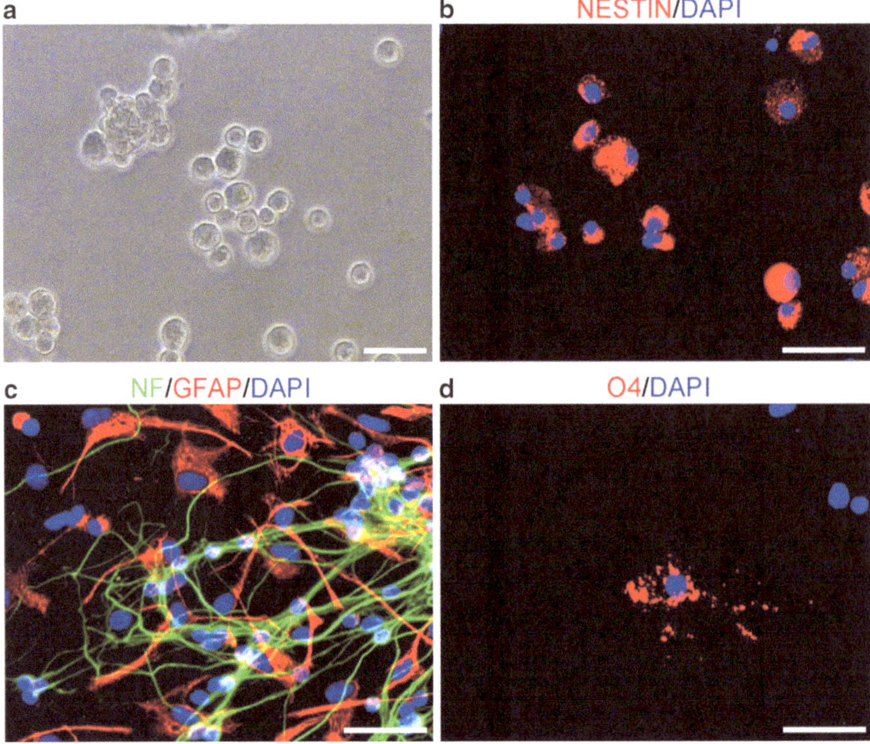

Fig. 1. Morphology and characterization of cerebellar neural progenitors. (**a**) Representative phase contrast photomicrograph of a cerebellar neural progenitor cell culture maintained in *growth medium* for 10 days. (**b**) Fluorescent photomicrograph of the specific immunoreactivity for nestin. Cells were replated onto poly-D-lysine-coated chamber slides in *differentiation medium*. After 3 days, cells were fixed with 4% paraformaldehyde and stained with specific antibodies. (**c**, **d**) Fluorescent photomicrographs of neurofilament- (NF), glial fibrillary acidic protein- (GFAP), and O4-positive cells. DAPI staining shows the nuclei. *Scale bars* = 50 μm (adopted in part from ref. (16), with modifications).

a sterile plastic pipette. Rinse plates with fresh medium, add rinse to the tube, and centrifuge cell suspension at 1,000 × *g* for 7 min.

2. To wash cells from growth factors, remove supernatant and resuspend cells in 10 mL of *differentiation medium*. Centrifuge at 1,000 × *g* for 7 min.
3. Remove the supernatant and resuspend cells in 1 mL of *differentiation medium*. Using a P200 Pipetman, gently triturate the pellet (30–40×) (see Note 15).
4. Count viable cells by trypan blue exclusion. Resuspend cells in the appropriate volume of culture medium (containing 10% fetal bovine serum) and plate them at the density of 1×10^4 cells/cm^2 onto poly-D-lysine-coated chamber slides.
5. Allow the cells to differentiate for 3–5 days and analyze the phenotype by immunocytochemical determination of cell-type-specific markers for neurons, astrocytes, and oligodendrocytes.

3.4. Immunocytochemical Staining for Characterization of Differentiation

All staining procedures are carried out at room temperature except where indicated. Washing steps are done for 5 min each:

1. Remove culture medium and carefully wash cultures twice with PBS.
2. Fix cells for 10 min in 4% paraformaldehyde at room temperature and wash three times with PBS (see Note 16).
3. Block nonspecific staining by incubation for 1 h in PBS containing 5% normal donkey serum (Jackson ImmunoResearch Laboratories) and 0.1% Triton X-100 (blocking solution) at room temperature (see Note 17).
4. Incubate with primary antibodies diluted in blocking solution for 2 h at room temperature or overnight at 4°C (see Note 18).
5. Wash cells three times with PBS and incubate for 1 h at room temperature in the dark with species-specific secondary antibodies (conjugated to the desired fluorophores) diluted in blocking solution (see Note 18).
6. Wash cells three times in PBS, incubate in PBS containing 0.1 μg/mL DAPI for 2 min (for nuclear counterstain), and then coverslip in Fluoromount mounting medium (Sigma-Aldrich) (see Note 19).
7. Analyze cell phenotype by confocal or fluorescence microscopy.

4. Notes

1. Dissecting solutions (trypsin and trypsin inhibitor solutions) must be prepared on the day of dissection.
2. Heat-inactivated fetal bovine serum: Heat inactivation of serum is important to destroy heat-labile complement. Degradation of these proteins by raising the temperature of the serum to 56°C for 30 min will ensure that antibody binding will not lyse the cells. Heat inactivation may change the appearance of the fetal bovine serum and result in precipitation. This is normal and does not indicate that the serum has been compromised:
   - Thaw the bottle of fetal bovine serum overnight at room temperature.
   - Immerse the serum bottle in the 56°C water bath for 30 min. Swirl the bottle every 5 min to ensure proper mixing.
   - Allow serum to cool for 1 h at room temperature, aliquot into 50-mL sterile tubes, and store at –20°C.
3. This fixative solution should be made up fresh. Paraformaldehyde is toxic and may be carcinogenic; therefore, prepare the solution

under a fume hood and wear gloves and mask. Disposal should follow chemical safety rules.

4. All the antibodies should be freshly diluted just before use. Do not store them as diluted solutions.
5. For a good cell yield, use at least one large litter of rats (12 or more). Rats are generally sacrificed by cervical dislocation prior to decapitation or by decapitation alone (this depends on the animal care protocols of your institution).
6. In Subheading 3.1, step 1 can be performed at the lab bench. To prevent contamination, use sterile instruments (autoclaved using autoclave bags and opened only when needed) and solutions and take precautions to maintain sterility, including the wearing of gloves. Speed is also important.
7. The Petri dish should be maintained on ice for the duration of dissection.
8. Chop the tissues with a rapid chopping motion, moving the dish in a circular fashion so that all tissue pieces are evenly cut.
9. Do not use the vacuum since there is a risk of sucking out the tissue.
10. Normally, 20–30 strokes are sufficient, but perform a second trituration of 15 strokes in case undissociated tissue is still visible. During triturating, always avoid foaming and bubbles.
11. Initially try a 1:5 dilution. The number of viable cells yielded by this protocol, using a single postnatal day 7 rat cerebellum, is 10×10^6 cells. Approximately 5–10% of plated cells form progenitor cell aggregates.
12. Always avoid foaming and bubbles.
13. Feed cells with fresh medium (replace half of the medium) every 5–6 days.
14. Isolated cerebellar neural progenitors can be subcultured for three passages.
15. Trituration into a single cell suspension is critical in the differentiation protocol. A drop of cell suspension should be added to a Petri dish to check that the suspension is free from large aggregates before plating. Trituration in trypsin/EDTA should be avoided as it damages the cell membrane and can cause adhesion defects.
16. Fixed cells can be stored in PBS at 4°C for about a week or can be processed immediately for immunocytochemistry.
17. Make blocking solution with serum from the species in which the secondary antibodies were raised. Since blocking solution contains serum, it can be contaminated. Blocking solution is stable for 1 week at 4°C; if the solution becomes cloudy, it has become contaminated and should not be used.

18. Primary and secondary antibody concentrations should always be titrated for the particular experiment, to obtain the best possible staining.
19. Fluoromount stabilizes the fluorescence well for long periods. Refrigerated, Fluoromuont-mounted slides kept in the dark will retain their fluorescence for months.

References

1. Gage FH (2000) Mammalian neural stem cells. *Science* 287, 1433–1438
2. Taupin P, and Gage FH (2002) Adult neurogenesis and neural stem cells of the central nervous system in mammals. *J Neurosci Res* 69, 745–749
3. Kempermann G, Wiskott L, and Gage FH (2004) Functional significance of adult neurogenesis. *Curr Opin Neurobiol* 14, 186–191
4. Morshead CM, Reynolds BA, Craig CG, McBurney MW, Staines WA, Morassutti D, et al (1994) Neural stem cells in the adult mammalian forebrain: a relatively quiescent subpopulation of subependymal cells. *Neuron* 13, 1071–1082
5. Palmer TD, Takahashi J, and Gage FH (1997) The adult rat hippocampus contains primordial neural stem cells. *Mol Cell Neurosci* 8, 389–404
6. Palmer TD, Markakis EA, Willhoite AR, Safar F, and Gage FH (1999) Fibroblast growth factor-2 activates a latent neurogenic program in neural stem cells from diverse regions of the adult CNS. *J Neurosci* 19, 8487–8497
7. Lie DC, Dziewczapolski G, Willhoite AR, Kaspar BK, Shults CW, and Gage FH (2002) The adult substantia nigra contains progenitor cells with neurogenic potential. *J Neurosci* 22, 6639–6649
8. Weiss S, Dunne C, Hewson J, Wohl C, Wheatley M, Peterson AC, et al (1996) Multipotent CNS stem cells are present in the adult mammalian spinal cord and ventricular neuroaxis. *J Neurosci* 16, 7599–7609
9. Klein C, Butt SJ, Machold RP, Johnson JE, and Fishell G (2005) Cerebellum- and forebrain-derived stem cells possess intrinsic regional character. *Development* 132, 4497–4508
10. Altman J, and Bayer SA (1997) Development of the cerebellar system: in relation to its evolution, structure, and functions. CRC Press, Boca Raton, Florida, USA
11. Fujita S, Shimada M, and Nakamura T (1966) H^3-Thymidine autoradiographic studies on the cell proliferation and differentiation in the external and internal granular layers of the mouse cerebellum. *J Comp Neurol* 128, 191–208
12. Altman J (1969) Autoradiographic and histological studies of postnatal neurogenesis. III. Dating the time of production and onset of differentiation of cerebellar microneurons in rats. *J Comp Neurol* 137, 433–458
13. Rakic P (1971) Neuron-glia relationship during cell migration in developing cerebellar cortex: A Golgi and electron microscopic study in *Macacus* Rhesus. *J Comp Neurol* 141, 283–312
14. Abraham H, Tornoczky T, Kosztolanyi G, and Seress L (2001) Cell formation in the cortical layers of the developing human cerebellum. *Int J Dev Neurosci* 19, 53–62
15. Lee A, Kessler JD, Read TA, Kaiser C, Corbeil D, Huttner WB, et al (2005) Isolation of neural stem cells from the postnatal cerebellum. *Nat Neurosci* 8, 723–729
16. Zusso M, Debetto P, Guidolin D, Barbierato M, Manev H, and Giusti P (2008) Fluoxetine-induced proliferation and differentiation of neural progenitor cells isolated from rat postnatal cerebellum. *Biochem Pharmacol* 76, 391–403
17. Sottile V, Li M, and Scotting PJ (2006) Stem cell marker expression in the Bergmann glia population of the adult mouse brain. *Brain Res* 1099, 8–17

Chapter 5

Culture of Rodent Cortical and Hippocampal Neurons

Laura Facci and Stephen D. Skaper

Abstract

Neurons cultured from rodent central nervous system tissues represent an important tool in the study of neurodegenerative disease mechanisms and neuroregenerative processes, including the survival- and axon growth–promoting properties of neurotrophic factors. This chapter presents a detailed protocol for the preparation of rat and mouse cortical and hippocampal neuron cell cultures using either embryonic or postnatal tissue with enzymatic digestion.

Key words: Cortex, Hippocampus, Neurons, Cell culture, Rat, Mouse, Embryonic, Newborn

1. Introduction

The study of central nervous system diseases has been greatly facilitated by the use of ex vivo models which are based on the culture of neurons from discrete brain areas, for example, cortex, hippocampus, and striatum. Procedures have been developed to allow for the survival and development of cultured central neurons largely free of unwanted glial cell elements, in particular the now widely used B27/Neurobasal formulation (1). Such cultures allow one to investigate in detail the molecular events involved in amyloid β-peptide toxicity, excitatory amino acid toxicity, oxidative stress, the survival-promoting effects of growth factors and cell adhesion molecule mimetics, kinase signaling pathways, and electrophysiological behaviors (2–13). The following chapter describes a protocol for the culture of cortical and hippocampal neurons from embryonic and newborn rodent (rat, mouse) tissues, based on the B27/Neurobasal serum-free medium.

Stephen D. Skaper (ed.), *Neurotrophic Factors: Methods and Protocols*, Methods in Molecular Biology, vol. 846, DOI 10.1007/978-1-61779-536-7_5, © Springer Science+Business Media, LLC 2012

2. Materials

2.1. Equipment and Labware

1. Stereo dissecting microscope (backlighting of stage is preferred) with fiber optic light source
2. Laminar flow cabinet for dissections
3. Laminar flow biological safety cabinet (CL2)
4. Humidified, water-jacketed culture incubator at 37°C and 5% CO_2/95% air
5. Water bath set at 37°C
6. Dissecting tools (Fine Science Tools)
7. Bench centrifuge to accommodate 15- and 50-mL tubes
8. 15- and 50-mL Polypropylene plastic centrifuge tubes (sterile)
9. 10-cm ∅ Sterile tissue culture dishes
10. 0.22-μm Filters (Millipore)
11. Neubauer hemocytometer (Fisher Scientific)
12. 96-Well poly-D-lysine-treated tissue culture plates (BD Biosciences)

2.2. Reagents

1. Hank's balanced salt solution, without $CaCl_2$ and $MgCl_2$, sterile (HBSS) (Invitrogen)
2. Neurobasal medium (Invitrogen)
3. Neurobasal-A medium (Invitrogen)
4. B27 supplement, 50×, with antioxidants (Invitrogen)
5. B27 supplement, 50×, without antioxidants (Invitrogen)
6. Fetal calf serum (FCS) (Invitrogen) (see Note 1)
7. L-Glutamine (200 mM stock), sterile, for cell culture (Invitrogen)
8. Sodium pyruvate (100 mM stock), sterile, for cell culture (Sigma)
9. Penicillin/streptomycin, 10,000 U/mL penicillin + 10,000 μg/mL streptomycin (100× stock), sterile, for cell culture (Invitrogen)
10. $MgCl_2$, 1 M, sterile (Sigma)
11. KCl, 3 M, sterile (Sigma)
12. HEPES, 1 M, sterile (Sigma)
13. Cytosine β-D-arabinofuranoside
14. Papain digestion kit (Worthington Biochemicals)

2.3. Culture Media and Other Solutions

1. *HBSS for dissection*. To 490 mL of calcium- and magnesium-free HBSS, add 2.25 g of D-glucose, 5 mL of 100-mM sodium pyruvate solution, and 5 mL of 1-M HEPES solution. Filter-sterilize.

2. *Embryonic neurons.* Neurobasal medium containing B27 supplement (1×), 2 mM L-glutamine, 100 U/mL penicillin, and 100 μg/mL streptomycin: To 500 mL Neurobasal medium, add 10 mL 50× B27 supplement, 5 mL 200-mM glutamine, and 5 mL 10,000-U/mL penicillin–10,000 μg/mL streptomycin. Complete medium can be stored at +4°C for up to 2 months.
3. *Neonatal neurons.* Neurobasal-A medium containing B27 supplement (1×), 1 mM sodium pyruvate, 2 mM L-glutamine, 25 mM KCl, 5 mM $MgCl_2$, penicillin (100 U/mL), and streptomycin (100 μg/mL): To 500 mL Neurobasal-A medium, add 10 mL 50× B27 supplement, 5 mL 100-mM sodium pyruvate, 5 mL 200-mM glutamine, 2 mL 1-M $MgCl_2$, 3.3 mL 3-M KCl, and 5 mL 10,000-U/mL penicillin–10,000 μg/mL streptomycin. Complete medium can be stored at +4°C for up to 2 months.
4. *Cytosine β-D-arabinofuranoside (araC) stock solution (10 mM).* Prepare araC as a 10-mM stock solution in phosphate buffered saline, filter-sterilize, and store aliquots at −20°C (good for up to 6 months). Once thawed, the aliquot tube can be kept at 4°C for 1 month (see Note 2).

3. Methods

3.1. Dissection of Rat Embryonic Cortex and Hippocampus

1. Rat cultures are routinely performed using animals at 18 days of gestation. Euthanize the pregnant rat (CD strain, Sprague Dawley) following appropriate national and/or institutional guidelines for animal sacrifice (see Note 3).
2. Place the animal dorsal side down and wipe thoroughly with a tissue soaked in 70% ethanol. Make an incision through the skin along the midventral line from the sternum to the pubis.
3. Make an incision through the belly wall muscles. While pushing away the muscle, pick up one of the uterine horns with forceps and cut the attachment of the uterine horns to the dorsal body wall.
4. Place the uterine horns in a 10-cm ∅ Petri dish containing cold HBSS for dissection.
5. Remove the second uterine horn in a similar manner.
6. Remove the embryos from the uterine segments and place in a new Petri dish containing ice-cold HBSS for dissection.
7. Decapitate embryos and store heads in HBSS (for dissection) on ice until/during dissection.
8. Remove brain from head by cutting the scalp along the midbrain line with fine scissors.

9. Scoop the brain out using a small spatula.
10. Place the brain in a 10-cm ∅ Petri dish containing fresh, ice-cold HBSS for dissection.
11. Remove and discard the cerebellum, brainstem, and optic nerve stumps.
12. Separate the two cerebral hemispheres and remove and discard the meninges from the cortex (see Note 4).
13. Using angled forceps, scoop the white matter from the underside of the cortex and the striatum which lies below this, in direct contact with the cortex.
14. The hippocampus, which appears as a banana-shaped area, is removed and transferred to a 15-mL centrifuge tube containing HBSS for dissection; place tube on ice (see Note 5).
15. Transfer the cortices to a 10-cm ∅ Petri dish containing fresh, ice-cold HBSS for dissection. Using a fine forceps, gently pince the tissue in thirds. Use a Pasteur pipette to transfer the pieces to a conical 15-mL sterile polypropylene tube.
16. Store on ice until all the brains have been processed.

3.2. Papain Digestion of Dissected Cortical and Hippocampal Tissue and Preparation of a Single Cell Suspension

1. The papain dissociation kit contains all needed reagents mentioned below except HEPES.
2. Add 2 mL of 1-M HEPES buffer to 100 mL of EBSS (vial 1 in kit). This will still be referred to as EBSS.
3. Solution C: Add 32 mL of EBSS to the albumin–ovomucoid inhibitor mixture (vial 4 in kit, containing 10 mg ovomucoid and 10 mg albumin/mL) and allow to dissolve.
4. Solution A: Add 5 mL of EBSS to the papain vial (vial 2 in kit, containing 20 U of papain/mL in 1 mM L-cysteine with 0.5 mM EDTA).
5. Solution B: Add 500 μL of EBSS to the DNase vial (vial 3 in kit, giving 2,000 U DNAse/mL) and mix gently.
6. Solution D: To solution A, add 250 μL of solution B.
7. Remove excess HBSS from the cortical tissue and add 5 mL of solution D.
8. Incubate at 37°C for 15 min with occasional gentle agitation.
9. Vigorously triturate tissue with a long (9-in.) Pasteur pipette until the cell suspension is devoid of visible clumps.
10. Although not usually necessary, small clumps of tissue remaining can be removed by passing the suspension over a 70-μm mesh filter (Falcon-BD Biosciences), having first inserted the strainer into the mouth of a 50-mL conical centrifuge tube.
11. Centrifuge the cell suspension (or filtrate) at 200 × *g* for 5 min.

12. While the cells are centrifuging, in a 15-mL tube, mix 2.7 mL EBSS with 0.3 mL solution C and add 0.15 mL solution B ("EBSS mix").
13. Place 5 mL solution C in a second 15-mL conical centrifuge tube.
14. Following centrifugation, discard the supernatant and resuspend the cell pellet in the above "EBSS mix."
15. Using a 5-mL pipette, carefully layer the cell suspension on top of the 5 mL ovomucoid solution.
16. Centrifuge the ovomucoid gradient at 150×*g* for 5 min.
17. Discard the supernatant and resuspend the cell pellet as required in Neurobasal medium for embryonic or neonatal cells.
18. Optimal cell attachment and survival can be obtained by precoating the poly-D-lysine culture surface with culture medium containing 10% (v/v) FCS. It is sufficient to expose the surface to FCS-containing medium for 1 h at 37°C. The FCS-containing medium should be removed just prior to cell plating.
19. Count cells (see Chap. 3) and dilute as required. For a 96-well plate, recommended density is 35,000 cells/well in 0.1 mL medium.
20. Average yields are $5–6 \times 10^6$ cells and $6–8 \times 10^5$ cells/rat embryonic day 18 cortex and hippocampus, respectively (see Note 6).
21. In our experience, such cultures can be maintained for up to 1 month (see Note 7).

3.3. Preparation of Newborn (Rat, Mouse) Cortical Neuron Cell Cultures

1. Use newborn pups (both sexes) within 24 h of birth. It is our experience that neuron cell survival diminishes markedly if the tissue is older than 1 day postnatal.
2. Euthanize the pups strictly following appropriate institution ally approved guidelines.
3. The dissection differs in some respects from that used for embryos. Cerebral hemispheres are dissected out following standard techniques and anatomical landmarks.
4. Following sacrifice and decapitation, place heads on a disinfectant wipe and spray with 70% ethanol.
5. Make an incision under the skin along the midline of the dorsal surface of the head and peel back the skin.
6. Make a second incision close to the midline along the exposed dorsal surface (the cut should be along the complete anteroposterior axis of the lateral ventricles).
7. Splay open the hemisphere along the cut surface and continue cut along the dorsal surface of the hemisphere.

8. Remove the cerebral hemispheres (minus cerebellum) and peel off meninges.
9. Place the hemispheres into a separate dish of cold HBSS for dissection and repeat for the other hemisphere and remaining pups.
10. The hippocampus and striatum are dissected away but can be saved if needed.
11. Papain dissociation of the tissue, cell counting, and plating are carried out as described earlier for embryonic tissue, *except that Neurobasal-A medium is used.*
12. The following day, add an equal volume of plating medium containing 20-μM araC (10 μM final; diluted 1:500 from the 10-mM stock). araC is added to retard growth of nonneuronal cell elements, especially astrocytes. The period of gliogenesis occurs around this time developmentally, which is why glial contamination is less of an issue when culturing cells from embryonic tissue (and hence araC is not used).
13. These cultures are generally used within 7–8 days and do not survive as long as those derived from embryonic tissues.

4. Notes

1. Heat inactivation of FCS is recommended to destroy heat-labile complement. Thaw the bottle of serum in advance using a 37°C water bath. Next, immerse the serum bottle in the water bath after reequilibrating to 56°C and leave for 30 min. Swirl the bottle occasionally to ensure proper mixing. Allow the serum to cool to room temperature, aliquot into 50-mL tubes, and store at −20°C.
2. Cytosine arabinoside is an antimitotic and potential carcinogen. Usage should be restricted to fume cupboards and biological safety cabinets. Avoid weighing out powder, or if not possible, then weigh out in a suitable closed environment. Culture media can be disposed of by normal routes. Stock/concentrated solutions should be enclosed within a leak-proof container and put into a burn bin for disposal.
3. It is preferable to flame-sterilize dissecting instruments by dipping in a solution of 70% ethanol and passing through a flame. However, if using a horizontal laminar flow cabinet for sterile work, it is possible with care to carry out the dissection with instruments swiped using a disinfectant wipe (e.g., Clinet 101 from Interscience, although this type of product can be obtained from any supplier of lab products).

4. Removal of meninges is a critical step, as their inclusion can lead to cultures contaminated with nonneuronal cell elements. We find that immobilizing the hemisphere with one pair of fine forceps while using a second pair of forceps to peel away the meninges works well. In particular, it is suggested to start at one end rather than in the middle of the tissue.
5. The hippocampal area is best visualized using a stereo dissecting microscope with illumination from below. This permits the hippocampal area to be easily distinguished from the cortical area due to differences in tissue contrast, with the hippocampus appearing darker.
6. An analogous procedure can be used to prepare cortical and hippocampal cell cultures from mouse embryonic tissue. However, optimal cell survival is achieved with tissue derived from embryonic day 15 fetuses. This is because mouse central nervous system development is precocious with respect to rat.
7. When attempting to maintain cultures for more than 10 days, the following is recommended: Evaporation can be an issue. If using 96-well plates, it is best to avoid plating cells in the outermost wells, that is, leave a "frame" in which sterile water is added. This will minimize volume loss. If larger wells are being used, for example, 24-, 12-, or 6-well plates, periodically check medium volume in the wells and top up with sterile tissue culture grade water if needed. Lastly, medium should be replenished weekly, by removing one half the volume and adding back a like volume of fresh plating medium.

Acknowledgment

L. Facci was supported by Fondazione CARIPARO "Progetto Dottorati di Ricerca" Anno 2009.

References

1. Brewer GJ (1995) Serum-free B27/neurobasal medium supports differentiated growth of neurons from the striatum, substantia nigra, septum, cerebral cortex, cerebellum, and dentate gyrus. J Neurosci Res 42, 674–683
2. Hussain I, Harrison DC, Hawkins J, Chapman T, Marshall I., Facci L, et al (2011) TASTPM mice expressing APP and PS-1 mutant transgenes are sensitive to γ-secretase modulation and Aβ42 lowering by GSM-10h. Neurodegenerative Dis 8, 15–24
3. Massa SM, Yang T, Xie Y, Shi J, Bilgen M, Joyce JN, et al (2010) Small molecule BDNF mimetics activate TrkB signaling and prevent neuronal degeneration in rodents. J Clin Invest 120, 1774–1785. doi: 10.1172/JCI41356
4. Edwards D, Das M, Molnar P, and Hickman JJ (2010) Addition of glutamate to serum-free culture promotes recovery of electrical activity in adult hippocampal neurons in vitro. J Neurosci Methods 190, 155–163
5. Baptista MS, Melo CV, Armelão M, Herrmann D, Pimentel DO, Leal G, et al (2010) Role of the proteasome in excitotoxicity-induced cleavage of glutamic acid decarboxylase in cultured hippocampal neurons. PLoS One 5, e10139

6. Franceschini D, Guisti P and Skaper SD (2006) MEK inhibition exacerbates ischemic calcium imbalance and neuronal cell death in rat cortical cultures. Eur J Pharmacol 553, 18–27
7. Skaper SD, Facci L, Williams G, Williams E-J, Walsh FS, and Doherty P (2004) A dimeric version of the short N-cadherin binding motif HAVDI promotes neuronal cell survival by activating an N-cadherin/fibroblast growth factor receptor signalling cascade. Mol Cell Neurosci 26, 17–23
8. Hughes JP, Staton PC, Wilkinson MG, Strijbos PJLM, Skaper SD, Arthur JSC, et al (2003) Mitogen and stress response kinase-1 (MSK1) mediates excitotoxic induced death of hippocampal neurones. J Neurochem 86, 25–32
9. Skaper SD, Facci L, Kee WJ, and Strijbos PJLM (2001) Potentiation by histamine of synaptically-mediated excitotoxicity in cultured hippocampal neurons: a possible role for mast cells. J Neurochem 76, 47–55
10. Chapman GA, Moores K, Harrison D, Campbell CA, Stewart BR, and Strijbos, PJ (2000) Fractalkine cleavage from neuronal membranes represents an acute event in the inflammatory response to excitotoxic brain damage. J Neurosci 20, RC87
11. Skaper SD, Ancona B, Facci L, Franceschini D, and Giusti P (1998) Melatonin prevents the delayed death of hippocampal neurons induced by enhanced excitatory neurotransmission and the nitridergic pathway. FASEB J 12, 725–731
12. Skaper SD, Leon A, and Facci L (1991) Death of cultured hippocampal pyramidal neurons induced by pathological activation of N-methyl-D-aspartate receptors is reduced by monosialogangliosides. J Pharmacol Exp Ther 259, 452–457
13. Choi DW (1987) Ionic dependence of glutamate toxicity. J Neurosci 7, 369–379

Chapter 6

Amyloid β-Peptide Neurotoxicity Assay Using Cultured Rat Cortical Neurons

Laura Facci and Stephen D. Skaper

Abstract

Neurons cultured from rodent central nervous system tissues represent an important tool in the study of neurodegenerative disease mechanisms and neuroregenerative processes, including the survival- and axon growth-promoting properties of neurotrophic factors. This chapter presents a detailed protocol for the preparation of rat and mouse cortical and hippocampal neuron cell cultures, using either embryonic or postnatal tissue.

Key words: Amyloid, Cortical neurons, Neurotoxicity, Lactate dehydrogenase assay, Luminescence assay, Cell culture, Rat, Embryonic

1. Introduction

A major pathological hallmark of Alzheimer's disease is the presence of extracellular senile plaques and intracellular neurofibrillary triangles in the brain (1, 2). The main protein components of senile plaques are amyloid-β (Aβ) peptides predominantly of 40 or 42 amino acids in length (Aβ40, Aβ42) which are generated following sequential cleavage of the amyloid precursor protein by β-site amyloid precursor protein cleaving enzyme 1 (β-secretase) and γ-secretase (3). A number of findings indicate the longer Aβ42 peptide to be the key pathogenic species in Alzheimer's disease. Aβ42 has a greater tendency to aggregate than Aβ40 and acts as the seed for Aβ deposition in the brain (4, 5). Aβ42 oligomers are neurotoxic to cell lines and neurons and can inhibit long-term potentiation in hippocampal slices (6, 7). In addition, a number of mutations in presenilin-1 that cause early onset familial Alzheimer's disease have been shown to specifically elevate production of Aβ42 (8).

Stephen D. Skaper (ed.), *Neurotrophic Factors: Methods and Protocols*, Methods in Molecular Biology, vol. 846, DOI 10.1007/978-1-61779-536-7_6, © Springer Science+Business Media, LLC 2012

Ex vivo models utilizing rodent-derived central nervous system neurons, in particular rat and mouse cortical and hippocampal neurons, have found widespread application in identifying small-molecule nonpeptide compounds with neuroprotective activity against Aβ42-induced death (9–13), as well as peptidomimetics of neurotrophic factors (14) and neurotrophins themselves. Compounds with this protective action would be potentially disease-modifying and thus highly desirable as therapeutics for Alzheimer's disease. This chapter describes a protocol for quantifying Aβ42-induced death (15, 16) of cultured rat embryonic cortical neurons, based on the measurement of either lactate dehydrogenase (LDH) release into the culture medium (colorimetric assay) or the intracellular content of ATP (luminescence assay).

2. Materials

2.1. Equipment and Labware

1. Laminar flow biological safety cabinet (CL2).
2. Humidified, water-jacketed culture incubator at 37°C and 5% CO_2/95% air.
3. Water bath set at 37°C.
4. 96-well poly-D-lysine treated tissue culture plates (BD Biosciences).
5. Microplate reader capable of reading absorbance at 490 nm.
6. Wallac 1450 MICROBETA PLUS Liquid Scintillation Counter (TopCount) or similar instrument for luminescence readout.
7. 3-mL glass weighing vials with cap.
8. Bench centrifuge to accommodate 15- and 50-mL tubes.
9. 15- and 50-mL polypropylene plastic centrifuge tubes, sterile.

2.2. Reagents

1. Hank's balanced salt solution, with $CaCl_2$ and $MgCl_2$, sterile (HBSS) (Invitrogen).
2. Neurobasal medium (Invitrogen).
3. B27 supplement, 50×, with antioxidants (Invitrogen).
4. B27 supplement, 50×, without antioxidants (Invitrogen).
5. Fetal calf serum (FCS) (Invitrogen) (see Note 1).
6. L-Glutamine (200 mM stock), sterile, for cell culture (Invitrogen).
7. Sodium pyruvate (100 mM stock), sterile, for cell culture (Sigma).
8. Penicillin/streptomycin, 10,000 U/mL penicillin + 10,000 μg/mL streptomycin (100× stock), sterile, for cell culture (Invitrogen).

9. HEPES, 1 M, sterile (Sigma).
10. Human β-amyloid peptide 1–42 (Aβ42) (California Peptide Research).
11. Hexafluoroisopropanol (HFIP) (Sigma).
12. Dimethylsulfoxide (DMSO) for cell culture (Sigma).
13. CytoTox 96® nonradioactive cytotoxicity assay kit (Promega).
14. CellTiter-Glo® luminescent cell viability assay kit (Promega).
15. ATP standard solution, supplied as 100 mM stock (BioVision).

2.3. Culture Medium

1. Neurobasal medium containing B27 supplement (1×), 2 mM L-glutamine, 100 U/mL penicillin, and 100 μg/mL streptomycin: To 500 mL Neurobasal medium, add 10 mL 50× B27 supplement, 5 mL 200 mM glutamine, and 5 mL 5,000 U/mL penicillin to 5,000 μg/mL streptomycin. Complete medium can be stored at +4°C for up to 2 months.

2.4. Aβ42 Solution

1. Weigh the desired amount of Aβ42 into a 3-mL glass weighing vial (see Notes 2 and 3).
2. In a CL2 cabinet, add the required volume of HFIP to the Aβ42 (use 400 μL HFIP per mg of Aβ42) (see Note 4).
3. Briefly mix and leave at room temperature for 10 min to allow peptide to fully dissolve (the solution should be clear at this stage).
4. Evaporate the HFIP under a gentle stream of air, for example, by attaching a Pasteur pipette to the tube delivering the air stream. *Perform this operation under a fume cupboard.*
5. At this point, a film of peptide should be visible around the base of the vial.
6. Resuspend the dried Aβ42 peptide in tissue culture grade DMSO 1 mg peptide in 44 μL DMSO, ensuring that the DMSO is pipetted around the base of the eppendorf tip where the peptide film is visible.
7. This produces a solution of 5-mM peptide, to which is now added 0.1% trifluoroacetic acid (73 μL/mg peptide).
8. The Aβ42 peptide solution is now diluted directly into pre-warmed (37°C) culture medium, to give a concentration of 80 μM (taking into account the volume contributed by DMSO and trifluoroacetic acid).
9. The cortical neuron cell cultures receive one-fourth volume of the Aβ42 solution (80 μM) in medium, to yield a final peptide concentration of 20 μM in the cultures.

3. Methods

3.1. Treatment of Neuronal Cell Cultures with Test Agent(s)

1. Cortical (or hippocampal) neurons prepared from embryonic rat (or mouse) are plated in 96-well poly-D-lysine-coated wells, 30,000 cells per well in 0.1 mL of Neurobasal/B27 plating medium (see Chapter 5 for cell preparation protocol).
2. *Day 2* in culture: Add to each well 0.1 mL of Neurobasal medium supplemented with antioxidant-free B27. The medium formulation is identical to that described in Subheading 2.3, except that B27 with antioxidants is substituted for by the antioxidant-free version.
3. *Day 6* in culture: Remove medium from the wells and add 50 μL Neurobasal/B27 medium (antioxidant-free).
4. Next, add to the well 25 μL of the same medium containing test compound at 4× the desired final concentration (0.1% DMSO contributed by compound).
5. Incubate cells with compound for 30 min (or longer, as appropriate to target) at 37°C.
6. Next, add to each well (except controls) 25 μL of Aβ42 solution (80 μM) to give a final concentration of 20 μM (0.4% DMSO contribution). Total final concentration of DMSO in all wells from test compound + amyloid is 0.5% (v/v).
7. Controls receive the same solutions but without Aβ42 (DMSO is maintained at 0.5%).
8. Return the plate(s) to the 5% CO_2 incubator at 37°C for 3 days.

3.2. Lactate Dehydrogenase Release Assay Protocol (CytoTox 96® Nonradioactive Cytotoxicity Assay)

The assay is composed of two component part numbers—G1781 and G1782—and is sufficient to process 5-×96-well plates.

Contents

G1781

- G179A: substrate mix (lyophilized diaphorase, lactate, and NAD^+), five vials.
- G180A: assay buffer (Tris-buffered tetrazolium dye (INT-chloride) and Triton X-100), 60 mL.

G1782

- G181A: LDH-positive control (slurry in ammonium sulfate, pH 6.0), 25 μL.
- G182A: lysis solution (10×), 3 mL.
- G183A: stop solution, 65 mL.

The assay procedure is as follows:

1. Remove 12 mL of assay buffer and promptly store the unused portion at −20°C.

2. Add the 12 mL of assay buffer (room temperature) to a bottle of substrate mix. Once resuspended, protect the solution from direct light and use immediately.
3. Dilute the LDH-positive control (bovine heart LDH, 800 U/mL) according to the scheme shown in the table below:

#	LDH (mU/mL)	μL	Medium (μL)
B	160	2 μL (*)	10,000
C	112	2,000 (B)	857
D	80	2,000 (C)	800
E	64	2,000 (D)	500
F	48	2,000 (E)	667
G	32	2,000 (F)	1,000
H	16	2,000 (G)	2,000
I	8	2,000 (H)	2,000
0	0	0	1,000

Add 50 μL of LDH standard-curve sample according to the template illustrated below. The stock solution should be prepared fresh each time.

	1	2	3	4	5	6	7	8	9	10	11	12
A	0	0	0	I	I	I	H	H	H	G	G	G
B												
C												
D												
E												
F												
G												
H	F	F	F	E	E	E	D	D	D	C	C	C

4. Collect from each well the full amount (100 μL) of culture medium and set aside.
5. Transfer 50 μL of sample medium to a flat bottom clear microtiter clear 96-well plate.
6. Lyse cells by adding to each well 100 μL culture medium containing 1% lysis solution (provided in the kit). Leave in the 37°C incubator for at least 10 min (see Note 5).
7. Transfer 10 μL of cell lysate to the flat-bottom 96-well plate. Add 40 μL culture medium to each of these wells.
8. Add 50 μL reconstituted substrate to all wells.

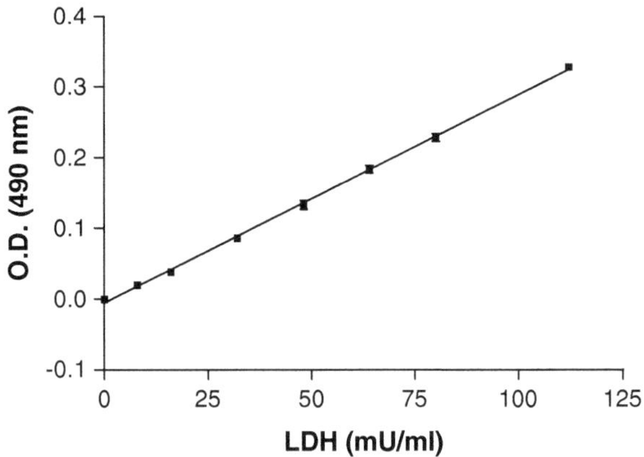

Fig. 1. Sample LDH standard curve.

9. Cover the plate with aluminum foil to protect from light.
10. Incubate at room temperature (20–23°C) for 30 min.
11. Add 50 μL of stop solution to each well (final volume –150 μL).
12. Read samples in a plate reader at A_{490} within 1 h of stopping the assay (see Note 6).
13. Calculate LDH values from the standard curve (see Fig. 1 for a "representative" example) (see Note 7).

3.3. ATP Cell Content Assay Protocol (CellTiter-Glo® Luminescent Cell Viability Assay)

1. Allow CellTiter-Glo® substrate and buffer to come to room temperature (see Notes 8 and 9).
2. Starting with the 100 mM ATP stock solution, prepare a working stock solution of 1 mM ATP using Neurobasal medium with B27 supplement (no antioxidants).
3. Prepare serial dilutions of ATP in the above culture medium, covering the range 0.003–2 μM. A 96-well template may be used, analogous to that described above for LDH assay but modified to accommodate the concentration range of ATP standard dilutions.
4. Add 100 μL per well of ATP standards (using empty wells but in the same 96-well plate with cells).
5. Add to each well (cells or ATP standards) 100 μL of CellTiter-Glo solution.
6. Allow at least 10 min for cell to lyse and come to equilibrium.
7. Read plate in a Wallac 1450 MICROBETA PLUS liquid scintillation counter (TopCount) or similar instrument for luminescence.
8. Calculate ATP values from the standard curve (see Fig. 2 for a "representative" example).

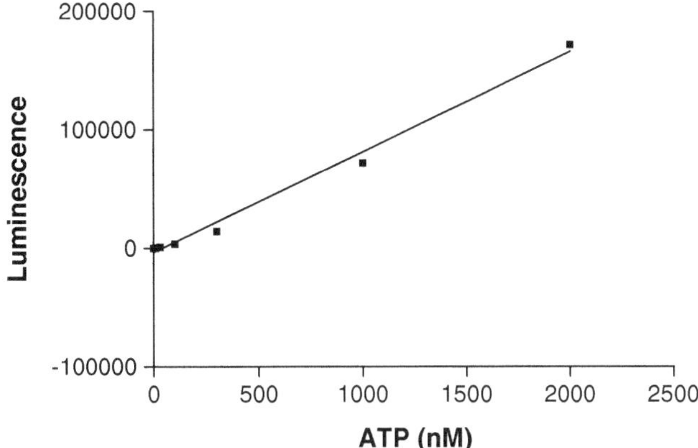

Fig. 2. Sample ATP standard curve.

4. Notes

1. Heat inactivation of fetal calf serum is recommended to destroy heat-labile complement. Thaw the bottle of serum in advance using a 37°C water bath. Next, immerse the serum bottle in the water bath after reequilibrating to 56°C and leave for 30 min. Swirl the bottle occasionally to ensure proper mixing. Allow the serum to cool to room temperature, aliquot into 50-mL tubes, and store at −20°C.
2. We have tested Aβ42 from different suppliers and find that California Peptide Research gives the most consistent results from batch to batch, and is easiest to dissolve in HFIP compared to Aβ42 from other sources (e.g., Sigma, Calbiochem, Bachem). If one intends to perform these assays on a routine basis (e.g., structure-activity screening of compounds rather than a "one off") we suggest requesting a 1 mg sample from the company to test for neurotoxicity. If satisfactory, a large amount of that peptide batch should be purchased and set aside.
3. Aβ42 (at least that from California Peptide) is supplied lyophilized and is rather hygroscopic. It should be stored at −20°C, desiccated, and under vacuum. The parent bottle should be allowed to warm to room temperature before opening.
4. We recommend purchasing HFIP in small quantities, for example, as 25-mL bottles from Sigma-Aldrich. Once opened, the bottle should be used for no more than 2 weeks and then discarded if not consumed. The HFIP solvent absorbs moisture from the air, and this will interfere with the proper solubilization of Aβ42.

5. If sufficient lysis solution is not available, you can use 0.5% Triton X-100.
6. The presence of bubbles, especially large ones, may cause spurious absorbance readings. These can be eliminated by using a hypodermic needle to pierce and break the bubble. For smaller bubbles, one method which works quite well is to first heat the needle tip, then touch the hot tip to the bubble(s).
7. A test compound which is cytotoxic per se can cause a false-positive result (indicating an apparent neuroprotective effect). This is especially true when the test compound causes a rapid release of LDH, which is then degraded over the 3-day incubation period. This leads to a false low value of LDH released and an apparent "neuroprotective" effect. To minimize the risk of such false positives, we strongly recommend measuring the LDH content of both the culture medium and in the remaining cells, and calculating LDH release as a percent of total (extra- and intracellular) LDH in the culture well.
8. For long-term storage, it is best to keep CellTiter-Glo® substrate and buffer at –20°C.
9. Reconstituted CellTiter-Glo® reagent can be stored at +4°C with approximately 5% reduction in luminescence intensity or 4 days at +4°C with approximately a 20% decrease in intensity. The reconstituted reagent may be frozen at –20°C for storage. Ten freeze-thaw cycles decrease the luminescence intensity by only 5–10%. The frozen reagent should be thawed in a 37°C water bath prior to assay.

Acknowledgments

L. Facci was supported by Fondazione CARIPARO Progetto Dottorati di Ricerca' Anno 2009.

References

1. Glenner GG, and Wong CW (1984) Alzheimer's disease: initial report of the purification and characterization of a novel cerebrovascular amyloid protein. *Biochem Biophys Res Comm* 120, 885–890
2. Goedert M, Wischik CM, Crowther RA, Walker JE, and Klug A (1988) Cloning and sequencing of the cDNA encoding a core protein of the paired helical filament of Alzheimer disease: identification as the microtubule-associated protein tau. *Proc Natl Acad Sci USA* 85, 4051–4055
3. Selkoe DJ (1994) Normal and abnormal biology of the beta-amyloid precursor protein. *Annu Rev Neurosci* 17, 489–517
4. Jarrett JT, Berger EP, and Lansbury PT Jr (1993) The carboxy terminus of the beta amyloid protein is critical for the seeding of amyloid formation: implications for the pathogenesis of Alzheimer's disease. *Biochemistry* 32, 4693–4697
5. McGowan E, Pickford F, Kim J, Onstead L, Eriksen J, Yu C, et al (2005) Aβ42 is essential for parenchymal and vascular amyloid deposition in mice. *Neuron* 47, 191–199

6. Dahlgren KN, Manelli AM, Stine Jr WB, Baker LK, Krafft GA, and LaDu MJ (2002) Oligomeric and fibrillar species of amyloid-β-peptides differentially affect neuronal viability. *J Biol Chem* 277, 32046–32053
7. Walsh DM, Klyubin I, Fadeeva JV, Cullen WK, Anwyl R, Wolfe MS, et al (2002) Naturally secreted oligomers of amyloid β protein potently inhibit hippocampal long-term potentiation in vivo. *Nature* 416, 535–539
8. Scheuner D, Eckman C, Jensen M, Song X, Citron M, Suzuki N, et al (1996) Secreted amyloid beta-protein similar to that in the senile plaques of Alzheimer's disease is increased in vivo by the presenilin 1 and 2 and APP mutations linked to familial Alzheimer's disease. *Nat Med* 2, 864–870
9. Bozyczko-Coyne D, O'Kane TM, Wu ZL, Dobrzanski P, Murthy S, Vaught JL, et al (2001) CEP-1347/KT-7515, an inhibitor of SAPK/JNK pathway activation, promotes survival and blocks multiple events associated with Abeta-induced cortical neuron apoptosis. *J Neurochem* 3, 849–863
10. Hussain I, Harrison DC, Hawkins J, Chapman T, Marshall I, Facci L, et al (2011) TASTPM mice expressing APP and PS-1 mutant transgenes are sensitive to γ-secretase modulation and Aβ42 lowering by GSM-10 h. *Neurodegener Dis* 8, 15–24
11. Yang T, Knowles JK, Lu Q, Zhang H, Arancio O, Moore LA, et al (2008) Small molecule, non-peptide $p75^{NTR}$ ligands inhibit Aβ-induced neurodegeneration and synaptic impairment. *PLoS One* 3, e3604
12. Liu F, Gong X, Zhang G, Marquis K, Reinhart P, and Andree TH (2005) The inhibition of glycogen synthase kinase 3β by a metabotropic glutamate receptor 5 mediated pathway confers neuroprotection to Aβ peptides. *J Neurochem* 95, 1363–1372
13. Chen J, Zhou Y, Mueller-Steiner S, Chen LF, Kwon H, Yi S, Mucke L, et al (2005) SIRT1 protects against microglia-dependent amyloid-β toxicity through inhibiting NF-κB signaling. *J Biol Chem* 280, 40364–40374
14. Massa SM, Yang T, Xie Y, Shi J, Bilgen M, Joyce JN, et al (2010) Small molecule BDNF mimetics activate TrkB signaling and prevent neuronal degeneration in rodents. *J Clin Invest* 120, 1774-1785. doi: 10.1172/JCI41356
15. Pym LJ, Buckingham SD, Tsetlin V, Boyd CA, and Sattelle DB (2007) The Aβ1-42M35C mutated amyloid peptide Aβ1-42 and the 25-35 fragment fail to mimic the subtype-specificity of actions on recombinant human nicotinic acetylcholine receptors (α7, α4β2, α3β4). *Neurosci Lett* 427, 28–33
16. Wogulis M, Wright S, Cunningham D, Chilcote T, Powell K, and Rydel RE (2005) Nucleation-dependent polymerization is an essential component of amyloid-mediated neuronal cell death. J Neurosci 25, 1071–1080

Chapter 7

Culture of Neonatal Rodent Microglia, Astrocytes, and Oligodendrocytes from Cortex and Spinal Cord

Stephen D. Skaper, Carla Argentini, and Massimo Barbierato

Abstract

The protocol described in this chapter covers the preparation and culture of enriched populations of microglia, astrocytes, and oligodendrocytes from the cortex and spinal cord of neonatal rat and mouse. The procedure is based on the enzymatic digestion of tissue, followed by the culture of a mixed glial cell population which is then utilized as the starting point for the isolation, via differential attachment, of the different cell types.

Key words: Microglia, Astrocytes, Oligodendrocytes, Cortex, Spinal cord, Rat, Cell culture

1. Introduction

Glia, which outnumber neurons by as much as 10 to 1 in some regions of the human brain, were at one time thought to function solely by providing physical support and housekeeping for the neurons. We now know that during brain development, glia guide migrating neurons to their destinations and instruct them to form synapses. In the adult brain, glia talk back to neurons, releasing neurotransmitters and other signals that regulate synaptic strength. The emerging realization is that glia may have key roles in central nervous system disorders from neuropathic pain and epilepsy to neurodegenerative disorders such as Alzheimer's—and may even contribute to schizophrenia, depression, and other psychiatric disorders. The brain's glia can be divided into three principle types: microglia (the brain's resident macrophages), astrocytes, and oligodendrocytes. Each glial cell type displays distinct properties and behaviors. For example, oligodendrocytes, the glia that form a fatty myelin sheath around the axons of neurons in the brain and spinal cord (1), are sensitive to injury from glutamate-induced oxidative

Stephen D. Skaper (ed.), *Neurotrophic Factors: Methods and Protocols*, Methods in Molecular Biology, vol. 846, DOI 10.1007/978-1-61779-536-7_7,

stress (2, 3) and excitotoxicity (4, 5) and are also capable of producing mature amyloid β-peptide (6). Microglia, depending on the context in which they occur, can provide both neurotrophic support (7) and neurotoxins under pathological conditions (8–10). Astrocytes, which are the most abundant glial cell type in the brain, are known to synthesize and release low (11) and high (12) molecular weight neurotrophic agents, as well as reactive nitrogen oxide species (13), glutamate, and pro-inflammatory cytokines (14). Cultured astrocytes have also proven useful for the study of cellular differentiation signals (15) and gene transcription (16). Clearly, enriched populations of these glial cell types are of considerable use in the study of nervous system development, injury, and therapeutic targets. The following chapter presents a protocol for the isolation and culture of rodent central nervous system glia.

2. Materials

2.1. Equipment and Labware

1. Stereo dissecting microscope (backlighting of stage is preferred) with fiber-optic light source.
2. Horizontal flow cabinet for dissections.
3. Laminar flow biological safety cabinet (CL2).
4. Humidified, water-jacketed culture incubator at 37°C and 5% CO_2/95% air.
5. Water bath set at 37°C.
6. Dissecting tools (Fine Science Tools (InterFocus Ltd) are very high quality for the price).
7. Bench centrifuge to accommodate 15- and 50-mL tubes.
8. Orbital shaker with enclosed, temperature controlled chamber.
9. 10-cm ∅ sterile tissue culture dishes (any supplier).
10. 15- and 50-mL polypropylene plastic centrifuge tubes (sterile).
11. Sterilin 10-cm ∅ sterile plastic petri dishes (Bibby-Sarstedt).
12. Tissue culture flasks (T-75; 75 cm^2) with canted neck and 0.2 μm vented cap (Corning) or without vented cap (BD Falcon).
13. Cell scrapers, 25-cm handle, 1.8-cm blade (BD Falcon).
14. 0.22-μm filters (Millipore Corporation).

2.2. Reagents

1. Phosphate-buffered saline (pH 7.4, sterile) (Invitrogen).
2. Dulbecco's modified Eagle's medium (DMEM) (+4.5 g/L glucose, L-glutamine, pyruvate) (Invitrogen).
3. N2 supplement (Invitrogen).
4. Trypsin inhibitor (Sigma).

5. DNAse (Type I) (Sigma).
6. L-15 medium (+L-glutamine, L-amino acids) (Invitrogen).
7. Trypsin inhibitor, type I from soybean (Sigma).
8. Papain (Worthington (Lorne)).
9. Bovine serum albumin, Cohen fraction V (Sigma).
10. T3 (3,3′,5-triiodo-L-thyronine, sodium salt), cell-culture-tested (Sigma).
11. T4 (L-thyroxine sodium salt pentahydrate), γ-irradiated, cell-culture-tested (Sigma).
12. Penicillin/streptomycin (100× stock), sterile, for cell culture (Invitrogen).
13. Gentamicin (50 mg/mL stock), sterile, for cell culture (Sigma).
14. Fetal calf serum (Invitrogen).
15. Poly-L-lysine (MW 70,000–150,000), sterile, for cell culture (Sigma P-6282).
16. L-Glutamine (200 mM stock), sterile, for cell culture (Invitrogen).
17. Sodium pyruvate (100 mM stock), sterile, for cell culture (Sigma).
18. HEPES (1 M), sterile for cell culture (Sigma).
19. L-Leucine methyl ester HCl (Sigma).

2.3. Culture Media and Other Solutions

1. *T3*: Dissolve 4 mg of T3 in 10 mL of 1 N NaOH (1,000× stock, 400 μg/mL), filter-sterilize, aliquot, and store at −20°C.
2. *T4*: Dissolve 4 mg of T4 in 10 mL of tissue culture grade water (1,000× stock, 400 μg/mL), filter-sterilize, aliquot, and store at −20°C.
3. *DNase*: Dissolve 40 mg of DNase in 10 mL of L-15 medium (100× stock, 4 mg/mL), filter-sterilize, aliquot, and store at −20°C.
4. *L-Cysteine*: Dissolve 24 mg of L-cysteine in 1 mL of L-15 medium (100× stock). Prepare fresh each time.
5. *Trypsin inhibitor*: Dissolve 100 mg of trypsin inhibitor in 1 mL (100× stock) of L-15 medium. Filter-sterilize, aliquot, and store at −20°C.
6. *Bovine serum albumin*: Dissolve 5 mg of bovine serum albumin in 1 mL (100× stock) of L-15 medium. Filter-sterilize, aliquot, and store at −20°C.
7. *0.15 M borate buffer pH 8.4*: Dissolve 28.6 g of sodium borate ($Na_2B_4O_7$, 10 H_2O) in 500 mL water (pH will be ~9.2). Adjust pH to 8.4 with 5 N HCl. Filter-sterilize and store at 4°C (up to 6 months).

8. *Plating medium/growth (maintenance) medium*: To a 500-mL bottle of DMEM, add 50 mL fetal calf serum (10% final), 5 mL stock penicillin/streptomycin (100 U/mL penicillin + 100 μg/mL streptomycin final), and 0.5 mL of gentamicin stock (50 μg/mL final).
9. *Sato medium (for oligodendrocytes)*: To 500 mL DMEM, add 5 mL N2 stock (1:100), 0.5 mL T3 stock (1:1,000), 0.5 mL T4 stock (1:1,000), 5 mL penicillin/streptomycin stock (1:100), and 2.5 mL fetal calf serum (0.5% final concentration). Store at +4°C for up to 2 months. Note: the defined medium supplement contains components which have been isolated from human sources (e.g., transferrin), and their hazard is not known.

2.4. Preparation of Substratum

1. Dissolve a 5-mL bottle of poly-L-lysine (MW 70,000–150,000) in 5 mL 0.15 M borate buffer (pH 8.4). This 1-mg/mL stock solution can be stored at +4°C for up to 2 months.
2. Dilute the 1-mg/mL stock solution of poly-L-lysine 1:100 (1 mL in 99 mL sterile water) and add 10 mL to each of the T-75 culture flasks. Rock the flask to cover the entire culture surface and place flask in the 5% CO_2 incubator (37°C) overnight.
3. The following day aspirate the poly-L-lysine coating solution and leave the flask(s) under the CL2 cabinet uncapped to dry. This will usually take several hours. Once dried, the poly-L-lysine-coated flasks can be stored at +4°C for 1 month.

3. Methods

3.1. Dissection of Rat Cortical Tissue

1. Rat cultures are routinely performed using animals at postnatal day 1–2. Following appropriate national and institutional guidelines for animal sacrifice, pregnant rats (CD strain, Sprague Dawley) are euthanized (see Note 1).
2. Dissect out cerebral hemispheres following standard techniques and anatomical landmarks. The procedure is applicable to either rat or mouse pups.
3. Make an incision under the skin along the midline of the dorsal surface and peel the skin back.
4. Make a second incision close to the midline along the exposed dorsal surface (the cut should be along the complete antero-posterior axis of the lateral ventricles).
5. Splay open the hemisphere along the cut surface and continue the cut along the dorsal surface of the hemisphere.

6. Remove the cerebral hemispheres (minus cerebellum and brain stem) and peel off meninges. The latter step is best accomplished by using two fine-tipped #5 Dumont forceps, using one pair to anchor the hemisphere while peeling off the meninges with the second pair. (see Note 2).
7. Place the dissected hemisphere into a separate 10-cm ∅ dish of cold L-15 medium.
8. Repeat this procedure for the other hemisphere and remaining pups.
9. The hippocampal area can be dissected away and retained or discarded, depending on need.

3.2. Enzymatic Digestion of Cortical Tissue and Preparation of Cell Suspension for Plating

1. Grip a straight-edge razor blade with a hemostat, dip in 70% ethanol, and pass through a flame to sterilize.
2. Mince the collected cortical tissue with the razor blade.
3. Collect the minced tissue with L-15 and transfer to a 15-mL centrifuge tube.
4. Centrifuge collected cortical tissue for 30 s (200 × *g*) to compact the minced tissue.
5. While the tissue is centrifuging, prepare the following: 3 mL of L-15 medium containing 140 μL stock papain solution (as purchased), 30 μL of 100× stock L-cysteine solution, and 30 μL of 100× DNase stock solution. Filter-sterilize before use.
6. Remove L-15 medium from the tube with cortical tissue pellet, and add the L-15 solution containing papain, DNase, and L-cysteine.
7. Incubate tissue for 60 min in a 37°C water bath with occasional swirling.
8. During this time, prepare the ovomucoid solution: to 3 mL of L-15 medium add 30 μL of 100× stock DNase solution, 30 μL of 100× stock bovine serum albumin solution, and 30 μL of 100× stock trypsin inhibitor solution.
9. Upon completion of the enzyme incubation step, remove the supernatant and replace with 1.5 mL ovomucoid solution (see Note 3).
10. Incubate for 2 min in a 37°C water bath.
11. Centrifuge at 250 × *g* for 3 min.
12. Remove the supernatant and add 1.5 mL fresh ovomucoid solution.
13. Triturate 20–25 times with a long (9-in.) cotton-plugged Pasteur pipette (see Note 4).
14. Add 4 mL plating medium and centrifuge at 200 × *g* for 5 min.

15. Using a 5-mL tissue culture pipette, resuspend the cell pellet in 3 mL plating medium. Add an additional amount of plating medium such that the final volume is equivalent to 2 mL × number of T-75 flasks to be used (e.g., 20 mL final for 10 flasks).
16. Add 18 mL plating medium to each flask, followed by 2 mL of cell suspension. Normally, we seed cultures at a ratio of 1.5 brains/flask (see Note 5).
17. The following day, replace the culture medium with 15 mL fresh plating (growth) medium.
18. Twice per week, replace one-half the volume of culture medium with an equal volume of fresh growth medium.
19. The mixed glial cell cultures are incubated for 10–14 days, after which time, confluence will have been reached and they can be taken for harvesting different cell populations (astrocytes, microglia, oligodendrocytes) (see Note 6).

3.3. Isolation of Oligodendrocyte Precursors and Microglia

1. Remove 3 mL medium from each T-75 flask; this should leave about 12 mL of medium.
2. As the culture medium is bicarbonate-buffered, to prevent pH excursions, close the flask cap (or cover with parafilm if the cap is of the filter type).
3. Place the flasks on an orbital shaker fitted with a temperature-controlled chamber. Flasks can be fastened to the shaker platform using double-sided adhesive tape, or taped directly to the platform using high-strength industrial ducting tape.
4. Shake the flasks for 2 h at 200 cycles/min (37°C).
5. Remove the culture medium (containing mainly microglia) from the flasks and transfer to 10-cm ∅ Sterilin plastic petri dishes. A volume of 20–25 mL/dish is best (see Notes 7 and 8).
6. Add 13 mL maintenance medium to the flasks and place on the orbital shaker overnight at 200 cycles/min.
7. Place the Sterilin dishes in the 5% CO_2 incubator (37°C) and leave for 45–60 min.
8. Aspirate the medium on the Sterilin plates and add 6 mL fresh maintenance medium.
9. Use a cell scraper to remove the attached microglia from the Sterilin dishes. Transfer the medium with cells to a 50-mL centrifuge tube. Rinse the dishes with 5 mL maintenance medium and transfer to the collection tube.
10. Resuspend the microglia cell pellet in maintenance medium and plate into poly-L-lysine-coated multiwall plates as dictated by the experimental design. Typically, we plate 10^5 cells in a 96-well plate (equivalent to 3.5×10^5 cells/cm^2). For cell counting

procedure, see Chap. 3. Microglia obtained are ≥97% pure. Cultures will maintain their viability for approximately 10 days.

11. Collect culture medium from the overnight shaking of flasks and transfer to Sterilin dishes as in step 5 above.
12. Place the Sterilin dishes in the 5% CO_2 incubator (37°C) and leave for 45–60 min.
13. Collect the medium from this last step (containing mainly non-adherent oligodendrocytes) and centrifuge at 200 × *g* for 5 min. The Sterilin dishes contain a limited number of microglia and can be discarded.
14. Resuspend the oligodendrocyte cell pellet in Sato medium, count, and dilute in Sato medium to the desired cell density for plating. Cultures prepared in this manner contain approximately 93% oligodendrocytes and 3% cells positive for the astrocytic marker glial fibrillary acidic protein. The remaining cells do not stain by any of these antibodies, including markers specific for microglia (7). The oligodendrocytes can be cultured for up to 10 days before loss of viability begins.
15. Oligodendrocytes are normally plated on a poly-D-lysine-coated tissue culture plastic surface: this substratum is prepared as described for coating with poly-L-lysine, except that the 1 mg/mL poly-D-lysine stock solution in borate buffer is diluted 1:20 for coating.

3.4. Isolation of an Enriched Population of Astrocytes

1. The T-75 flasks remaining after shaking to recover oligodendrocytes and microglia are used as a source of highly enriched astrocytes.
2. Aspirate the medium from the flask and rinse the cell monolayer with 10-mL sterile phosphate-buffered saline.
3. Add to the flask 3 mL of 0.25% trypsin/EDTA solution, rock the flask to spread the trypsin solution over the entire monolayer, and then remove by aspiration. This will leave a thin film of trypsin solution over the cells.
4. Incubate the flask(s) at 37°C for 10 min.
5. Tap the flask against the palm of the hand to dislodge cells and add 5 mL maintenance medium per flask; the fetal calf serum in the medium inactivates the trypsin.
6. Rinse the flask with the medium added, collect cells, and transfer to a 15- or 50-mL centrifuge tube, as needed.
7. Pellet cells by centrifugation at 200 × *g* for 5 min, re-suspend pellet in maintenance medium and count. The typical cell yield is $4–5 \times 10^6$ cells/T-75 flask. On this basis, dilute the cell suspension to $1–1.5 \times 10^6$ cells/mL for ease of counting. Astrocytes thus obtained are ≥95% pure.

8. If needed, dilute further the cell suspension for plating into poly-L-lysine-coated culture vessels are dictated by the experimental design. Typically, we plate 5×10^4 cells in a 96-well plate (equivalent to 1.75×10^5 cells/cm^2) (see Note 9).

3.5. Preparation of Spinal Cord Glia

1. The procedure is essentially as that described for cortical tissue, except for the dissection.
2. Sacrifice pups by decapitation.
3. Place the body face down on a paper towel. Spray 70% ethanol over the back.
4. Remove the back skin with surgical scissors to expose the vertebral column.
5. Entering from the anterior side of the body, make a cut through the midline as far as the tail.
6. Using a fine pair of surgical scissors trim back, at a 45° angle, the tissue on either side of the spinal canal (being careful not to disrupt the spinal cord itself).
7. Carefully strip off the spinal cord with a fine pair of forceps and transfer to a 10-cm Ø dish of cold L-15 medium.
8. Repeat this procedure for the remaining pups.
9. Peel off the meninges using fine tipped forceps, and transfer the tissue pieces to a 15-mL centrifuge tube. It is not necessary to mince the tissue as for cortex; the spinal cord is much smaller and will already have been fragmented to some extent in the course of removing the meninges.
10. Centrifuge the collected tissue for 30 s ($200 \times g$) to compact.
11. Remove L-15 medium from the tube and add the L-15 solution containing papain, DNase, and L-cysteine.
12. Incubate tissue for 60 min in a 37°C water bath with occasional swirling.
13. Proceed with tissue dissociation as for cortex, except the final 5-min centrifugation (Subheading 3.2, step 14). Doing so reduces the risk of losing material, which is limited in comparison to cortex.
14. Following dissociation, plate the spinal cord cell suspension in medium and distribute in poly-L-lysine-coated T-75 flasks, five spinal cords per flask.
15. For harvesting microglia, we find that spinal cord mixed glia cultures mature more slowly than do the cortical cultures. As such, optimal recovery of microglia is achieved after 10–11 days in vitro. Isolation of separate populations of microglia, astrocytes, and oligodendrocytes is carried out as described for cortical cultures.

4. Notes

1. When preparing cortical glia cultures from mouse, we recommend using pups at 6–7 days of age. The yield of microglia and oligodendrocytes is superior to that obtained with younger animals.
2. We find that removal of meninges is facilitated by using a dissecting microscope with backlighting (i.e., under the stage, so that the tissue is illuminated from beneath). In this way, meninges have a different contrast from the underlying cortex, which makes it easier to delineate. This principle is valid also for delineating the hippocampal region from the cortex.
3. If there is no clear boundary between the incubation solution and tissue, you can centrifuge for 1 min at $200 \times g$ to achieve a better separation.
4. The number of up-and-down strokes to achieve complete dispersal of the tissue can vary; the value given here is normally sufficient but can be increased if tissue pieces are still evident. If necessary, and to clear the suspension of any remaining small clumps, this may be passed through a cell strainer (70 μm, BD Falcon) placed over the mouth of a 50-mL centrifuge tube.
5. The choice of flask type, i.e., with or without vented cap, is largely immaterial. The advantage of the vented cap is that it eliminates the risk of forgetting to leave the cap loose, to allow for CO_2 exchange in the incubator. The disadvantage is that one needs to remember to cover the cap with parafilm before placing on the orbital shaker (and remove when returning to the incubator).
6. The time to reach maturity can vary between labs, which may be a reflection of differences in reagent source, batch number (e.g., serum, enzymes), and animal (sub)strain. In our former lab, microglia could be harvested after 10–14 days following setting up of the initial mixed glial culture. In our present lab, this timing appears to be accelerated, with optimal recovery of microglia (numbers, vitality) after 7–8 days. We would suggest that investigators determine the best time in their hands.
7. Sterilin plastic petri dishes appear to work best for collecting microglia by differential attachment. We have used other types (sources) of bacterial plastic petri dishes, but observed that the microglia recovered from these had a lower vitality than when using Sterilin.
8. The efficiency of the differential plating step over Sterilin plastic petri dishes depends on the volume in the dish and time of incubation. We find that a volume of 20–25 mL/dish is optimal, when combined with an incubation time of 45–60 min.

Larger volumes tend to decrease efficiency, as do shorter (<30 min) incubation times. Incubation times longer than 1 h can increase the percent of nonmicroglial cellular elements (hence inpurities).

9. Residual microglia can be removed by treating the astrocyte monolayers with medium containing 50 mM L-leucine methyl ester for 60 min (17). This medium is then removed and replaced with fresh maintenance medium and the cultures incubated for at least one day prior to use. L-Leucine methyl ester, a lysosomotropic agent originally used to selectively destroy macrophages (18), has also been employed to deplete microglia from neural cell cultures including astrocytes (19) and oligodendrocytes (20).

References

1. Schwab ME, and Schnell L (1991) Channeling of developing rat corticospinal tract axons by myelin-associated neurite growth inhibitors. *J Neurosci* 11, 709–721
2. Rosin C, Colombo S, Calver AA, Bates TE, and Skaper SD (2005) Dopamine D2 and D3 receptor agonists limit oligodendrocyte injury caused by glutamate oxidative stress and oxygen/glucose deprivation. *Glia* 52, 336–343
3. Rosin C, Bates TE, and Skaper SD (2004) Excitatory amino acid induced oligodendrocyte cell death in vitro: receptor-dependent and-independent mechanisms. *J Neurochem* 90, 1173–1185
4. Matute C, Alberdi E, Domercq M, Perez-Cerda F, Perez-Samartin A, and Sanchez-Gomez MV (2001) The link between excitotoxic oligodendroglial death and demyelinating diseases. *Trends Neurosci* 24, 224–230
5. Gallo V, and Ghiani CA (2000) Glutamate receptors in glia: new cells, new inputs and new functions. *Trends Pharmacol Sci* 21, 252–258
6. Skaper SD, Evans NA, Soden PE, Rosin C, Facci L, and Richardson JC (2009) Oligodendrocytes are a novel source of amyloid peptide generation. *Neurochem Res* 34, 2243–2250 (doi: 10.1007/s11064-009-0022-9)
7. Trang T, Beggs S, Wan X, and Salter MW (2009) P2X4-receptor-mediated synthesis and release of brain-derived neurotrophic factor in microglia is dependent on calcium and p38-mitogen-activated protein kinase activation. *J Neurosci* 29, 3518–3528
8. Zhang P, Hatter A, and Liu B (2007) Manganese chloride stimulates rat microglia to release hydrogen peroxide. *Toxicol Lett* 173, 88–100
9. Culbert AA, Skaper SD, Howlett DR, Evans NA, Facci L, Soden PE, et al (2006) MAPKAP kinase 2 deficiency in microglia inhibits pro-inflammatory mediator release and resultant neurotoxicity: Relevance to neuroinflammation in a transgenic mouse model of Alzheimer's disease. *J Biol Chem* 281, 23658–23667
10. Skaper SD, Facci L, Culbert A, Evans NA, Chessell I, Davis JB, et al (2006) $P2X_7$ receptors on microglial cells mediate injury to cortical neurons in vitro. *Glia* 54, 234–242
11. Varon S, Skaper SD, Barbin G, Selak I, and Manthorpe M (1984) Low molecular weight agents support survival of cultured neurons from the central nervous system. *J Neurosci* 4, 654–658
12. Shen LH, Li Y, and Chopp M (2010) Astrocytic endogenous glial cell derived neurotrophic factor production is enhanced by bone marrow stromal cell transplantation in the ischemic boundary zone after stroke in adult rats. *Glia* 58, 1074–1081
13. Skaper SD, Facci L, and Leon A (1995) Inflammatory mediator stimulation of astrocytes and meningeal fibroblasts induces neuronal degeneration via the nitridergic pathway. *J Neurochem* 64, 266–276
14. Miller G (2005) The dark side of glia. *Science* 308, 778–781
15. Facci L, Skaper SD, Favaron M, and Leon A (1988) A role for gangliosides in astroglial cell differentiation in vitro. J Cell Biol 106, 821–828
16. Gabellini N, Facci L, Milani D, Negro A, Callegaro L, Skaper S.D, et al (1991) Differences in induction of c-fos transcription by cholera toxin-derived cyclic AMP and Ca^{2+} signals in astrocytes and 3 T3 fibroblasts. *Exp Cell Res* 194, 210–217

17. Hamby ME, Uliasz TF, Hewett SJ, and Hewett JA (2006) Characterization of an improved procedure for the removal of microglia from confluent monolayers of primary astrocytes. *J Neurosci Methods* 150, 128–137
18. Thiele DL, Kurosaka M, and Lipsky PE (1983) Phenotype of the accessory cell necessary for mitogen-activated T and B cell responses in human peripheral blood: delineation by its sensitivity to the lysosomotropic agent, L-leucine methyl ester. *J Immunol* 131, 2282–2290
19. Giulian D, Vaca K, and Corpuz M (1993) Brain glia release factors with opposing actions upon neuronal survival. *J Neurosci* 13, 29–37
20. Hewett JA, Hewett SJ, Winkler S, and Pfeiffer SE (1999) Inducible nitric oxide synthase expression in cultures enriched for mature oligodendrocytes is due to microglia. *J Neurosci Res* 56, 189–198

Chapter 8

Central Nervous System Neuron-Glia Co-culture Models

Stephen D. Skaper and Laura Facci

Abstract

Glial cell activation plays an important role in the pathogenesis of various neurodegenerative disorders. This article presents a protocol for the preparation of cultures consisting of rat embryonic cortical neurons grown in the presence of cortical microglia, in which the glia are present in physical contact with the neurons or separated by a semi-permeable membrane barrier.

Key words: Neurons, Microglia, Astrocytes, Cortex, Co-culture, Rat, Cell culture

1. Introduction

Our appreciation of inflammation's role in neurodegenerative disease pathogenesis has increased rapidly in recent years. Neuroinflammation in disorders such as Alzheimer's disease has previously been viewed as an epiphenomenon, with inflammation occurring when damaged neurons provoke an activation response from glia. Accumulating evidence, however, is now challenging this earlier perspective and points to a more active role of neuroinflammation in pathophysiology onset and progression. In the central nervous system, glial cells (microglia, astroglia, and oligodendroglia) serve supportive and nutritive roles for neurons; in the healthy brain, they often respond to stress and insults by transiently upregulating inflammatory processes. These processes are kept in check by other endogenous anti-inflammatory and neuroprotective responses that return the brain to homeostasis. Otherwise "normal" glial functions can sometimes result in a more severe and chronic neuroinflammatory cycle that actually promote or propagate neurodegenerative disease (1, 2). The delicate balance in this homeostasis can be disturbed, resulting in disease or exacerbation of initiating factors that result in disease (i.e., the neuroinflammation hypothesis)

Stephen D. Skaper (ed.), *Neurotrophic Factors: Methods and Protocols*, Methods in Molecular Biology, vol. 846, DOI 10.1007/978-1-61779-536-7_8, © Springer Science+Business Media, LLC 2012

(3). A corollary of the neuroinflammation hypothesis is that suppression of neurotoxic products by glial activation will result in neuroprotection (4, 5). Indeed, activated microglia can cause neuronal cell death in vitro (6–12), and inhibiting the former can limit neuron injury (cf. (5, 10–12)). Therefore, controlling glial activation and, hence, inflammation is of great therapeutic interest in mitigating neuronal cell death in different neurological disorders. This chapter describes a protocol for preparing cultures of rat embryonic cortical neurons grown in the presence of cortical microglia, either as co-cultures in which the two cell populations are in physical contact or with the use of a transwell insert system whereby glia are cultured on a semi-permeable membrane in close proximity to the underlying monolayer of neurons.

2. Materials

2.1. Equipment and Labware

1. Laminar flow biological safety cabinet (CL2).
2. Humidified, water-jacketed culture incubator at 37°C and 5% CO_2/95% air.
3. Water bath set at 37°C.
4. Dissecting tools (Fine Science Tools are very high quality for the price).
5. Bench centrifuge to accommodate 15- and 50-mL tubes.
6. Orbital shaker with enclosed, temperature-controlled chamber.
7. 10-cm ∅ sterile tissue culture dishes.
8. 15-mL and 50-mL polypropylene plastic centrifuge tubes (sterile).
9. Sterilin 10-cm ∅ sterile petri plastic dishes (Bibby-Sarstedt).
10. Tissue culture flasks (T-75; 75 cm^2) with canted neck and 0.2-μm vented cap (Corning) or without vented cap (BD Falcon).
11. Cell scrapers, 25-cm handle, 1.8-cm blade (BD Falcon).
12. 0.22-μm filters (Millipore Corporation).
13. Cell culture inserts for 24-well plate, PET (polyethylene terephthalate) transparent membrane, 0.4-μm pore size, BD Falcon™ (BD Biosciences).
14. Tissue culture plate, 24-well plate, BD Falcon™. Tissue-culture-treated polystyrene, flat bottom, with low-evaporation lid (BD Biosciences).

2.2. Reagents

1. Hank's balanced salt solution (HBSS), with $CaCl_2$ and $MgCl_2$ (Invitrogen).
2. Neurobasal medium (Invitrogen).

3. Phenol red-free Neurobasal medium (Invitrogen).
4. B27 supplement, 50×, with antioxidants (Invitrogen).
5. B27 supplement, 50×, without antioxidants (Invitrogen).
6. Dulbecco's modified Eagle's medium (DMEM), containing 4.5 g/L glucose, L-glutamine, and pyruvate (Invitrogen).
7. Fetal calf serum (FCS) (Invitrogen) (see Note 1).
8. L-Glutamine (200 mM stock), sterile, for cell culture (Invitrogen).
9. Sodium pyruvate (100 mM stock), sterile, for cell culture (Sigma).
10. Penicillin/streptomycin (100× stock), sterile, for cell culture (Invitrogen).
11. Gentamicin (50 mg/mL stock), sterile, for cell culture (Sigma).
12. Poly-D-lysine hydrobromide, γ-irradiated, cell culture–tested, mol. wt. 70,000–150,000 (Sigma).
13. 3-(4,5-dimethylthiazol-2-yl)-2,5-diphenyl tetrazolium bromide (MTT) (Sigma).
14. CytoTox® non-radioactive cytotoxicity assay kit for lactate dehydrogenase (LDH) (Promega).

2.3. Culture Media and Solutions

1. *Culture medium for neurons*: Neurobasal medium containing B27 supplement (1×), 2 mM L-glutamine, 100 U/mL penicillin, and 100 μg/mL streptomycin: To 500 mL Neurobasal medium, add 10 mL 50× B27 supplement, 5 mL 200 mM glutamine, and 5 mL 10,000 U/ml penicillin to 10,000 μg/mL streptomycin. Complete medium can be stored at +4°C for up to 2 months.
2. *Culture medium for microglia*: To a 500-mL bottle of DMEM, add 50 mL FCS (10% final), 5 mL stock penicillin/streptomycin (100 U/mL penicillin + 100 μg/mL streptomycin final), and 0.5 mL of gentamicin stock (50 μg/ml final).
3. *0.15 M borate buffer pH 8.4*: Dissolve 28.6 g of sodium borate ($Na_2B_4O_7 \cdot 10H_2O$) in 500 mL water (pH will be ~9.2). Adjust pH to 8.4 with 5 N HCl. Filter-sterilize and store at 4°C (up to 6 months).

2.4. Preparation of Poly-D-Lysine Substratum

1. Dissolve a 5 mL bottle of poly-D-lysine (MW 70,000–150,000) in 5 mL 0.15 M borate buffer (pH 8.4). This 1 mg/mL stock solution can be stored at +4°C for up to 2 months (see Note 2).
2. Dilute the 1-mg/mL stock solution of poly-D-lysine 1:10 (1 in 9 mL sterile water) and add 0.5 mL to each well of a 48-well plate or 1 mL to each well of a 24-well plate. Place the plate(s) in the 5% CO_2 incubator (37°C) overnight.

3. The following day aspirate the poly-D-lysine coating solution and leave the open plate(s) under the CL2 cabinet to dry. The poly-D-lysine-coated plate(s) can be stored at +4°C for 1 month.

3. Methods

3.1. Neuron-Glia Co-cultures: Physical Contact Model

1. Prepare primary cerebral cortical neuron cell cultures from gestational day 18 fetuses of time-mated CD rats (Charles River) (see Chap. 5).
2. Re-suspend the final cell pellet in Neurobasal medium containing 2% B27 supplements, 1 mM sodium pyruvate, 2 mM L-glutamine, and penicillin (100 units/mL)/streptomycin (100 μg/mL).
3. Plate cells in poly-D-lysine-coated 48-well plates, 1×10^5 cells/well in 0.35 mL of the same medium. Place cultures in a humidified incubator at 37°C in a 5% CO_2/95% air atmosphere.
4. Expose the poly-D-lysine-coated culture wells overnight to medium containing 10% FCS prior to cell seeding.
5. After 2 days, add to each culture well 0.35 mL of plating medium, but containing B27 supplement *without* antioxidants. Neuron cell cultures cells are normally used on day 6 for experiments.
6. Culture microglia from the cortex of newborn CD (Charles River) rat pups, as described in Chap. 7.
7. Collect microglia from the 10-cm Ø Sterilin plates by mechanically scraping into culture medium for glia (see Chap. 7).
8. Pellet microglia by centrifugation ($200 \times g$ for 5 min) and re-suspend the cell pellet in Neurobasal medium containing 2% B27 supplements (without antioxidants), 1 mM sodium pyruvate, 2 mM L-glutamine, and penicillin/streptomycin. Use 1 mL medium for each Sterilin plate of microglia collected, and re-suspend pellet using a long (9-in.) Pasteur pipette attached to an automatic pipettor. Eight or nine up-and-down strokes are usually sufficient to fully disperse the cell pellet. Determine the cell density in the suspension by counting with a hemocytometer (See Chap. 3).
9. To prepare the co-cultures, dilute the microglia cell suspension to 285,000 cells/mL, using the same medium for suspension of the starting cell pellet.
10. Take the 48-well plate of neurons set up 6 days ago and remove 0.35 mL of medium. Now add to each well 0.35 mL of the

microglia cell suspension, which gives a neuron to microglia ratio of 1:1. Return the plate to the 5% CO_2 incubator.

11. After incubation for 30 min, wash the plate with 1 mL of Neurobasal/B27 (antioxidant-free) medium (point 8 above) to remove non-adherent microglia and debris, then add to each well 0.7 mL of the same medium. Return the plate to the 5% CO_2 incubator; wait several hours before commencing treatments.
12. Co-cultures are usually utilized the same day. Of course, the treatment regimen will be determined by the experimental plan one wishes to carry out. One example may be to treat with an agent (lipopolysaccharide) which is known to stimulate microglia to release neurotoxic factors, and then add potential anti-inflammatory drugs to test for neuroprotection. However, description of such material is beyond the scope of this chapter (see Note 3).

3.2. Assessment of Neurotoxicity in Physical Contact Neuron-Microglia Co-cultures

1. LDH release into the culture medium was used as a measure of cell death and has been utilized previously to quantify neuronal cell injury in microglia-neuron co-cultures (8, 9).
2. LDH activity in the cell culture supernatants is determined after 72 h, using the CytoTox® non-radioactive cytotoxicity assay kit.
3. Collect 100 μL of medium from each well of the co-cultures for assay. Use 50 μL of this sample for assay, and proceed with the assay as described in Chap. 6 (see Note 4).
4. Express values of LDH release as milliunits/mL (see Notes 5 and 6).
5. Establish separate cultures of microglia without neurons in poly-D-lysine-coated 48-well plates (100,000 cells/well) in 0.7 mL Neurobasal/B27 (antioxidant-free) at the time microglia are added to the neuron cultures. Incubate these microglia cultures for 72 h in the presence of stimuli, and then measure LDH release from the microglia alone ± stimuli.
6. Subtract the microglia only LDH values obtained from the LDH values obtained for the neuron-microglia co-cultures receiving the same treatments (see Note 7). The purpose of this step is to correct for the contribution of microglia LDH release.

3.3. Neuron-Glia Co-cultures: Transwell Insert Model

1. This model allows one to culture glia on a porous membrane, which is then placed in a well containing an established monolayer of neurons. Application of a stimulus to the glia layer allows for glia-derived agents to diffuse across the membrane and act on the underlying neurons. The insert with glia can then be removed, permitting morphological or biochemical analysis of the neurons. Figure 1 shows a typical commercially

Fig. 1. A typical cell culture insert chamber.

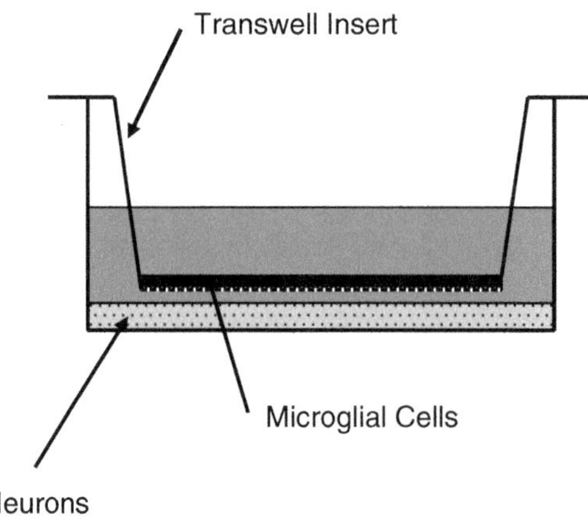

Fig. 2. Schematic representation of the transwell insert neuron-glia co-culture model.

available insert chamber, and Fig. 2 illustrates schematically the principal of this co-culture model.

2. Prepare primary cerebral cortical neuron cell cultures from gestational day 18 fetuses of time-mated CD rats (Charles River) (see Chap. 5).
3. Re-suspend the final cell pellet in Neurobasal medium containing 2% B27 supplements, 1 mM sodium pyruvate, 2 mM L-glutamine, and penicillin (100 units/mL)/streptomycin (100 μg/mL).

4. Plate cells in poly-D-lysine-coated 24-well plates, 2×10^5 cells/well in 0.5 mL of the same medium. Place cultures in a humidified incubator at 37°C in a 5% CO_2/95% air atmosphere.
5. Expose the poly-D-lysine-coated culture wells overnight to medium containing 10% FCS prior to cell seeding.
6. After 2 days, add to each culture well 0.5 mL of plating medium, but containing B27 supplement *without* antioxidants. Neuron cell cultures cells are normally used on day 6 for experiments.
7. Culture microglia from the cortex of newborn CD (Charles River) rat pups (see Chap. 7).
8. Collect microglia from the Sterilin plates by mechanically scraping into culture medium for microglia. This should be done on the day prior to intended transfer of inserts to wells containing neurons (the neuron cultures will be 5 days old at this point) (see Chap. 5).
9. Pellet microglia by centrifugation ($200 \times g$, 5 min) and re-suspend cell pellet in culture medium for microglia. Use 1 mL medium for each Sterilin plate of microglia collected, and re-suspend pellet using a 9-in. Pasteur pipette. Determine cell density of the suspension (see Chap. 3).
10. To prepare the co-cultures, dilute the microglia cell suspension to 187,500 cells/mL, using the same medium for suspension of the starting cell pellet.
11. To an empty 24-well plate, add 0.8 mL/well of culture medium for microglia.
12. Using flame-sterilized forceps, transfer an empty cell culture insert to each well of the above 24-well plate.
13. Add to each insert 0.4 mL of microglia suspension containing 75,000 cells.
14. The following day, wash the inserts containing microglia once with 0.5 mL of Neurobasal medium containing 2% B27 supplements (without antioxidants), 1 mM sodium pyruvate, 2 mM L-glutamine, and penicillin/streptomycin.
15. Transfer the inserts to the 24-well plate of cortical neurons (set up 6 days ago), without changing the medium in the well. The porous membrane allows free diffusion of molecules. The distance between the neuron monolayer and microglia on the insert membrane is 1 mm. Return the plate to the 5% CO_2 incubator.
16. The insert co-cultures are now ready for use. The treatment regimen will, of course, be determined by the experimental plan one wishes to carry out. For example, one can treat the upper (microglia-containing) chamber with amyloid β-peptide, using a concentration sufficient to stimulate microglia without causing injury to the underlying neurons (10), in the presence

or absence of potential anti-inflammatory drugs to test for neuroprotection. Generally, we terminate the experiment 3 days after addition of the microglia stimulus (see Note 8).

3.4. Assessment of Neurotoxicity in Transwell Insert Co-cultures

1. At the end of the incubation period, remove the microglia-containing inserts and evaluate neuronal cell viability by the colorimetric method based upon conversion of 3-(4,5-dimethylthiazol-2-yl)-2,5-diphenyltetrazolium bromide (MTT) by mitochondrial dehydrogenases to a blue formazan product.
2. Add to each well 400 μL of phenol red-free Neurobasal medium containing 2% B27 supplements (without antioxidants), 1 mM sodium pyruvate, 2 mM L-glutamine, and penicillin/streptomycin, plus 0.15 mg/mL MTT (which represents a 1:20 dilution of the 3 mg/mL stock). We find that this MTT concentration gives optimal color yield (under these conditions).
3. Incubate the MTT-containing cultures for 1 h at 37°C. This incubation time gives optimal color yield (under these conditions).
4. Examination of the cultures under a light microscope will show that viable cells contain blue crystals. This is the formazan reaction product.
5. Remove the culture medium by aspiration, and add 400 μL of dimethyl sulfoxide per well. This is to dissolve the reaction product. Gently rock the plate to mix the blue reaction product uniformly with the solvent. This should take only 1–2 min.
6. Transfer 100 μL of the colored dimethyl sulfoxide solution from each well to the well of a clear polystyrene plastic 96-well microplate (any brand will do). Read the plate on a microplate (ELISA) reader. We use a SpectraMax M2 microplate reader from Molecular Devices, although there are a number of quality plate readers on the market. For optimal results, the plate should be read at a wavelength of 570 nm (test wavelength), followed by 630 nm (reference wavelength). The difference (A_{570}–A_{630}) is used as the final absorbance value generated by the sample. Correcting for optical imperfections in the microplates by subtracting A_{630} is recommended but is not an essential procedure.
7. Normalize the absolute MTT values by scaling to the mean of cultures which have not been treated with test compound (defined as 100%).
8. Cell vitality should be routinely confirmed by morphological observation using a light microscope with phase contrast optics (see Note 9).

4. Notes

1. Heat inactivation of fetal calf serum is recommended to destroy heat-labile complement. Thaw the bottle of serum in advance, using a 37°C water bath. Next, immerse the serum bottle in the water bath after re-equilibrating to 56°C and leave for 30 min. Swirl the bottle occasionally to ensure proper mixing. Allow the serum to cool to room temperature, aliquot into 50-mL tubes and store at −20°C.
2. The molecular weight of the poly-D-lysine used is critical. Optimal culture of neurons on poly-D-lysine is a function of cell type and should be determined by the experimenter. We find that the poly-D-lysine used here (Sigma P-6407) works best for cortical, hippocampal, and striatal neuron cell cultures. A molecular weight lower or higher than this will result in cell clumping, which is most undesirable.
3. Neuron-microglia co-cultures ("physical contact") as described are not limited to the use of rat cortical neurons but can also be carried out using hippocampal neurons, or cortical, or hippocampal neurons prepared from mouse embryonic tissues. In addition, one can make use of microglia derived from mouse, rather than rat cortex. These variations can be of particular utility if one wishes to examine the influence of gene target deletion. We find that this system performs equally well even when using "heterogeneous" cell populations, for example, co-culturing rat cortical neurons and mouse microglia, or vice versa.
4. As a starting point, we recommend using 50 μL of culture medium for the LDH assay. If the value obtained is outside the standard curve, dilute the culture medium sample twofold in culture medium and repeat the assay.
5. Values of absolute LDH release will most likely vary from experiment to experiment, even for control (untreated) cultures. This is not unexpected, given that microglia in culture are never in a truly "resting" state. Their tonic level of activation will influence the absolute LDH values obtained. We find that data can be normalized to achieve a reasonable consistently across experiments by expressing values as a percent of the control value for each individual trial.
6. Our historical experience in using these neuron-microglia co-cultures is that in a minority of preparations, (about 1 in 10) the microglia are in a "highly" activated state per se, and the neurons will degenerate in the presence of the microglia without addition of any stimulus. We recommend observing the co-cultures microscopically for appearance of the neurons prior to commencing treatments; neuronal cell injury (if present) will be evident morphologically by disintegration of the neurite network and cell bodies.

7. The purpose of this step is to correct for the contribution of microglia LDH release. In our experience, most experimental treatments (e.g., lipopolysaccharide, anti-inflammatory drugs) have little or no influence on this LDH release. However, it is recommended to do in every case, as the data thus generated will represent a more rigorous analysis. This procedure will double the number of microglia cells needed for an experiment. One can also perform the same using neuron-only cultures. In our experience, this is not necessary as LDH release from neurons in the absence of microglia is virtually below the limit of detection.
8. Determination of the optimal time of exposure of neurons to the injury-inducing effects of activated microglia is empirical. Using, for example, lipopolysaccharide to stimulate the microglia, degeneration of the underlying neurons is largely complete by 3 days. With incubation times of less than 2 days, neuronal cell injury in our hands was not evident.
9. A limitation of the insert system is the following: The "footprint" of the culture insert is smaller than the area of the underlying well (approximately 50%). When microscopically examining the neuron monolayer after removing the microglia-containing insert, one will notice that damage to neurons (degeneration of neurite network and cell bodies) tends to be limited to the area immediately under the insert. Possibly, the activated microglia release neurotoxic agents with a limited diffusion or lifetime, although we have not explored this. One thus needs to keep in mind that the measure of neuronal cell viability loss obtained is an underestimate, as only those cells in the vicinity of the insert will "register" with the MTT assay.

Acknowledgments

L. Facci was supported by Fondazione CARIPARO Progetto Dottorati di Ricerca' Anno 2009.

References

1. Griffin WS, Sheng JG, Royston MC, Royston MC, Gentleman SM, McKenzie JE, et al (1998) Glial-neuronal interactions in Alzheimer's disease: the potential role of a "cytokine cycle" in disease progression. *Brain Pathol* 8, 65–72
2. Colton CA, and Wilcock DM (2010) Assessing activation states in microglia. *CNS Neurol Disord Drug Targets* 9, 174–191
3. Craft JM, Watterson DM, and Van Eldik LJ (2005) Neuroinflammation: a potential therapeutic target. *Expert Opin. Ther Targets* 9, 887–900
4. Skaper SD (2010) Microglia as a target for inflammatory processes and neuroprotective strategies. *Am J Neuroprotec Neuroregen* 2, 35–47
5. Skaper SD (2011) Ion channels on microglia: therapeutic targets for neuroprotection. *CNS Neurol Disord Drug Targets*, 10, 44–56
6. Chao CC, Hu S, Molitor TW, Shaskan EG, and Peterson PK (1992) Activated microglia mediate neuronal cell injury via a nitric oxide mechanism. *J Immunol* 149, 2736–2741

7. Flavin MP, Zhao G, and Ho LT (2000) Microglial tissue plasminogen activator (tPA) triggers neuronal apoptosis in vitro. *Glia* 29, 347–354
8. Golde S, Chandran S, Brown GC, and Compston A (2002) Different pathways for iNOS-mediated toxicity in vitro dependent on neuronal maturation and NMDA receptor expression. *J Neurochem* 82, 269–282
9. Parvathenani LK, Tertyshnikova S, Greco CR, Roberts SB, Robertson B, and Posmantur R (2003) $P2X_7$ mediates superoxide production in primary microglia and is upregulated in a transgenic mouse model of Alzheimer's disease. *J Biol Chem* 278, 13309–13317
10. Culbert AA, Skaper SD, Howlett DR, Evans NA, Facci L, Soden PE, et al. (2006) MAPKAP kinase 2 deficiency in microglia inhibits pro-inflammatory mediator release and resultant neurotoxicity: Relevance to neuroinflammation in a transgenic mouse model of Alzheimer's disease. *J Biol Chem* 281, 23658–23667
11. Skaper SD, Facci L, Culbert A, Evans NA, Chessell I, Davis JB, et al. (2006) $P2X_7$ receptors on microglial cells mediate injury to cortical neurons in vitro. *Glia* 54, 234–242
12. Jana A, and Pahan K (2010) Fibrillar amyloid-β-activated human astroglia kill primary human neurons via neutral sphingomyelinase: implications for Alzheimer's disease. *J Neurosci* 30, 12676–12689

Chapter 9

Culture and Characterization of Rat Mesencephalic Dopaminergic Neurons

Stephen D. Skaper, Giulia Mercanti, and Laura Facci

Abstract

Dopaminergic neuronal cell degeneration is the principal characteristic feature of the neuropathology of Parkinson's disease. Cultures of mesencephalic neurons are widely used as a source of dopaminergic neurons for the study of mechanisms implicated in dopaminergic cell death and for the evaluation of potential dopaminergic neuroprotective agents, including neurotrophic factors. This chapter presents a detailed protocol for the preparation of rat mesencephalic cell cultures and their application to evaluating the neuroprotective action of brain-derived neurotrophic factor.

Key words: Mesencephalic, Ventral tegmentum, Rat, Brain-derived neurotrophic factor, Dopamine uptake, Tyrosine hydroxylase, Parkinson's disease

1. Introduction

Parkinson's disease is the most common neurodegenerative movement disorder, whose neuropathological hallmarks are characterized by progressive and profound loss of neuromelanin-containing dopaminergic neurons in the substantia nigra pars compacta with the presence of eosinophilic intracytoplasmic, proteinaceous inclusions termed Lewy bodies and dystrophic Lewy neurites in surviving neurons (1, 2). Brain-derived neurotrophic factor (BDNF) enhances the survival of and differentiation of dopaminergic neurons in vitro (3, 4), and the intranigral transplantation of epigenetically induced BDNF-secreting human mesenchymal stem cells in a dopamine neurotoxin model in rats rescued some neurological deficits (5). In addition, BDNF is neuroprotective against a number of insults (6–8). Mesencephalic dopaminergic neurons in vitro respond also to basic fibroblast growth factor (9) and their striatal target cells (10). This chapter describes in detail a protocol for culturing

Stephen D. Skaper (ed.), *Neurotrophic Factors: Methods and Protocols*, Methods in Molecular Biology, vol. 846,
DOI 10.1007/978-1-61779-536-7_9, © Springer Science+Business Media, LLC 2012

rat embryonic mesencephalic cells and identifying the dopaminergic neuron component. The response of the dopamine cells to BDNF will also be presented, to demonstrate an application of these cultures.

2. Materials

2.1. Equipment and Labware

1. Stereo dissecting microscope (backlighting of stage is preferred) with fiber optic light source.
2. Laminar flow cabinet for dissections.
3. Laminar flow biological safety cabinet (CL2).
4. Humidified, water-jacketed culture incubator at 37°C and 5% CO_2/95% air.
5. Water bath set at 37°C.
6. Dissecting tools (Fine Science Tools (InterFocus Ltd.) are very high quality for the price).
7. Bench centrifuge to accommodate 15- and 50-mL tubes.
8. 10-cm ∅ sterile tissue culture dishes (any supplier).
9. Glass coverslips (12 mm ∅ #1, Menzel-Gläser, Menzel GmbH, Braunschweig, Germany) (see Note 1).
10. 15- and 50-mL polypropylene plastic centrifuge tubes (sterile).
11. 0.22-μm filters (Millipore).

2.2. Culture Media, Supplements, and Additives

1. Hank's balanced salt solution (HBSS, with $CaCl_2$ and $MgCl_2$) (Invitrogen).
2. Dulbecco's modified Eagle's medium, high glucose with GlutaMAX I, and sodium pyruvate (Invitrogen).
3. Neurobasal medium (Invitrogen).
4. B27 supplement (with antioxidants, 50×) (Invitrogen).
5. B27 supplement (without antioxidants, 50×) (Invitrogen).
6. N2 supplement (100×) (Invitrogen).
7. Fetal calf serum (Invitrogen).
8. L-Glutamine (200 mM, sterile, for cell culture) (Invitrogen).
9. Non-essential amino acids (100×, sterile) (Invitrogen).
10. Penicillin/streptomycin (100×, sterile) (Invitrogen).
11. Poly-D-lysine (MW 30,000–70,000, sterile, for cell culture) (Sigma).
12. [^{3}H]Dopamine (DA) (Amersham Biosciences).
13. [γ-^{14}C]Aminobutyric (GABA) (Amersham Biosciences).

14. Mazindol (Sigma).
15. L-2,4-Diaminobutyric acid·2HCl (Sigma).
16. Anti–tyrosine hydroxylase (TH) rabbit polyclonal antibody (Abcam).
17. Donkey polyclonal to rabbit IgG-H&L (Rhodamine) (Abcam).
18. 2,4,5-Trihydroxyphenylalanine (TOPA) (Sigma).
19. BDNF (human, recombinant) (Peprotech).

2.3. Culture Media and Other Solutions

1. *Medium 1*: To 500 mL of Neurobasal medium, add B27 (10 mL of a 50×, containing antioxidants), 25 μM L-glutamate final (0.5 mL of a 1,000× stock in H_2O), 0.5 mM L-glutamine final (1.25 mL of a 200 mM stock), 1 mM sodium pyruvate final (5 mL of a 100 mM stock), and 100 U/mL penicillin/100 μg/mL streptomycin final (5 mL of a 100× stock). Store at 4°C for 1 month.
2. *Medium 2*: To 500 mL of Neurobasal medium, add B27 (10 mL of a 50×, without antioxidants), 0.5 mM L-glutamine final (1.25 mL of a 200 mM stock), 1 mM sodium pyruvate final (5 mL of a 100 mM stock), and 100 U/mL penicillin/100 μg/mL streptomycin final (5 mL of a 100× stock). Store at 4°C for 1 month (see Note 2).
3. *Medium 3*: To 500 mL of Dulbecco's modified medium high glucose with GlutaMAX I and sodium pyruvate, add 5% (v/v) fetal calf serum final, 1× non-essential amino acids (5 mL of a 100× stock), and 50 U/mL penicillin/50 μg/mL streptomycin final (2.5 mL of a 100× stock). Store at 4°C for 1 month (see Note 3).
4. *Medium 4*: To 500 mL of Hank's balanced salt solution (with $CaCl_2$ and $MgCl_2$), add 1 mM sodium pyruvate final (5 mL of a 100 mM stock), 10 mM HEPES final (5 mL of a 1 M stock), 0.035% $NaHCO_3$ final (2.3 mL of a 7.5% stock), and 100 U/mL penicillin/100 μg/mL streptomycin final (5 mL of a 100× stock). Can be stored for 3 months at 4°C.
5. *Krebs-Ringer buffer (pH 7.4)*: 16 mM $NaH_2PO_4 \cdot H_2O$ (2.20 g/L), 16 mM Na_2HPO_4, (2.27 g/L), 119 mM NaCl (6.95 g/L), 0.35 g 4.7 mM KCl (0.35 g/L), 0.26 g 1.8 mM $CaCl_2 \cdot 2H_2O$ (0.26 g/L), 1.2 mM $MgSO_4$ (0.14 g/L), mM EDTA·$Na_2 \cdot 2H_2O$ (0.48 g/L), and 5.6 mM D-glucose (1.00 g/L).
6. *Borate buffer (pH 8.4)*: Dissolve 28.6 g of sodium borate ($Na_2B_4O_7 \cdot 10H_2O$) in 500 mL water (pH will be ~9.2). Adjust pH to 8.4 with 5 N HCl. This gives a concentration of 0.15 M. Filter-sterilize and store at 4°C (up to 6 months).

3. Methods

3.1. Isolation of Mesencephalic Cells

1. Euthanize pregnant rats (CD strain, Sprague Dawley) following appropriate national and institutional guidelines. Under aseptic conditions, remove the whole brain from the rat embryos (gestational age 14–15 days) and collect in a petri dish (10 cm ∅) containing ice-cold Medium 4. Remove the blood vessels and meninges.
2. After having collected all brains, dissect out the ventral two-thirds of the mesencephalon. The dissected regions include dopaminergic neurons in the substantia nigra and the ventral tegmental area but not the noradrenergic neurons in the locus ceruleus.
3. When all cortices have been dissected, mechanically dissociate the tissue in 2 mL of Neurobasal medium containing 10% (v/v) fetal calf serum, using a series of three flame-polished Pasteur pipettes of decreasing bore size (see Notes 4 and 5).
4. Count the cell suspension (see Chapter 3). A typical yield is 10^6 viable cells per embryo.
5. Use a poly-L-lysine substratum for culture. The method of preparation follows that described in Chapter 3 (see Note 6).
6. Centrifuge the cell suspension ($200\times g$, 5 min), remove the supernatant, and resuspend the cell pellet in prewarmed (37°C) Medium 1. Plate at 1,100–2,500 cells/mm^2.
7. After 5 days in culture, replace 1/2 of the plating medium with Medium 2 (this may not be necessary—check pH).
8. Replace 1/2 of the culture medium twice a week with Medium 2, until the cultures are used.

3.2. Mesencephalic Neuron-Glia Cultures

1. Follow the procedure detailed in Subheading 3.1 up to and including step 5. Resuspend the cell pellet in Medium 3, count for cell number, and adjust cell density 1×10^6 cells/mL (see Note 7).
2. Plate cells on poly-L-lysine-coated tissue culture plastic surfaces at a density of 2.5×10^5 cells/cm^2.
3. Change culture medium every second day.
4. During the course of treatment, the culture medium is not supplemented with amino acids to avoid the cytotoxicity of glutamate.
5. The composition for major cell types in the culture can be estimated by visual counting of immunostained cells, using indirect immunofluorescence (see Chapter 20 for protocol). Seven-day-old cultures will "typically" contain the following cellular composition: ~45% astrocytes (glial fibrillary acidic protein$^+$,

1:400, Sigma), ~40% neurons (βIII-tubulin$^+$, 1:200, Upstate), ~1% dopaminergic neurons (tyrosine hydroxylase$^+$, 1:200, Chemicon), and the remainder being microglia (MAC-1$^+$, 1:200, Serotec). Information in parentheses refers to primary antibody, suggested dilution, and source. This culture composition is consistent with that described by others (11, 12).

3.3. High-Affinity [^{3}H]DA Uptake

1. Initially seed mesencephalic cells (either as neuron-enriched cultures or neuron-glia cultures) in poly-L-lysine-coated 24-well plates, 5×10^5 cells per well.
2. At the desired culture age (days in vitro; will depend on the experiment being performed), wash the cells twice with 1 mL of Krebs-Ringer buffer.
3. Incubate each culture well with 1 μCi (~50 nM) [^{3}H]DA in Krebs-Ringer buffer (0.4 mL/well) for 15 min at 37°C.
4. Determine nonspecific uptake of DA in parallel wells receiving both (^{3}H)DA and 10 μM mazindol (1:1,000, from a 10 mM stock in H_2O), an inhibitor of neuron DA uptake (see Notes 8 and 9).
5. Wash the cells 3 times with ice-cold Krebs-Ringer buffer (1 mL/well).
6. Obtain blank values by incubating cells under the same conditions at 0°C.
7. Lyse the cells with 1 N NaOH (0.5 mL/well).
8. Mix the lysate with 3 mL of scintillation fluid (there are a number of brands on the market which accommodate aqueous samples).
9. Determine radioactivity with a liquid scintillation counter.
10. Calculate the specific DA uptake by subtracting the amount of radioactivity observed in the presence of mazindol from that observed in the absence of mazindol.

3.4. High-Affinity [^{14}C] GABA Uptake

1. Incubate mesencephalic cells at 37°C for 15 min with 0.1 μM [^{14}C]GABA in Krebs-Ringer buffer in the presence of 2 mM β-alanine (1:250 from a 0.5 M stock solution in H_2O), an inhibitor of glial high-affinity GABA uptake.
2. Verify high-affinity neuron-specific (^{14}C)GABA uptake using the inhibitor L-2,4-diaminobutyric acid (1 mM final, 1:500 from a 0.5 M stock in H_2O).
3. Washing and extraction procedures are as for (^{3}H)DA uptake.

3.5. Catecholamine Cytofluorescence

1. Dissociated mesencephalic neurons can be analyzed by catecholamine fluorescence using a glyoxylic acid technique.
2. Grow cells on 12-mm ∅ glass coverslips previously coated with poly-L-lysine (or poly-L-ornithine), the coverslips having been

placed within the individual wells of a 24-well plate. Plating density is the same as for (^{3}H)DA uptake measurement.

3. At the desired culture age, rinse the cells once with ice-cold Krebs-Ringer buffer (1 mL/well).
4. Incubate cells with ice-cold 2% glyoxylic acid solution in Krebs-Ringer buffer (pH 7.4) for 5 min (see Note 10).
5. As an index of dopaminergic specificity, parallel cultures can be preincubated for 15 min with 10 μM mazindol, prior to fixation with glyoxylic acid.
6. Wash twice with Krebs-Ringer buffer to remove excess glyoxylic acid.
7. Dry the coverslips for 10 min under warm air.
8. Heat coverslips for 10 min at 95°C.
9. Mount coverslips on glass slides with liquid paraffin.
10. View under a fluorescence microscope equipped for catecholamine epifluorescence and phase-contrast optics.
11. Using prefixed coordinates, count the number of glyoxylic acid–positive neurons corresponding to at least 3% of the total surface area.

3.6. BDNF as a Neuroprotectant for Mesencephalic Dopaminergic Neurons

In the case of Parkinson's disease, it is intriguing that a substance closely related to the therapeutic agent 3,4-dihydroxydopamine (L-DOPA) acts as an excitatory agent. The 6-hydroxylated derivative of L-DOPA (2,4,5-trihydroxyphenylalanine or TOPA) in solution elicits both electrophysiological responses and neurotoxicity when applied to central nervous system neurons (13, 14), including mesencephalic-derived dopaminergic neurons (15). This paradigm can be utilized to show the neuroprotective action of BDNF:

1. Seed mesencephalic cell dissociates into poly-L-lysine-coated 24-well plates, without (for (^{3}H)DA and (^{14}C)GABA uptake) or with 13-mm ∅ glass coverslips previously coated with poly-L-lysine (for TH immunostaining). The plating density is 5×10^5 cells/well in all cases. Use culture medium as described in Subheading 3.1.
2. The day after cell plating, add BDNF to the cultures (1–50 ng/mL final) (see Note 11).
3. At the seventh day in culture, add TOPA to a final concentration of 10 μM (1:200 from a 2 mM stock—see Note 12). TOPA has LD_{50} values of 3.3 ± 0.5 and 3.8 ± 0.6 μM, respectively, for general cell viability (MTT oxidation) and dopaminergic cell survival (TH immunostaining) (15).

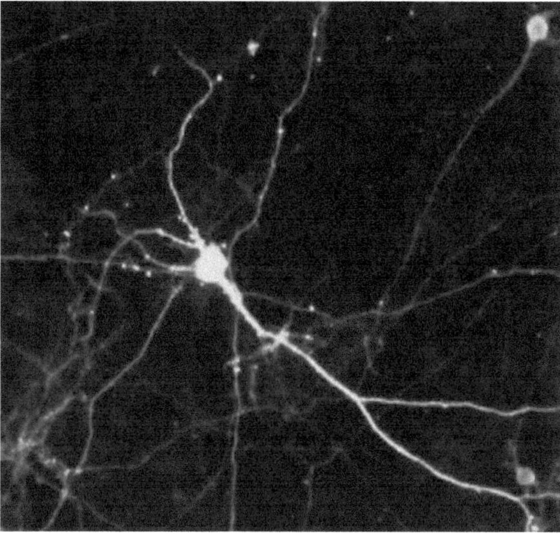

Fig. 1. Rat mesencephalic dopaminergic neurons at 7 days in culture, immunostained for tyrosine hydroxylase.

4. On the eighth day in culture (i.e., 24 h after TOPA addition), process the cultures for dopaminergic neuron survival, using tyrosine hydroxylase (TH) immunostaining and neurotransmitter uptake.
5. For TH immunostaining, drain cells of medium and fix by adding 1 mL cold (–20°C) methanol per well for 6 min.
6. Incubate with anti-TH antibody (60 min, 37°C) diluted 1:1,000 in PBS (pH 7.4) containing 0.5% BSA.
7. Wash twice with PBS (1 mL/well).
8. Stain with rhodamine-conjugated donkey anti-rabbit secondary antibodies (1:200 in PBS) for 30 min at 37°C.
9. Mount cover glasses as described in Chapter 20.
10. For observation, a photomicroscope equipped with rhodamine epifluorescence filters and phase-contrast optics will be required. Using prefixed coordinates, the number of TH^+ neurons, corresponding to at least 10% of the total surface area of the cover glass, should be counted. Only immunoreactive cells having round, smooth bodies and intact, non-fragmented neurites are scored as positive (see Fig. 1).
11. Figure 2 and Table 1 illustrate the results of a typical experiment, using this model and the described assay procedures.

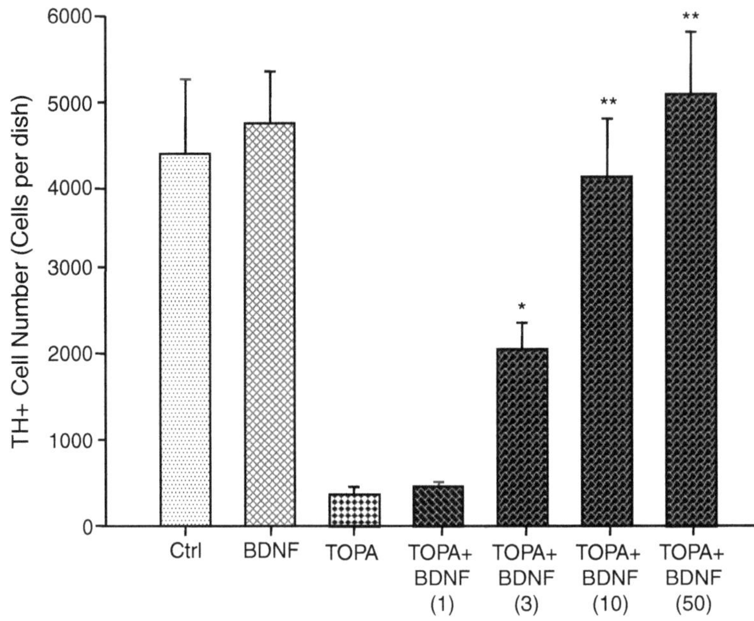

Fig. 2. BDNF protects cultured mesencephalic dopaminergic neurons from the neurotoxic effects of TOPA. After 24 h in vitro, cultures received BDNF (1–50 ng/mL). Cultures were kept until seven DIV, before being exposed to 10 μM TOPA for 24 h. At the end of the experiment (8 days in total), all cover glasses were processed for TH immunoreactivity, and the number of TH$^+$ cells was determined in each group (mean ± SD, $n=6$, 3 experiments). *Values in parentheses* indicate BDNF concentration (ng/mL). *$P<0.05$ or **$P<0.01$ vs. TOPA. The "TOPA + BDNF (10)" and "TOPA + BDNF (50)" groups were not significantly different from each other or from the no-treatment group (CTRL). Reprinted with permission from Wiley-Blackwell (15).

Table 1
Effect of BDNF on [^{3}H]dopamine and [^{14}C]GABA uptake in control and TOPA-treated mesencephalic cells

Treatment	Dopamine (pmol/well)	GABA (nmol/well)
5 None	50 ± 9	76 ± 11
BDNF	63 ± 14	86 ± 18
TOPA	6.9 ± 2.6*	9.3 ± 3.4*
BDNF/TOPA	43 ± 7**	15 ± 4

Mesencephalic cells were plated (5×10^5) in 25 mm diameter cluster wells (24-well plate) and treated with 50 ng/mL BDNF on day 1. At 7 days, TOPA, where indicated, was added to a final concentration of 10 μM. Twenty-four hours later, specific uptake of [^{3}H]dopamine and [^{14}C]GABA was determined

*$P<0.01$ vs. control ("none") or BDNF-treated cultures; **$P<0.01$ vs. TOPA-treated cultures ($n=6$). Reprinted with permission from Wiley-Blackwell (15)

4. Notes

1. The source of glass is very important for the coverslips: German suppliers are highly recommended, in particular the one listed here.
2. Neurobasal medium/B27 medium yields cultures that contain relatively few glial cells. Should a mixed neuron/glial cell culture be desired (approximately 30% glia), replace B27 supplement with N2 supplement of 10% fetal calf serum, and increase L-glutamine to 3 mM.
3. Heat inactivation of fetal calf serum is recommended to destroy heat-labile complement. Thaw the bottle of serum in advance, using a 37°C water bath. Next, immerse the serum bottle in the water bath after re-equilibrating to 56°C and leave for 30 min. Swirl the bottle occasionally to ensure proper mixing. Allow the serum to cool to room temperature, aliquot into 50-mL tubes, and store at –20°C.
4. When mechanically dissociating the mesencephalic tissue, it is important to use medium with protein (e.g., fetal calf serum). Alternatively, one can use Neurobasal medium with 1% (w/v) bovine serum albumin. The presence of protein protects cells from damage caused by trituration.
5. It is preferable to use a long (9 in.) Pasteur pipette for dissociation, rather than a shorter (e.g., 6 in.) pipette. The dissociation of tissue occurs mainly as a result of the shear force within the narrow bore (lower) part of the pipette, and the longer pipette facilitates the process.
6. Poly-L-ornithine (30,000–70,000 MW) can be substituted for poly-L-lysine.
7. In some cases, one may find that fetal calf serum is cytotoxic for neurons. It is advisable to test several batches of serum from different suppliers before initiating a series of experiments making use of neuron-glia cultures.
8. Benztropine (0.5 μM) can also be used to specifically block (^{3}H)DA uptake.
9. As additional controls, (^{3}H)DA uptake can be assessed in the presence of either desmethylimipramine (0.5 μM) or fluoxetine (1 μM), drugs which block the transport of catecholamines in noradrenergic terminals and amines in serotoninergic nerve terminals, respectively.
10. Alternatively, cultures can be fixed in 2% glyoxylic acid/0.5% paraformaldehyde in Krebs-Ringer bicarbonate buffer (pH 7) for 15 min.

11. BDNF is supplied as a solid. When reconstituting, it is important to use a solution with carrier protein, for example, PBS with 0.1% BSA. Neurotrophins (NGF, BDNF, NT-3, NT-4) have a very high isoelectric point (which means they are quite basic) and will stick to glass and plastic (especially polystyrene) surfaces. We recommend reconstitution in the container provided by the supplier to a concentration between 25 and 100 μg/mL. Aliquot this stock into small Eppendorf tubes (10–20 μL/tube) and store at −20°C for up to 6 months. Once thawed, an aliquot can be safely kept in the refrigerator for up to 4 weeks. *Do not subject BDNF to repeated freeze-thaw cycles.*
12. When preparing solutions of TOPA, the following are recommended: (1) in 1 N HCl, up to 50 mg/mL. Solutions should be freshly prepared and protected from exposure to light. (2) In water, up to 3 mg/mL. Dissolve in oxygen-free boiled water containing 0.1% sodium metabisulfite or other antioxidant. Solutions should be freshly prepared and protected from exposure to light.

Acknowledgments

L. Facci was supported by "Fondazione CARIPARO Progetto Dottorati di Ricerca" Anno 2009.

References

1. Forno LS (1996) Neuropathology of Parkinson's disease. *J Neuropathol Exp Neurol* 55, 259–272
2. Gelb DJ, Oliver E, and Gilman S (1999) Diagnostic criteria for Parkinson disease. *Arch Neurol* 56, 33–39
3. Hyman C, Hofer M, Barde Y-A, Juhasz M, Yancopoulos GD, Squinto SP, et al. (1991) BDNF is a neurotrophic factor for dopaminergic neurons in the substantia nigra. *Nature* 350, 230–232
4. Knüsel B, Winslow JW, Rosenthal A, Burton LE, Seid DP, Nikolics K, et al (1991) Promotion of central cholinergic and dopaminergic neuron differentiation by brain-derived neurotrophic factor but not neurotrophin-3. *Proc Natl Acad Sci USA* 88, 961–965
5. Somoza R, Juri C, Baes M, Wyneken U, and Rubio FJ (2010) Intranigral transplantation of epigenetically-induced BDNF-secreting human mesenchymal stem cells: implications for cell-based therapies in Parkinson s disease. *Biol Blood Marrow Transplant* 16, 1530–1540
6. Wilkins A, Kemp K, Ginty M, Hares K, Mallam E, and Scolding N. Human bone marrow-derived mesenchymal stem cells secrete brain-derived neurotrophic factor which promotes neuronal survival in vitro. *Stem Cell Res* 2009 Mar 27. (Epub ahead of print)
7. Wang R, Li YH, Xu Y, Li YB, Wu HL, Guo H, et al (2010) Curcumin produces neuroprotective effects via activating brain-derived neurotrophic factor/TrkB-dependent MAPK and PI-3K cascades in rodent cortical neurons. *Prog Neuropsychopharmacol Biol Psychiatry* 34, 147–153
8. Wu CL, Chen SD, Yin JH, Hwang CS, and Yang DI (2010) Erythropoietin and sonic hedgehog mediate the neuroprotective effects of brain-derived neurotrophic factor against mitochondrial inhibition. *Neurobiol Dis* 40, 146–154
9. Prochiantz A, Di Porzio U, Kato A, Berger B, and Glowinski J (1979) In vitro maturation of mesencephalic dopaminergic neurons from mouse embryos is enhanced in presence of

their striatal target cells. *Proc Natl Acad Sci USA* 76, 5387–5391

10. Ferrari G, Minozzi M-C, Toffano G, Leon A, and Skaper SD (1989) Basic fibroblast growth factor promotes the survival and development of mesencephalic neurons in culture. *Dev Biol* 133, 140–147
11. Qin L, Liu Y, Wang T, Wei SJ, Block ML, Wilson B, et al (2004) NADPH oxidase mediates lipopolysaccharide-induced neurotoxicity and proinflammatory gene expression in activated microglia. *J Biol Chem* 279, 1415–1421
12. Gao HM, Liu B, and Hong JS (2003) Critical role for microglial NADPH oxidase in rotenone-induced degeneration of dopaminergic neurons. *J Neurosci* 23, 6181–6187
13. Olney JW, Zorumski CF, Stewart GR, Price MT, Wang G, and Labruyere J (1990) Excitotoxicity of L-DOPA and 6-OH-DOPA: Implications for Parkinson's and Huntington's diseases. *Exp Neurol* 108, 269–272
14. Rosenberg PA, Loring R, Xie Y, Zaleskas V, and Aizenman E (1991) 2,4,5-Trihydroxyphenylalanine in solution forms a non N-methyl-D-aspartate glutamatergic agonist and neurotoxin. *Proc Natl Acad Sci USA* 88, 4865–4869
15. Skaper SD, Negro A, Facci L, and Dal Toso R (1993) Brain-derived neurotrophic factor selectively rescues mesencephalic dopaminergic neurons from 2,4,5-trihydroxyphenylalanine-induced injury. *J Neurosci Res* 34, 478–487

Chapter 10

Preparation of Adult Spinal Cord Motor Neuron Cultures Under Serum-Free Conditions

Jose V. Montoya-Gacharna, Jhon Jairo Sutachan, Wai Si Chan, Alexandra Sideris, Thomas J.J. Blanck, and Esperanza Recio-Pinto

Abstract

Spinal cord motor neuron cultures are an important tool for the study of mechanisms involved in motor neuron survival, degeneration and regeneration, volatile anesthetic-induced immobility, motor neuron disorders such as amyotrophic lateral sclerosis or spinal muscular atrophy as well as in spinal cord injury. Embryonic spinal cord motor neurons derived from rats have been successfully cultured; unfortunately, the culture of adult motor neurons has been problematic due to their short-term survival. Recently, by using a cocktail of target-derived factors, neurotrophins (brain-derived neurotrophic factor and glial cell line-derived neurotrophic factor) and a permeable cyclic adenosine monophosphate analog, we have established a reproducible protocol for long-term cultures of healthy and functional adult motor neurons (Exp Neurol 220:303–315, 2009). Here, we now describe in detail the steps that we used for the optimization of the process of isolation and maintenance of adult rat ventral horn motor neurons in vitro.

Key words: Motor neurons, Spinal cord, Brain-derived neurotrophic factor, Glial cell line-derived neurotrophic factor, Cyclic AMP, Conditioned medium, Adult rat

1. Introduction

Embryonic spinal cord motor neuron cultures have been a very important research tool that has provided insights on the mechanisms involved in their physiology (1–6) and in several motor neuron diseases such as amyotrophic lateral sclerosis (6–8). However, the biology of embryonic motor neurons differs from their adult counterpart. Several physiological changes occur in motor neurons after birth (9–13) making an in vitro model of adult spinal cord motor neurons desirable. For a long time, it has been known that embryonic spinal cord motor neurons (SCMNs) require target-derived factors to survive in vitro (14–25). Interestingly, we recently found that adult rat spinal cord motor neurons survive and become

Stephen D. Skaper (ed.), *Neurotrophic Factors: Methods and Protocols*, Methods in Molecular Biology, vol. 846,
DOI 10.1007/978-1-61779-536-7_10,

electrically functional in the presence of soluble muscle-derived factors but also require the presence of two neurotrophins, brain-derived neurotrophic factor (BDNF) and glial cell line-derived neurotrophic factor (GDNF), and a cell-permeable cyclic adenosine monophosphate analog (cAMP) (26). Adult mouse motor neurons can also survive for 24 h in the presence of soluble muscle-derived factors (25) and for longer times with a mixture of muscle-conditioned medium, BDNF, GDNF, ciliary neurotrophic factor, and cAMP (data not published). This supports the view that mature SCMNs are strongly dependent on their interactions with neurotrophins, cAMP, and soluble factors from their target tissue. A similar trophic dependence of mature SCMNs has also been described by Das et al. (27, 28). Adult rat SCMNs survive in the presence of a complex cocktail of neurotrophins (27) and they reach electrical functionality in the presence of neurotransmitters (28).

Here, we describe a practical and detailed protocol for the isolation and maintenance of adult SCMNs derived from rats (Figs. 1 and 2). This protocol requires rapid isolation and dissection of the spinal cord at low temperature, separation of myelin and purification of the neuronal cell fraction with a density gradient according to Brewer and Torricelli (29) (Fig. 2), cell attachment on poly-D-lysine (PDL) substrate, and maintenance with a serum-free neuronal culture medium (Neurobasal-A with B-27 supplement) containing myocyte-conditioned medium generated from culturing C2C12 cells (mouse myoblast cell line), BDNF, GDNF, and cAMP. The adult SCMNs obtained using this protocol (Fig. 3) express proteins and transcription factors selective to SCMNs such as the non-phosphorylated neurofilament H, the transcription factors HB9 and Islet-1, and the enzyme choline acetyltransferase responsible for synthesis of the neurotransmitter acetylcholine (26). These SCMNs also express the major synaptic protein synaptophysin (Fig. 4) (data not published) that has been involved in regulating activity-dependent synapse formation (30). Cultured SCMNs are also electrically active: They express outward K^+ currents as early as 1 h after plating, and, fast and slow inward currents become detectable after 1 week in culture (26). Action potentials can also be evoked (26). Calcium transients have been detected after 1 week of culture (Fig. 5) (data not published) using calcium imaging techniques. This in vitro system provides rapid and efficient serum-free cultures of adult SCMNs and hence represents a valuable tool for future studies on motor neuron survival, synapse formation, anesthetic-induced immobility, motor neuron diseases, and spinal cord injury.

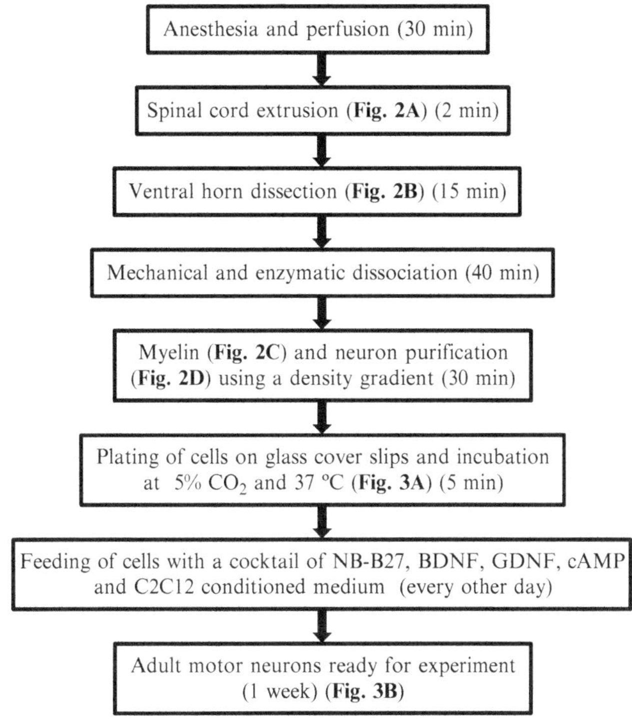

Fig. 1. Flowchart protocol for the isolation and maintenance of adult rat spinal cord motor neurons in vitro. Correlate the procedure with Fig. 2.

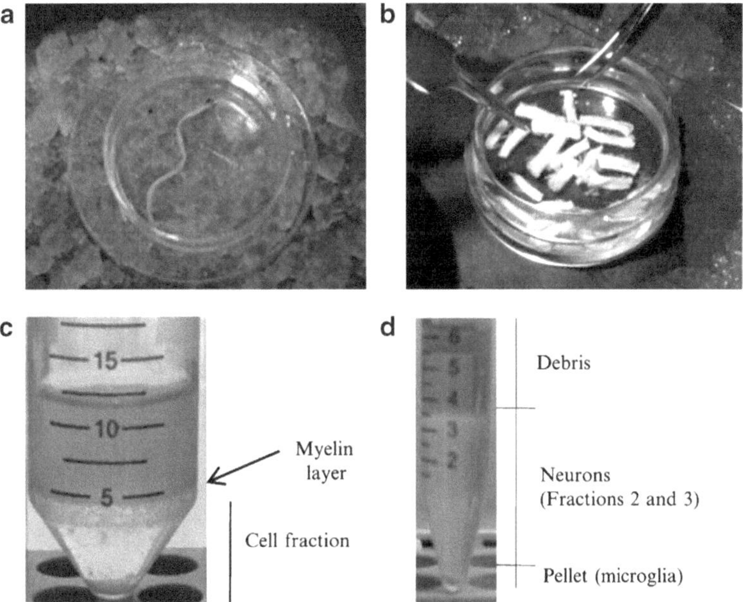

Fig. 2. Isolation and purification of spinal cord neurons. (**a**) Spinal cord obtained by hydraulic spinal cord extrusion (modified from Meikle and Martin (34)). (**b**) Ventral horn segments in neuronal isolation medium. (**c**) Density gradient for myelin removal (modified from Brewer (33)). (**d**) Density gradient for neuron purification (modified from Brewer and Torricelli (29)).

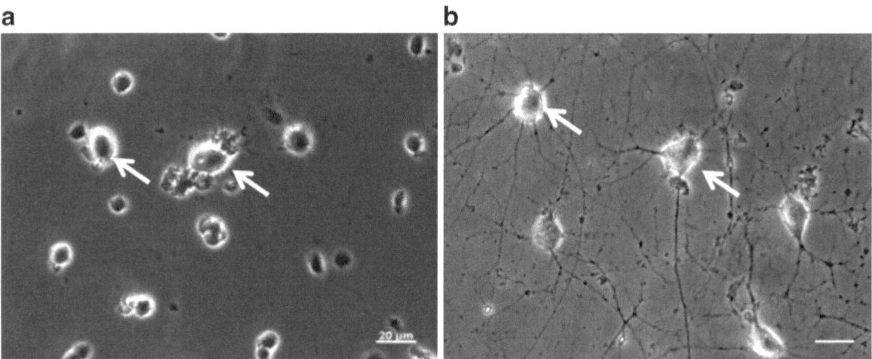

Fig. 3. Adult rat spinal cord cultures 1 h (**a**) and 7 days (**b**) after plating. Cells were maintained with a cocktail of GDNF, BDNF, cAMP, and C2C12 myocyte-conditioned medium (26). The *arrows* indicate large phase bright cells which are considered morphologically as motor neurons. These neurons display extensive neurites in culture. Bar = 20 µm.

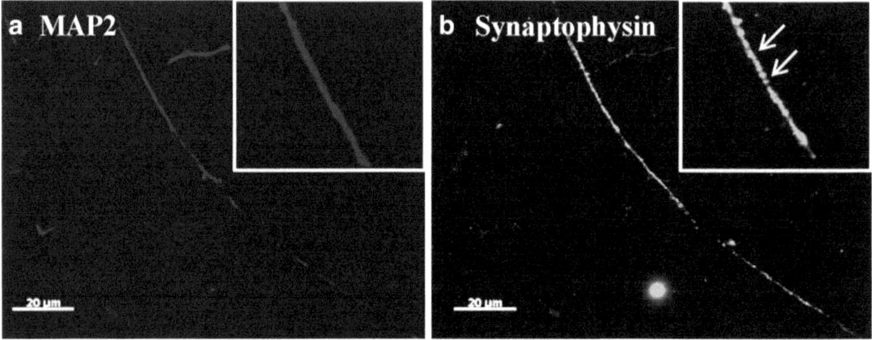

Fig. 4. Spinal cord neurons express a major synaptic protein, synaptophysin in vitro (7 days post plating). Cell cultures were co-immunostained with microtubule-associated protein 2 (**a**) and synaptophysin (**b**). Synaptophysin puncta (*arrows*) are found in the microtubule-associated protein 2-positive neurites. *Inset*: Neurites at higher magnification. Bar = 20 µm.

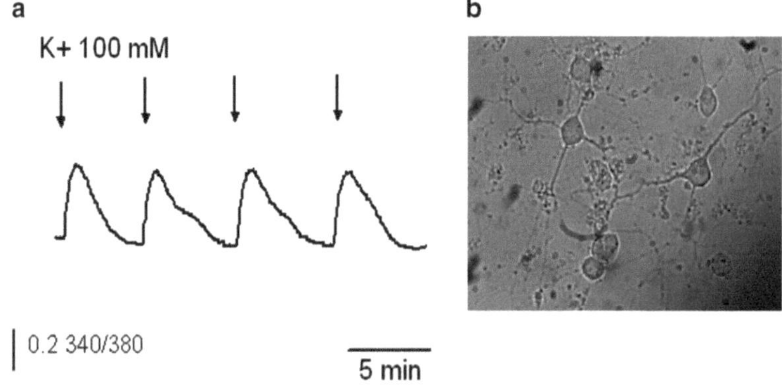

Fig. 5. KCl-evoked calcium cytoplasmic transients in dissociated adult rat spinal neurons (14 days in vitro). (**a**) The SCMNs cultures were loaded with the calcium indicator Fura 2-AM (5 µM) (Molecular Probes) and stimulated with 100 mM KCl. The excitation wavelengths for Fura (338 and 380 nm) were alternately (every 0.55 s) applied and the emitted fluorescence was filtered with the fluorescence barrier filter BA 475 nm and collected with an intensified CCD camera (ATTO Instruments, Rockville, MD, USA). (**b**) Phase bright light image of the spinal cord culture after KCl depolarization (40×).

2. Materials

Prepare all solutions using 0.22-μm filtered double distilled water. Every solution and medium in contact with cell cultures should be equilibrated at least 1 h in the cell incubator at 5% CO_2 and 37°C unless specified otherwise. Follow all waste disposal regulations.

2.1. C2C12 Myocyte-Conditioned Medium

1. C2C12 mouse myoblast cell line (CRL-1772, ATCC, VA). C2C12 is a sub-clone (31) of the mouse myoblast cell line established by Yaffe and Saxel (32).
2. C2C12 cell proliferation medium: Dulbecco's modified Eagle's medium (DMEM), 10% fetal calf serum and 1% penicillin/streptomycin. Store at 4°C.
3. C2C12 differentiation medium: DMEM, 1% penicillin/streptomycin supplemented with 10% horse serum. Store at 4°C.
4. Cryoprotectant medium: DMEM, 10% fetal calf serum supplemented with 5% (v/v) dimethylsulfoxide.
5. 0.25% Trypsin-0.53 mM EDTA. Prepare 5 mL aliquots in 15-mL conical tubes. Store at −20°C.

2.2. Intracardiac Animal Perfusion

1. Adult Sprague–Dawley rats (70–100 days old) (see Note 1).
2. A dose of 80 mg/kg ketamine (Ketaset® 100 mg/mL; Fort Dodge Animal Health, IA) and 12 mg/kg xylazine (AnaSed® 20 mg/mL; Shenandoah, IA) is used per animal. Store anesthetics at room temperature in a locked cabinet.
3. Artificial cerebrospinal fluid solution (ACSF): 0.4 mM dextran, 15 mM glucose, 3 mM KCl, 125 mM sucrose, 2.1 mM HEPES, 1.2 mM KH_2PO_4, 125 mM glycerol, 1.3 mM $MgSO_4$, and 26 mM $NaHCO_3$. Store at 4°C. Use within a month. On the day of the spinal cord isolation, bubble the ACSF solution for 45 min with 95% O_2/5% CO_2 on ice. Adjust the pH to 7.4 on ice with 1 N NaOH (see Note 2).

2.3. Spinal Cord Cell Isolation

1. Neuronal isolation medium (NIM): Hibernate A Medium (Brain Bits, IL), 2% B27 supplement (Gibco, CA, 50×), 0.5 mM Glutamax, 100 U/mL penicillin, and 100 μg/mL streptomycin (Gibco, CA) (see Note 3). Store at 4°C.
2. 2% DNAse I (Worthington, NJ) stock solution in NIM: Use within 1–2 weeks. For long-term storage in solution follow the manufacturer's instructions. Store at −20°C.
3. Enzyme digestion solution (EDS): Prepare the day of neuron isolation. Pre-warm 10 mL of NIM at 37°C and dissolve 20 mg of papain (36 U/mL) (Worthington, NJ) (see Note 4). Add

100 μL of 2% DNAse stock solution to 9,900 μL of papain-NIM solution. Filter with a 0.22-μm membrane into a 15-mL conical tube and place on ice.

4. Density gradient for myelin separation: Prepare OptiPrep™ 1.32 (Axis-Shield, Oslo) stock solution: Mix homogeneously 4.95 mL OptiPrep with 5.05 mL Hibernate A with 2% B27 in a 15-mL conical tube (modified from Brewer, (33)) (Figs. 1 and 2c). Store at 4°C. Prepare *6% OptiPrep 1.32 layer*: Mix homogeneously 2 mL Optiprep 1.32 stock solution with 8 mL Hibernate A with 2% B27 in a 15-mL conical tube (modified from Brewer (33)). Store at 4°C (see Note 5).
5. Density gradient for neuronal cell purification according to Brewer and Torricelli (29) (see Note 6) (Figs. 1 and 2d).
6. Motor neuron basal medium: Neurobasal-A medium, 2% B27 supplement, 0.5 mM glutamine, 100 U/mL penicillin, and 100 μg/mL streptomycin. Prepare a total volume of 100 mL per spinal cord isolation.
7. PDL solution: 50 μg/mL PDL (molecular weight 70,000–150,000, Sigma, MO) dissolved in sterile water (see Note 7). Store 1 mL PDL aliquots at −20°C.
8. Acid-washed glass cover slips: 24-h acid wash (1 N HCl) of 12 mm ∅ glass cover slips (Fisher HealthCare, TX). Rinse cover slips with double distilled water at least five times. Air dry cover slips for 3 h and sterilize them by autoclaving.
9. PDL-coated glass cover slips: Add 500 μL of PDL solution to cover slips placed on 2 × 2Nunc® wells dishes (Invitrogen, CA) for 3 h. Remove the solution and rinse with 1 mL of sterile water three times. Air dry the cover slips for 2–3 h under a sterile hood.

2.4. Motor Neuron Feeding Medium

1. Concentrated stock solutions of growth factors (made under sterile conditions): 1.1 μg/mL BDNF (Invitrogen, CA) and 0.4 μg/mL GDNF (Invitrogen, CA) in 0.1% bovine serum albumin (manufacturer's recommendation) (Sigma, MO) (see Note 8). Store at −20°C for up to 6 months.
2. Concentrated stock solution of the cell-permeable cAMP analog: 25 mg/mL 8-(4 chlorophenylthio) cyclic adenosine-3′, 5′-monophosphate sodium salt (cAMP) (Sigma, MO) in double distilled water. Store at −20°C.
3. Motor neuron feeding medium (MFM): Motor neuron basal medium, 30% of C2C12 cell-conditioned medium, 1 ng/mL BDNF, 0.1 ng/mL GDNF, and 125 μM cAMP.

3. Methods

Every medium or solution in contact with the cell culture should be equilibrated in a controlled 5% CO_2 incubator at 37°C for 1 h.

3.1. Preparation of C2C12 Conditioned Medium

1. Prepare C2C12 myocyte-conditioned medium at least 2 weeks before motor neuron isolation. Handle following the biosafety level I containment protocol.
2. Add 14 mL of cold C2C12 proliferation medium into a 15-mL conical tube. Leave another empty 15-mL tube next to it.
3. Rapidly thaw C2C12 frozen cells by adding 1 mL of C2C12 proliferation medium to the aliquot of cells.
4. Pipette up and down and place that 1 mL of cell suspension into the empty conical tube. Continue this process until all C2C12 differentiation medium has been used.
5. Centrifuge the cell suspension at 1,315 × *g* for 5 min to obtain a pellet.
6. Fill up two 12.5-cm^2 flasks with 4 mL of DMEM containing 10% FCS.
7. After centrifugation carefully decant the supernatant and resuspend the pellet in 2 mL of C2C12 proliferation medium.
8. Add 1 mL of this cell suspension to each 12.5-cm^2 flask to bring the final volume to 5 mL.
9. Place the cells in the incubator at 5% CO_2 and 37°C.
10. Feed the cells every other day and let grow to confluence (1 week).
11. After approximately 1 week, fill up six 75-cm^2 flasks with 15 mL of C2C12 proliferation medium and place them in cell culture incubator 5% CO_2 at 37°C to equilibrate.
12. Remove and discard culture medium from the C2C12 culture.
13. Incubate the cells with a pre-warmed trypsin–EDTA solution for 3 min at 37°C.
14. Add 4 mL of complete growth medium to stop the action of trypsin. Aspirate cells by gently pipetting.
15. Prepare a 1:10 dilution with proliferation medium for a total volume of 15 mL for each 75-cm^2 flask.
16. Expand C2C12 cells in proliferation medium until they reach 30% confluence (see Note 9).
17. After 30% confluence is reached, rinse the cells three times with motor neuron basal medium.
18. Add 15 mL of C2C12 differentiation medium (ATCC) to differentiate the myoblasts into myocytes.

19. Within 3 days, replace the existing medium with 30 mL motor neuron basal medium.
20. Collect C2C12 muscle-conditioned medium in 50-mL conical tubes after 2 days of culture.
21. Filter the conditioned medium with a 0.22-μm membrane and store at −80°C.

3.2. Intracardiac Animal Perfusion

1. Set up the perfusion pump system (Masterflex®, Cole Parmer, IL). Attach the perfusion needle and the tubing to the pump. Run double distilled water through the tubing to remove any residue. Fill up the tubing with cold ACSF and adjust the pump to a slow steady drip (20–40 mL/min). The volume of ACSF perfused is usually 200–300 mL per animal (see Note 10).
2. Anesthetize the rats with an intraperitoneal injection containing a mixture of ketamine and xylazine following standard protocols for anesthesia of small animals.
3. Test the level of anesthesia with a firm toe pinch. All surgical procedures should be done under deep anesthesia.
4. Place the animal on its back. Palpate the sternum and the inferior border of the last rib. These landmarks define the inferior aspect of the rib cage. Make a transverse incision with sharp scissors in the upper abdomen below the rib cage. Identify the diaphragm and make a transverse incision to access the thoracic cavity. Rapidly cut the ribs toward the head at the point of the anterior thoracic line. Open up the thoracic cavity. Hold the beating heart with forceps and insert the needle into the apex of the heart. Clamp the inserted needle near the entry point. Incise the right atrium and perfuse cold oxygenated ACSF through the heart. A volume of approximately 200–300 mL of ACSF is used per animal. The perfusion is completed when the liver changes color from a dark red to a gray color, and the extremities become white.

3.3. Spinal Cord Cell Isolation

Spinal cord cells are isolated following the protocol by Brewer and Torricelli (29) with some modifications.

1. Place the rat on its abdomen and decapitate it using a guillotine. Incise the dorsal skin longitudinally from head to tail. Make a transverse incision in the thigh to localize the femur. Fracture both femurs with scissors as well as the spinal column at the level of the sacrum, approximately 4 cm from the base of the rat's tail. Pull up the sacral vertebra and identify the spinal canal.
2. Place a Petri dish filled with cold NIM at the cervical/decapitated end of the rat. Then, insert a syringe filled with 10 mL of cold ACSF in the sacral canal and apply hydraulic pressure to extrude the spinal cord into the NIM (Figs. 1 and 2a). Quickly compress the plunger of the syringe and the spinal cord should

come out of the cervical end. Collect the spinal cord in cold NIM (34) (see Note 11).

3. Immerse the spinal cord in the cold NIM (Fig. 2a) (see Note 12).
4. The spinal cord should be immersed into a Petri dish containing cold NIM at all times. Using a stereoscope, obtain transversal sections of the spinal cord using a pair of iridectomy scissors (Figs. 1 and 2b).
5. Tease away the meninges using small forceps.
6. Cut the spinal cord longitudinally through the posterior median sulcus and the anterior median fissure to obtain two halves of cord (each one containing a dorsal and a ventral column).
7. Identify a darker gray central area, which is butterfly shaped in the cross-section of the spinal cord. The dorsal and ventral horns are easily identified under a stereoscope. Separate the dorsal and ventral columns by cutting longitudinally between both ventral horns.
8. Cut the ventral horns in small pieces (2–3 mm) for better cell dissociation.
9. Add 10 mL of EDS to a 15-mL conical tube. Transfer the ventral horns to the conical tube and incubate for 5 min at room temperature followed by 25 min at 30°C in a rotating incubator.
10. In the meantime, prepare 6 mL of 0.02% DNAse in EDS medium by adding 60 μL of 2% DNAse to 6 mL of EDS medium into a 15-mL conical tube.
11. Prepare three 2-mL fire-polished Pasteur pipettes (Fisher brand) of decreasing diameter (large, medium, and small). The Pasteur pipettes are prepared by flaming the pipette tips using an alcohol burner.
12. Centrifuge the tissue at 1,228 × *g* for 5 min. All the following steps (except for the centrifugation) should be done under a sterile hood, and sterile techniques should be followed.
13. Remove the supernatant, and add 2 mL of NIM with 0.02% DNAse to the pellet.
14. Triturate the tissue ten times with each of three Pasteur pipettes starting with the pipette with the largest opening.
15. After each trituration, the homogenate is allowed to settle for 1 min and the supernatant is transferred into a separate 15-mL conical tube.
16. Add an additional 2 mL of NIM to the precipitate and repeat the process for a total of three times.
17. First gradient for myelin separation: Add 5 mL of 6% OptiPrep 1.32 to a 50-mL conical tube. Layer the final cell suspension (6 mL) on top of the 6% OptiPrep 1.32 and centrifuge at

822 × *g* for 15 min at 4°C (Figs. 1 and 2c). Discard the supernatant (6% OptiPrep) and the myelin layer (white band) at the interface. Collect the cell-containing fractions (bottom layer and the pellet). Resuspend and bring the volume to 20 mL with NIM.

18. Filter the cell suspension through a 70-μm nylon mesh.
19. Centrifuge the cell suspension at 480 × *g* for 5 min at 4°C.
20. Resuspend the pellet with 6 mL of NIM and place on ice.
21. Second gradient for neuron separation: Prepare an OptiPrep 1.32 density gradient according to Brewer and Torricelli (29). Add the 6 mL cell suspension on top of this gradient (Figs. 1 and 2d).
22. Centrifuge the gradient at 1,027 × *g* (1,900 rpm Beckman JS-4.2 rotor) for 15 min at 4°C.
23. Collect the neuronal cell fractions (Fig. 10.2d) (fractions 2 and 3 according to Brewer and Torricelli (29) which contain both neurons and glia) in separate 15-mL conical tubes.
24. Pool the neuronal cell fractions and dilute them with 5 mL of NIM and centrifuge at 244 × *g* for 2 min at 4°C to remove the OptiPrep.
25. Resuspend each pellet separately with 150 μL of MFM. Pool both cell suspensions together. Mix them by softly flicking the bottom of the tube. Plate the cells with a 10-μL tip pipette. Place a volume of 10 μL of cell suspension (6,000 cells) in the middle of each 12-mm ∅ glass cover slip.
26. Allow the attachment of the cells for 1 h at 37°C and 5% CO_2.
27. Place 500 μL of pre-warmed MFM into each culture plate by resting the pipette tip on the side of the well.
28. Evaluate cell viability after dissociation by using the trypan blue exclusion test (see Note 13).
29. Leave the spinal cord cultures in the incubator at 5% CO_2 and 37°C (see Note 14).

3.4. Feeding of Motor Neuron Cultures

The cells are fed the day after plating, and every other day until the end of the experiment. Care must be taken to minimize the time the cells are outside of the incubator and exposed to ambient CO_2 levels, as it can cause cell death. The cells should not be out of the incubator for more than 3 min at a time.

To feed the cells:

1. Under the sterile hood, remove approximately 350 μL of MFM (see Note 14).
2. Carefully add 350 mL of MFM by resting the pipette tip on the side of the well to avoid detachment of the cells.

3. Place the cells in the incubator at 5% CO_2 and 37°C. Immunocytochemistry can be performed using several motor neuron cell markers (26) at different time points. Electrically functional SCMNs can be found after 1 week of culture (26) and can be measured at different time points thereafter.

4. Notes

1. The age of the animal is a factor inversely related to the survival of spinal cord MNs. Mostly we have evaluated the survival of SCMNs derived from 70- to 100-day-old rats.
2. We usually prepare 1 L of ACSF solution. The dextran should be the first chemical to be added to at least 500 mL of double distilled water due to its low solubility. The solution should also be filtered with 0.2-µm pore membrane and stored at 4°C. On the day of the perfusion, the ACSF solution is bubbled for at least 45 min with 95% O_2/5% CO_2 on ice and the pH adjusted to 7.4 just prior to perfusing the animal.
3. B27 supplement is light sensitive and should be handled according to the manufacturer's instructions. It should be kept frozen at −80°C indefinitely when not in use. Once thawed, it should not be frozen again. Thawed B27 expires within a month from the day it is opened and can be stored at 2–8°C in the dark. We have observed cases where the survival of SCMNs varied depending on the batch of B27 used. For a series of extended experiments we recommend ordering enough B27 of the same lot in order to decrease the variability of the results. Glutamax is stable in aqueous solution (Gibco). We recommend storing 1 mL aliquots of Glutamax at −20°C. NIM solution should be prepared 1 day before isolation. A volume of 250 mL of NIM is prepared for each rat. Store the solution in an aluminum foil covered bottle at 4°C.
4. The NIM should be warmed to 37°C before the papain is added. Mix the solution well on a vortex for 5 min and filter it with 0.22-µm membrane. As the papain powder is very light, it is best to wear a mask to avoid inhaling the enzyme during weighing. Store on ice until ready to use. This volume is sufficient for one spinal cord.
5. Density stock solutions can be stored at 4°C and used within 1 month. Mix well by vortexing before preparing the density gradient.
6. We recommend alternating between clear Hibernate A (without phenol red) and red hibernate A (with phenol red) when preparing the density gradients so as to facilitate the visualization of the different fractions.

7. We have also evaluated several substrates like collagen, poly-ornithine, laminin, and PDL. In our experience, PDL gives the best results. Others have shown that the cell growth-promoting organosilane substrate, N-1(3[trilnethoxysilyl] propyl)-diethylenetriainine, coated on a glass surface is necessary for the survival of these neurons (27, 28). PDL should not be used after one freeze–thaw cycle. The PDL-coated glass cover slips are prepared the day before cell isolation and can be stored at 4°C. PDL-coated cover slips may be used up to 2 days later.
8. For growth factors storage follow manufacturer's recommendations. Do not store in glass, use polypropylene vials (Invitrogen). Do not store diluted solutions and avoid adsorptive loss using at least 0.1% bovine serum albumin (Invitrogen). Avoid repeated freeze–thaw cycles of the stock solutions. In our studies, we found that some of the essential factors for the survival of dissociated SCMNs were cAMP and muscle-conditioned medium. However, other molecules may be involved in the survival of SCMNs. For instance, Das et al. (27, 28) showed that a more complex mixture of factors (BDNF, GDNF, neurotrophin-3, heparin sulfate, cardiotrophin-1, vitronectin, and acidic fibroblast growth factor) is necessary for the maintenance of adult SCMNs under their conditions.
9. Do not let the C2C12 cells become confluent because the myoblasts will be depleted (ATCC).
10. Make sure there are no bubbles in the tubing while perfusing as they can destroy the rat tissues.
11. A 16G 1 1/2 needle is inserted into the spinal canal to form a tight seal. The seal is important because otherwise the spinal cord will not be extruded. Additionally, it is necessary to make sure that approximately three quarters of the needle is inside the spinal canal. When inserting the needle, there should be a slight resistance. If it is too difficult to insert, stop, readjust the needle and/or make another transverse cut of the sacral spine if necessary. Never force the needle, because you will break the bone, and extrusion may become impossible at this point as a seal cannot be formed. On the other hand, be careful not to cut the spinal column too high (lumbar region) because if the needle slides inside the canal too easily, enough pressure cannot be sustained, and extrusion will not be successful.
12. Keep the time between decapitation and immersion of the spinal cord into the medium as short as possible (2 min) as the spinal cord neurons are very sensitive to ischemia. Most of our experiments were done with the cervical enlargement of the spinal cord; however, any segment of the cord can be used.
13. The viability of the SCMNs cultures is more than 90% (26) after plating.

14. The SCMN cells are quite sensitive to temperature and pH, and therefore it is essential to change the media quickly to minimize the exposure to colder temperature and ambient pH levels. If the cells remain outside for too long they are likely to detach. We recommend the use of 2×2 Nunc® well plates because they allow quick medium change. We also recommend resting the well plate on a tube rack that is placed in the hood (the night before, so it can be sterilized by the UV light) because this reduces the contact the cells would have to the relatively cold hood surface.

References

1. Zhang HM, Robinson N, Gomez-Curet I, Wang W, and Harrington MA (2009) Neuronal and network activity in networks of cultured spinal motor neurons, Neuroreport 20, 849–854
2. Streit J, Tscherter A, Heuschkel MO, and Renaud P (2001) The generation of rhythmic activity in dissociated cultures of rat spinal cord, Eur. J. Neurosci. 14, 191–202
3. Das M, Molnar P, Devaraj H, Poeta M, and Hickman JJ (2003) Electrophysiological and morphological characterization of rat embryonic motoneurons in a defined system, Biotech. Prog. 19, 1756–1761
4. Wallis DI, Elliott P, Foster GA, and Stringer BM (1998) Synaptic activity, induced rhythmic discharge patterns, and receptor subtypes in enriched primary cultures of embryonic rat motoneurones, Can J Physiol Pharmacol 76, 347–359
5. Fruns M, Krieger C, and Sears TA (1987) Identification and electrophysiological investigations of embryonic mammalian motoneurones in culture, Neurosci Lett 83, 82 88
6. Silani V, Braga M, Botturi A, Cardin V, Bez A, Pizzuti A, and Scarlato G (2001) Human developing motor neurons as a tool to study ALS, Amyotroph Lateral Scler Other Motor Neuron Disord 2 Suppl 1, S69–76
7. Bar PR (2000) Motor neuron disease in vitro: the use of cultured motor neurons to study amyotrophic lateral sclerosis, Eur. J Pharmacol 405, 285–295
8. Jackson M, Ganel R, and Rothstein JD (2002) Models of amyotrophic lateral sclerosis, Curr Protoc Neurosci Chapter 9, Unit 9 13.
9. Cameron WE, and Nunez-Abades PA (2000) Physiological changes accompanying anatomical remodeling of mammalian motoneurons during postnatal development, Brain Res Bull 53, 523–527
10. Carrascal L, Nieto-Gonzalez JL, Cameron WE, Torres B, and Nunez-Abades PA (2005) Changes during the postnatal development in physiological and anatomical characteristics of rat motoneurons studied in vitro, Brain Res Brain Res Rev 49, 377–387
11. Viana F, Bayliss DA, and Berger AJ (1994) Postnatal changes in rat hypoglossal motoneuron membrane properties, Neuroscience 59, 131–148
12. Berger AJ, Bayliss DA, Bellingham MC, Umemiya M, and Viana F (1995) Postnatal development of hypoglossal motoneuron intrinsic properties, Adv Exp Med Biol 381, 63–71
13. Nunez-Abades PA, and Cameron WE (1995) Morphology of developing rat genioglossal motoneurons studied in vitro: relative changes in diameter and surface area of somata and dendrites, J Comp Neurol 353, 129–142
14. Giller EL, Jr, Schrier BK, Shainberg A, Fisk HR, and Nelson PG (1973) Choline acetyltransferase activity is increased in combined cultures of spinal cord and muscle cells from mice, Science 182, 588–589
15. Bennett MR, Lai K, and Nurcombe V (1980) Identification of embryonic motoneurons in vitro: their survival is dependent on skeletal muscle, Brain Res 190, 537–542
16. Bennett MR, and Nurcombe V (1979) The survival and development of cholinergic neurons in skeletal muscle conditioned media, Brain Res 173, 543–548
17. Henderson CE, Huchet M, and Changeux JP (1984) Neurite-promoting activities for embryonic spinal neurons and their developmental changes in the chick, Dev Biol 104, 336–347
18. Berg DK, and Fischbach GD (1978) Enrichment of spinal cord cell cultures with motoneurons, J Cell Biol 77, 83–98

19. O'Brien RJ, and Fischbach GD (1986) Isolation of embryonic chick motoneurons and their survival in vitro, J Neurosci 6, 3265–3274
20. Dohrmann U, Edgar D, Sendtner M, and Thoenen H (1986) Muscle-derived factors that support survival and promote fiber outgrowth from embryonic chick spinal motor neurons in culture, Dev Biol 118, 209–221
21. Smith RG, Vaca K, McManaman J, and Appel SH (1986) Selective effects of skeletal muscle extract fractions on motoneuron development in vitro, J Neurosci 6, 439–447
22. Comella JX, Sanz-Rodriguez C, Aldea M, and Esquerda JE (1994) Skeletal muscle-derived trophic factors prevent motoneurons from entering an active cell death program in vitro, J Neurosci 14, 2674–2686
23. Arce V, Pollock RA, Philippe JM, Pennica D, Henderson CE, and deLapeyriere O (1998) Synergistic effects of schwann- and muscle-derived factors on motoneuron survival involve GDNF and cardiotrophin-1 (CT-1), J Neurosci 18, 1440–1448
24. Wiese S, Herrmann T, Drepper C, Jablonka S, Funk N, Klausmeyer A, Rogers M L, Rush R, and Sendtner M (2010) Isolation and enrichment of embryonic mouse motoneurons from the lumbar spinal cord of individual mouse embryos, Nat Protoc 5, 31–38
25. Eagleson KL, and Bennett MR (1983) Survival of purified motor neurones in vitro: effects of skeletal muscle-conditioned medium, Neurosci Lett 38, 187–192
26. Montoya G J, Sutachan JJ, Chan WS, Sideris A, Blanck TJ, and Recio-Pinto E (2009) Muscle-conditioned media and cAMP promote survival and neurite outgrowth of adult spinal cord motor neurons, Exp Neurol 220, 303–315
27. Das M, Bhargava N, Gregory C, Riedel L, Molnar P, and Hickman JJ (2005) Adult rat spinal cord culture on an organosilane surface in a novel serum-free medium, In vitro Cell Dev Biol 41, 343–348
28. Das M, Bhargava N, Bhalkikar A, Kang JF, and Hickman JJ (2008) Temporal neurotransmitter conditioning restores the functional activity of adult spinal cord neurons in long-term culture, Exp Neurol 209, 171–180
29. Brewer GJ, and Torricelli JR (2007) Isolation and culture of adult neurons and neurospheres, Nat Protoc 2, 1490–1498.
30. Tarsa L, and Goda Y (2002) Synaptophysin regulates activity-dependent synapse formation in cultured hippocampal neurons, Proc Natl Acad Sci USA 99, 1012–1016
31. Blau HM, Pavlath GK, Hardeman EC, Chiu CP, Silberstein L, Webster SG, Miller SC, and Webster C (1985) Plasticity of the differentiated state, Science 230, 758–766
32. Yaffe D, and Saxel O (1977) Serial passaging and differentiation of myogenic cells isolated from dystrophic mouse muscle, Nature 270, 725–727
33. Brewer GJ (1997) Isolation and culture of adult rat hippocampal neurons, J Neurosci Meth 71, 143–155
34. Meikle AD, and Martin AH (1981) A rapid method for removal of the spinal cord, Stain Technol 56, 235–237

Chapter 11

Rodent Retinal Ganglion Cell Cultures

Stephen D. Skaper

Abstract

Central neurons lose the ability for axonal regrowth during development and typically do not regenerate their axons following axotomy once they become mature unless given a growth-permissive environment, for example, a peripheral nerve graft. Retinal ganglion cells (RGCs) of the optic nerve represent a highly useful cell model for the study of neurotrophic factor responsiveness, although the presence of nonneuronal cells in the retina makes it difficult to interpret the direct effects of tested factors on RGCs. Cultures of purified RGCs thus represent an excellent tool for the study of optic nerve cell trophic responsiveness, in terms of both survival and axonal regeneration.

Key words: Retina, Ganglion cells, Rat, Neonatal, Embryonic, Culture, Antibody panning, Neurotrophic factors, Axonal regeneration

1. Introduction

Adult central nervous system (CNS) neurons typically do not regenerate their axons following axotomy, with the consequence that CNS injury results in permanent impairment. Studies have concluded that it is the CNS environment that is inhibitory to regenerative growth and that at least some of the adult CNS neurons do have an intrinsic ability to regenerate if given a permissive environment (1, 2). The optic nerve is a readily accessible CNS tract that contains glial cells (oligodendrocytes, astrocytes, and microglia) and only one type of axon, those arising from retinal ganglion cells (RGCs). Neurotrophic factors applied intraocularly or to the axotomized optic nerves enhance the survival of RGCs in adult rodents during early stages after injury. Such factors include brain-derived neurotrophic factor (BDNF) (3, 4), neutrophin-4/5 (5, 6), nerve growth factor (7, 8), ciliary neurotrophic factor (CNTF) (9, 10), and fibroblast growth factor (11, 12). Although

Stephen D. Skaper (ed.), *Neurotrophic Factors: Methods and Protocols*, Methods in Molecular Biology, vol. 846,
DOI 10.1007/978-1-61779-536-7_11,

RGCs can be studied as a flat-mounted preparation, as explants, or as dissociated retinal cultures, the presence of nonneuronal cells in the retina makes it difficult to interpret the direct effects of tested factors on RGCs. Cultures of purified RGCs thus represent an excellent tool for the study of optic nerve cell trophic responsiveness, in terms of both survival and axonal regeneration. Indeed, rat RGCs display age-dependent responsiveness to isolated neurotrophic factors, as well as Schwann cell–secreted factors (13). The protocol described here outlines a procedure for the purification of rat RGCs from a whole retina cell suspension (14). The proteolytic enzyme papain is used to dissociate rat retinal tissue (15–17). Purification is effected by the technique of antibody-mediated plate adhesion, "panning," developed by Wysocki and Sato (18). This procedure yields a highly pure (>99%) population of ganglion cells, with yields ranging from 25% to 50%.

2. Materials

2.1. Equipment and Labware

1. Stereo dissecting microscope with fiber optic light source.
2. Laminar flow cabinet for dissections.
3. Laminar flow biological safety cabinet (CL2).
4. Humidified, water-jacketed culture incubator at 37°C and 5% CO_2/95% air.
5. Water bath set at 37°C.
6. Dissecting tools (Fine Science Tools).
7. Benchtop centrifuge to accommodate 15- and 50-mL tubes.
8. 15- and 50-mL polypropylene plastic centrifuge tubes (sterile).
9. 12-mL test tubes, sterile.
10. 6-mL flat bottom tubes (Bijou), sterile.
11. 1.5-mL Eppendorf tubes.
12. 10-cm ∅ sterile tissue culture dishes.
13. 6-, 10-, and 15-cm ∅ sterile Petri culture dishes.
14. #21- and #23-gauge hypodermic needles.
15. 5- and 10-mL syringes.
16. Nitex mesh, 20 μm (Tetko HC3-20).
17. 0.22- and 0.45-μm filters (Millipore).
18. Neubauer hemocytometer (Fisher Scientific).
19. 24- and 96-well tissue culture plates.
20. 24- and 96-well poly-D-lysine treated tissue culture plates (BD Biosciences).

2.2. Reagents

1. Earle's balanced salt solution (EBSS) (Invitrogen).
2. EBSS containing sodium bicarbonate and phenol red but lacking Ca^{2+} and Mg^{2+} (CMF) (Invitrogen).
3. Dulbecco's phosphate-buffered saline (D-PBS) (Invitrogen).
4. Minimal essential medium with Earle's salts (MEM) (Invitrogen).
5. Dulbecco's modified Eagle's medium (DMEM), containing 4.5 g/L glucose, L-glutamine, and pyruvate (Invitrogen).
6. DMEM (American Type Culture Collection, ATCC).
7. Neurobasal medium (Invitrogen).
8. B27 supplement, 50×, with antioxidants (Invitrogen).
9. Fetal calf serum (FCS) (Invitrogen) (see Note 1).
10. Penicillin/streptomycin, 10,000 U/mL penicillin + 10,000 μg/mL streptomycin (100× stock), sterile, for cell culture (Invitrogen).
11. L-Glutamine (200 mM stock), sterile, for cell culture (Invitrogen).
12. Sodium pyruvate (100 mM stock), sterile, for cell culture (Sigma).
13. Papain (Worthington (Lorne)).
14. DNase, type I (Worthington (Lorne)).
15. Trypsin, 2.5% solution, sterile, for cell culture (Sigma).
16. Trypsin (0.25%)-EDTA, sterile, for cell culture (Invitrogen).
17. Poly-D-lysine (MW 70,000–150,000), sterile, for cell culture (Sigma).
18. Merosin, human (Invitrogen).
19. Bovine serum albumin (BSA), essentially fatty acid and globulin-free (Sigma).
20. Ovomucoid (trypsin inhibitor from chicken egg white, type III-O, free of ovoinhibitor) (Sigma).
21. L-Cysteine hydrochloride (Sigma).
22. Insulin, from bovine pancreas (Sigma).
23. Transferrin, human, partially iron saturated, tested for cell culture (Sigma).
24. Progesterone (Sigma).
25. Putrescine (Sigma).
26. Sodium selenite (Sigma).
27. T3 (3,3′,5-triiodo-L-thyronine, sodium salt), cell culture tested (Sigma).
28. Thyroxine (T4) (Sigma).
29. Affinity-purified goat anti-mouse IgG + IgM (H + L) (Jackson ImmunoResearch).

30. Affinity-purified goat anti-mouse IgM, mu-chain specific (Jackson ImmunoResearch).
31. Anti-rat-macrophage antiserum (Accurate/Axell).
32. RAN-2 hybridoma TIB-119™ (ATCC).
33. Thy 1.1 IgM hybridoma TIB-103™ (name T11D7e2) (ATCC).
34. BDNF, human, recombinant (PeproTech).
35. CNTF, human, recombinant (PeproTech).
36. Glial cell line–derived neurotrophic factor, human, recombinant (GDNF) (PeproTech).
37. Basic fibroblast growth factor (PeproTech).
38. Forskolin (Sigma).
39. Trypan blue stain 0.4%, for cell culture (Invitrogen).
40. 3-(4,5-dimethylthiazol-2-yl)-2,5-diphenyl tetrazolium bromide (MTT) (Sigma).
41. Dimethyl sulfoxide, for cell culture (Sigma).

2.3. Solutions for Tissue Dissociation

1. *Papain solution.* Immediately before the start of the dissection, prepare a papain solution by adding 165 units (for postnatal day 8 (P8) rats, for embryonic, much less is needed) of papain to 10 mL of EBSS (or P-PBS or MEM) in a 15-mL blue-top centrifuge tube. Next, add 100 μL of a 4 mg/mL solution of DNase. Dissolve the latter by placing the mixture, which is in a 15-mL centrifuge tube, in a test tube rack in a 37°C water bath. About 10 min before use, mix the papain solution with 2.4 mg of L-cysteine hydrochloride and pass through a 0.22-μM filter into a sterile scintillation vial or universal (Bijou) tube.
2. *Ovomucoid solution.* Dissolve 13 mg of ovomucoid and 13 mg of BSA in 13 mL of MEM, and then add 0.13 mL of a 4 mg/mL solution of DNase; adjust to pH 7.4 and filter-sterilize.
3. *Trypsin (0.125%).* Add 200 μL of a 2.5% trypsin stock to 4 mL of CMF. Be sure to prepare the trypsin aliquots immediately upon trypsin arrival; store at –20°C until use.
4. *Trypsin inactivating solution.* Combine 3 mL of FCS plus 10 mL DMEM.
5. *Sato 100× stock solution.* To 20 mL of Neurobasal medium, add the following (final concentrations in brackets): 200 mg transferrin (100 μg/mL), 200 mg BSA (100 μg/mL), 5 μL of stock progesterone solution (stock is 2.5 mg/100 μL ethanol) (60 ng/mL or 0.2 μM), 32 mg putrescine (16 μg/mL), 200 μL of stock sodium selenite (stock is 40 mg in 1 mL of 0.1 N NaOH plus 10 mL Neurobasal-A) (40 ng/mL), 10 μL of stock T3 (stock is 8 mg in 1 mL of 0.1 N NaOH, 30 ng/mL), and 10 μL of stock T4 (stock is 8 mg in 1 mL of 0.1 N NaOH, 40 ng/mL).

Aliquot the above 100× stock solutions (200 μL) and store at −20°C. The ethanol stocks for progesterone, sodium selenite, T3, and T4 should be prepared fresh each time a new batch of 100× Sato medium is made.

2.4. Sato Serum-Free Culture Medium

1. To 20 mL of Neurobasal medium (19), add the following: 200 μL Sato 100× stock solution, 400 μL 50× B27 supplement, 200 μL sodium pyruvate 100× stock, 200 μL glutamine 100× stock, 200 μL of 100× penicillin/streptomycin stock (stored at −20°C), and 20 μL *N*-acetyl cysteine 1,000× stock.
2. Next, add 200 μL of 100× insulin stock (stored at 4°C; stock must be prepared every 4–6 weeks—do not freeze). The 100× insulin stock (0.5 mg/mL) is prepared by adding 2 mg of insulin to 4 mL of tissue culture grade water containing 20 μL of 1 N HCl. Filter the stock through a 0.22-μm filter and store at 4°C.
3. Filter the above medium through a 0.22-μm filter. Before filtering the medium, filter about 5 mL of Neurobasal medium to rinse away traces of detergent in the filter. The quality of the filter matters—Millex-GV from Millipore is best. Falcon plastic is preferable, along with Monoject syringes.
4. Recommended trophic factors for mixed optic nerve cultures: insulin (5 μg/mL) plus CNTF (10 ng/mL). Do not filter the medium once the growth factors are added due to the risk of their sticking to the filter. In addition, forskolin (5 μM) can be included in the medium, as it promotes survival of mixed and purified cultures.
5. For purified RGCs, add BDNF (50 ng/mL), CNTF (10 ng/mL), and forskolin (5 μM final concentration; use a 1,000× stock prepared in dimethyl sulfoxide). Optional: Add basic fibroblast growth factor (10 ng/mL) for low-density culture.

3. Methods

3.1. Dissection

1. Normally, one to ten rats from a Sprague-Dawley or Long-Evans litter (Charles River) are used, typically between P8 and P10 (see Note 2).
2. Remove the retinas from the eye in situ. Sacrifice the animal by decapitation (following approved institutional protocols), and remove the skin overlying the eyes.
3. Pin back the head to a wax block and remove the anterior tissues of the eye with a #11 scalpel blade while the cornea is held with a pair of rat-tooth, microdissecting forceps. Remove

the lens and vitreous humor with forceps. Then, gently lift the retina away with a small spatula.

4. Store the retinas at room temperature in EBSS containing calcium and magnesium, pH 7.4, until retinas are removed from all animals.

3.2. Tissue Dissociation

1. Upon completion of the dissection, remove the EBSS with a sterile Pasteur pipette and replace with 2 mL of the papain solution (see Notes 3 and 4).
2. Transfer the retinas to the vial by pouring, and place the vial in a test-tube rack in the water bath at 37°C. Every 10 min or so, gently swirl the retinas. The incubation time with the papain solution is 30 min (see Note 5).
3. Remove the vial from the water bath and transfer the retinas (by pouring) to a 15-mL centrifuge tube. Allow the retinas to settle, remove the papain solution with a pipette, and rinse the tissue gently with 3 mL of ovomucoid solution. Allow the tissue pieces to settle and remove the rinse solution (see Note 6).
4. Triturate the tissue with a 1-mL pipette in the ovomucoid solution. The following steps are repeated ten times. Add 1 mL of ovomucoid solution, gently pull the tissue up into a 1-mL pipette and expel. Centrifuge the tissue dissociate for a few seconds at 200 × *g* to remove pieces of tissue from the cell suspension, and collect the supernatant. The retinal tissue is further triturated by adding 1 mL of fresh ovomucoid solution, making one pass through the 1-mL plastic pipette, spinning down the undissociated tissue, and collecting the supernatant. After repeating this trituration procedure 6–10 times, the retinas should be completely broken up. A cell count at this point should show a cell yield of about 20 million cells/retina (see Note 7).
5. Centrifuge the resulting retinal cell suspension at 400 × *g* for 10 min to separate the retinal cells from the ovomucoid solution.
6. Discard the supernatant and resuspend cells in 10 mL of MEM containing 0.05% (w/v) BSA. The BSA serves to slow the cells' settling time on the panning dishes. The cells are now ready for the panning step.

3.3. RGC Purification by Panning

3.3.1. Preparation of the Panning Dish

1. Incubate each of two 6-cm Petri dishes with 5 mL of 50 mM Trizma buffer, pH 9.5, with affinity-purified goat anti-mouse IgM, mu-chain specific (final concentration is 5–10 μg/mL) for 12 h at 4°C. To sterilize, pass the Tris solution through a filter into a sterile dish, and then add the azide-containing antibody directly to this; otherwise, much of the protein will be lost to the filter (see Notes 8 and 9).

2. Similarly, incubate two 15-cm Petri dishes with 15 mL of anti-rabbit IgG H+L (5–10 μg/mL in Tris buffer, pH 9.5). Wash each dish three times with 8 mL of D-PBS (see Note 10).
3. Incubate the 10-cm dishes with 10 mL of Thy 1.1 IgM monoclonal supernatant for at least 1 h at room temperature. The 15-cm dishes are not incubated with any other antibodies.
4. Remove the supernatant and wash the plates three times with D-PBS.
5. In order to prevent nonspecific binding of cells to the panning dish, 5–15 mL of MEM with BSA (0.2%) is placed on each plate for at least 20 min; this solution should remain on the plate until the plate is needed.

3.3.2. Panning Procedure

1. Incubate the retinal suspension in anti-rat-macrophage antiserum (1:100 in MEM) for 20 min, centrifuge, resuspend in MEM, and incubate on a 15-cm anti-rabbit IgG panning plate at room temperature for 45 min total.
2. After 20 min, swirl the plate to ensure access of all cells to the surface of the plate. If cells from more than eight retinas are being panned, at the end of 45 min, transfer the nonadherent cells to another 15-cm anti-rabbit IgG panning plate for a further 30 min.
3. Remove the nonadherent cells (in suspension) and filter through a 20-μm Nitex mesh to remove small clumps of cells. Sterilize the mesh before use (UV exposure for a few minutes is sufficient), prewet the mesh on the bottom side, then form a cone with the filter on top of a test tube, and pass 1 mL of MEM through it before using a Pasteur pipette to transfer the cell suspension through the filter.
4. Clumps could occur either because they were not broken into single cells during trituration or because they were clumped by the antimacrophage antiserum. The latter type of clumping represents the single greatest impediment to final RGC purity as they can bring down unwanted contaminating cell types.
5. Transfer the cell suspension from the previous step to the Thy1.1 panning dishes. The use of two 10-cm dishes improves the final yield, although the use of more than this is not helpful. Leave the cell suspension on the dish(es) for 1 h while swirling every 20 min.
6. Wash the dishes at least eight times with 6 mL of D-PBS (discard after each wash) and swirl rather vigorously each wash to dislodge nonadherent cells. Monitor the progress of nonadherent cell removal by observing the dishes under a light microscope. Washing is considered complete when only adherent cells remain on the dish.

3.3.3. Removing Adherent Cells from the Panning Dish

1. Incubate adherent cells on the panning dish with 0.125% trypsin solution for 10 min in a 37°C CO_2 incubator.
2. Remove the dish from the incubator and squirt off the cells with a 1-mL pipetman while they are still in the trypsin solution. Mark one point on the outside of the dish with marker. Squirt from this point, once around the dish, by squirting from the peripheral edge toward the center, rotate the dish, squirt again, etc., until the entire surface area has been covered once. Keep to a minimum the amount of squirting necessary to remove the cells (see Note 11).
3. Transfer the cell suspension from the dish to a 15-mL centrifuge tube. Add 1 mL of the trypsin inactivating solution to these cells and add another 4 mL to the dish. Again squirt the dish. Check quickly under the microscope to make sure most of the cells have been released (if not, this probably indicates that the trypsin has lost activity).
4. Collect this 4 mL and transfer to the centrifuge tube (total volume: 9 mL). Gently invert the tube several times to mix the cells and remove a 100 μL aliquot for counting. Pour the remainder of the trypsin inactivating solution into the tube with the cells to top up the tube. Centrifuge at 200×*g* for 5 min to collect the cells.
5. Resuspend the cells in Sato medium with growth factors for purified RGCs, mix a 20 μL aliquot with an equal volume of 0.4% trypan blue, count, and plate. Expected yield: 40,000 RGCs viable (trypan blue–excluding cells) per P8 rat pup (see Notes 12–15).
6. Culture the cells on a poly-D-lysine/merosin substrate. Dissolve a 5-mg bottle of poly-D-lysine in 5 mL of 0.15 M borate buffer, pH 8.4. Add 200 μL of this poly-D-lysine stock (1 mg/mL) to 20 mL of tissue culture grade sterile water. Use 0.2 mL poly-D-lysine per cm^2 surface area. Incubate with poly-D-lysine at room temperature (20–22°C) for 60 min. Merosin coating is done after rinsing off excess poly-D-lysine: Add a 2 μg/mL solution in Neurobasal medium and incubate at 37°C for several hours to overnight (if overnight, add penicillin/streptomycin) in a 5% CO_2 incubator. The cells can be plated at 5,000 cells per well of a 96-well plate (high density) or at low density on 6-cm tissue culture dishes, as desired.

3.4. Modifications for Purifying Mouse P8 RGCs Rather than Rat

1. Use CD1 mice or other Thy1.2-positive strains, between 8 and 10 days postnatal age.
2. Instead of using an IgM Thy1.1 antibody, use an IgM Thy 1.2 antibody for the final panning dish. For example, the F7D5 Thy1.2 antibody, which is a mouse monoclonal (Accurate Chemical Co., Cat No. MAS 731).

3. Add 5 mL of 0.2% BSA in D-PBS containing F7D5 ascites fluid at a dilution of 1:1,000 or 1:500.
4. All other steps remain the same.

3.5. Modifications for Purifying Embryonic Day 18 (E18) RGCs Rather than P8 RGCs

3.5.1. Dissection

1. Decapitate embryos with a pair of #5 microdissecting forceps.
2. Remove skins overlying the eyes with the same forceps.
3. Make an insertion on the cornea with a pair of #55 microdissecting forceps.
4. Peel off the cornea with another pair of #55 forceps to expose the lens.
5. Remove the lens with the #55 forceps.
6. Lift away the retina with a small spatula.

3.5.2. Dissociation

1. Use 50 U of papain in 10 mL of D-PBS (for 30 min, same length of time as for P8).
2. The yield of purified E18 RGCs will be substantially less than for P8 because they have not all been generated yet and because they express less Thy1.

3.5.3. Culture

The same Neurobasal serum-free Sato medium, use of B27 at 1:50 is key, as well as BDNF, which will only save about a quarter of the cells in any case. The E18 RGCs do not respond to the other trophic factors.

3.6. MTT Survival Assay

1. The MTT assay is frequently used to quantify neuronal cell survival. MTT reacts with mitochondrial dehydrogenases to produce a blue formazan product in living cells, but not in dying cells or their lytic debris (20, 21). Because of the low density of RGC cultures, MTT labeled cells are counted, rather than dissolving the reaction product in dimethyl sulfoxide and reading absorbance spectrophotometrically.
2. An MTT stock solution is prepared by dissolving MTT in D-PBS at 5 mg/mL. Sterilize by passage through a 0.22-µm filter. The stock solution can be stored at −20°C for up to 6 months.
3. Add 10 µL of the MTT stock to each 100-µL well of a 96-well plate. Return the plate to the 37°C incubator for at least 1 h (maximum 2 h). Viable cells with active mitochondria cleave the tetrazolium ring into a visible dark blue/black formazan reaction product (see Note 16).
4. Count the percentage of viable cells in each well by counting at least 200 cells per well. Ideally, at least three wells per test condition should be counted. Use an eyepiece reticle with a counting grid. Cells must be counted immediately after the MTT incubation to avoid the formation of large crystals of MTT and toxicity of MTT to the cells.

4. Notes

1. Heat inactivation of fetal calf serum is recommended to destroy heat-labile complement. Thaw the bottle of serum in advance, using a 37°C water bath. Next, immerse the serum bottle in the water bath after reequilibrating to 56°C and leave for 30 min. Swirl the bottle occasionally to ensure proper mixing. Allow the serum to cool to room temperature, aliquot into 50-mL tubes, and store at −20°C.
2. The dissociation procedure works on any age retina (E18–P12); purification using rats older than P12 has not been verified using this protocol (but see ref. (13)).
3. The papain should be obtained from Worthington.
4. Do not use L15 as it inhibits papain.
5. During the dissociation procedure, never expose the cells to glutamate, aspartate, or glutamine nor allow cells to be cooled lower below room temperature to avoid toxic effects on ganglion cells.
6. Ovomucoid itself does not inhibit the papain but probably a contaminant; the source of the ovomucoid matters.
7. In step 4 of the tissue dissociation, tissue pieces may be allowed to settle by gravity rather than using centrifugation.
8. Dishes used for panning must be Petri plastic, not tissue culture plastic treated. Falcon is a good brand, and others may work but should be tested first.
9. The time of the entire panning procedure can be significantly shortened by decreasing the time of all panning plate cell incubations by half without much loss in yield or purity.
10. When solutions are removed from the panning dishes during washes, they must be instantly replaced with fresh solution so that cells do not dry out.
11. Do not wash too vigorously, or you will wash off the RGCs.
12. Do not resuspend the final cell pellet in DMEM, because there is no BSA—you may lose cells on the wall of the pipette tip from nonspecific sticking.
13. DMEM should not be used, as the osmolarity is 335. Combined with the B27 supplement and other additions indicated, the final osmolarity of 360 becomes cytotoxic. Neurobasal osmolarity is 225, becomes about 260 with the various supplements/additions.
14. The B27 supplement seems to go off quickly once opened; it is recommended to buy the small size and replace the bottle of medium frequently.

15. Do not use F12 culture medium or combination with F12, as F12 contains amounts of iron which are cytotoxic for RGCs.
16. MTT is toxic. Handle with gloves; do not inhale, touch, or swallow.

Panning Tips

17. For panning, it is preferable to use antibodies that recognize surface antigens that are not cleaved by the papain or trypsin used to prepare the cell suspension. For this reason, antibodies to glycolipids (e.g., galactocerebroside) or gangliosides (e.g., A2B5) are ideal.
18. In some cases, when the panning antigen is a protein, the enzymes used do not completely cleave the antigen of interest, and so enzymes can still be used (e.g., RAN-2 and Thy1). In many cases, however, the enzymes completely destroy the antigen. It is still possible to pan using antibodies to this antigen, if the cell suspension is allowed to recover at 37°C for 1 h.
19. Always pan at room temperature, never at 37°C. The higher temperature allows activation of cell adhesion mechanisms and can result in adherence of contaminating cell types.
20. Always use at least two panning dishes; the first should be coated with an antibody to deplete an unwanted contaminating cell type or with an irrelevant antibody in order to make sure that microglia/macrophages are depleted (about 6% of cells in the developing optic nerve). They will not be successfully eliminated by panning for them on a dish that is not antibody coated.
21. An alternative for removing microglia/macrophages is to just coat the first dish with the lectin BSLI (Bandeiraea simplicifolia lectin I; this lectin also binds to brain endothelial cells).
22. When coating panning dishes, the first antibody (the goat anti-rabbit or anti-mouse) should always be affinity purified. Because unrelated proteins like BSA will occupy most binding sites on the Petri dish, monoclonal supernatants (which generally contain 10% FCS) cannot be put straight onto the dish unless you precoat with an affinity-purified anti-mouse antibody. However, an affinity-purified antibody (or a lectin) to the cell surface antigen of interest can be put onto the dish directly.
23. For panning, the second antibody (the one that actually binds to the cell surface antigen of interest) should be a monoclonal supernatant whenever possible. The reason for this is that the only mouse antibody type in a supernatant is the one of interest. Using a polyclonal antiserum risks that there will be many irrelevant antibodies to occupy (anti-rabbit) sites on the panning dish. Likewise, ascites fluid may contain numerous host-derived antibodies and will also occupy (anti-mouse) sites on the panning dish. In some cases, ascites from IgM-secreting

hybridoma cells can be used if an anti-mouse IgM, mu-chain-specific antibody is used to first coat the dish, as there do not seem to be many irrelevant IgM antibodies in ascites.

24. If there is no alternative and a polyclonal antiserum or ascites must be used, then it is still possible to use this antibody for panning. In this case, however, the panning dish must be coated with only one antibody (the anti-rabbit or anti-mouse antibody), and the cell suspension should be incubated with the rabbit or mouse antibody directly. In this way, the wanted antibodies will bind to the cell surface and the unwanted ones can be eliminated by spinning the cell suspension down and discarding the supernatant (and preferably washing the suspension one more time).

25. The most delicate step in panning is removing the tightly adhering purified cells from the panning dish. A saturating amount of antibody can be utilized on the first panning dishes used for depletion of contaminants from the cell suspension; however, the concentration of antibody on the final panning dish is crucial and must be titrated carefully. Cells that are too tightly bound will be difficult to release viably, while if too loosely bound, they will wash off (= poor yield). This is particularly crucial if the antigen is not destroyed by trypsin, as antibodies are not well cleaved by trypsin. In general, an optimal concentration of antibody causes about half of the adherent cells to be phase dark and flat (tightly adherent) and the other half to be phase bright and roundish (less adherent). This can be determined by a titration experiment in which the yield of viable cells is optimized; typically, supernatants on the final dish are diluted 1:20 and ascites about 1:2,000.

26. In general, if the antigen used to purify cells on the final panning dish is not trypsin sensitive, then the antibody used needs to be an IgM rather than an IgG. The IgG-bound cells are difficult to release while retaining viability, perhaps because of its high affinity for the antigen. If the only antibody available is an IgG and the antigen is not trypsin sensitive, it is still possible to purify the cells. In this case, the final dish must be coated with either protein A or protein G (or a mixture of both) rather than an anti-mouse IgG antibody. Protein A and protein G are easily cleaved by trypsin, and the cells are easily released.

References

1. David S, and Aguayo AJ (1981) Axonal elongation into peripheral nervous system "bridges" after central nervous system injury in adult rats. Science 214, 931–933
2. Gervasi NM, Kwok JC, and Fawcett JW (2008) Role of extracellular factors in axon regeneration in the CNS: implications for therapy. Regen Med 3, 907–923
3. Mansour-Robaey S, Clarke DB, Wang YC, Bray GM, and Aguayo AJ (1994) Effects of ocular injury and administration of brain-derived neurotrophic factor on survival and

regrowth of axotomized retinal ganglion cells. Proc Natl Acad Sci USA 91, 1,632–1,636

4. Mey J, and Thanos S (1993) Intravitreal injections of neurotrophic factors support the survival of axotomized retinal ganglion cells in adult rats in vivo. Brain Res 602, 304–317
5. Peinado-Ramon P, Salvador M, Villegas-Perez MP, and Vidal-Sanz M (1996) Effects of axotomy and intraocular administration of NT-4, NT-3, and brain-derived neurotrophic factor on the survival of adult rat retinal ganglion cells. A quantitative in vivo study. Invest Ophthalmol Vis Sci 37, 489–500
6. Clarke DB, Bray GM, and Aguayo AJ (1998) Prolonged administration of NT-4/5 fails to rescue most axotomized retinal ganglion cells in adult rats. Vision Res 38, 1,517–1,524
7. Carmignoto G, Maffei L, Candeo P, Canella R, and Comelli C (1989) Effect of NGF on the survival of rat retinal ganglion cells following optic nerve section. J Neurosci 9, 1,263–1,272
8. Lenzi L, Coassin M, Lambiase A, Bonini S, Amendola T, and Aloe L (2005) Effect of exogenous administration of nerve growth factor in the retina of rats with inherited retinitis pigmentosa. Vision Res 45, 1,491–1,500
9. Cui Q, Yip HK, Zhao RC, So KF, and Harvey AR (2003) Intraocular elevation of cyclic AMP potentiates ciliary neurotrophic factor-induced regeneration of adult rat retinal ganglion cell axons. Mol Cell Neurosci 22, 49–61
10. Leaver SG, Cui Q, Plant GW, Arulpragasam A, Hisheh S, Verhaagen J, et al (2006) AAV-mediated expression of CNTF promotes long-term survival and regeneration of adult rat retinal ganglion cells. Gene Ther 13, 1,328–1,341
11. Rios-Munoz W, Soto I, Duprey-Diaz MV, Blagburn J, and Blanco RE (2005) Fibroblast growth factor 2 applied to the optic nerve after axotomy increases Bcl-2 and decreases Bax in ganglion cells by activating the extracellular signal-regulated kinase signaling pathway. J Neurochem 93, 1,422–1,433
12. Unoki K, and LaVail MM (1994) Protection of the rat retina from ischemic injury by brain-derived neurotrophic factor, ciliary neurotrophic factor, and basic fibroblast growth factor. Invest Ophthalmol Vis Sci 35, 907–915
13. Ma CHM, and Taylor JSH (2010) Trophic responsiveness of purified postnatal and adult rat retinal ganglion cells. Cell Tissue Res 339, 297–310
14. Barres BA, Silverstein BE, Corey DP, and Chun LL (1988) Immunological, morphological, and electrophysiological variation among retinal ganglion cells purified by panning. Neuron 1, 791–803
15. Zhang XM, Li Liu DT, Chiang SW, Choy KW, Pang CP, Lam DS, et al (2010) Immunopanning purification and long-term culture of human retinal ganglion cells. Mol Vis 16, 2,867–2,872
16. Bader DR, MacLeish PR, and Schwartz EA (1978) Responses to light of solitary rod photoreceptors isolated from tiger salamander retina. Proc Natl Acad Sci USA 75, 3,507–3,511
17. Huettner JE, and Baughman RW (1986) Primary culture of identified neurons from the visual cortex of postnatal rats. J Neurosci 6, 3,044–3,060
18. Wysocki LJ, and Sato VL (1978) "Panning" for lymphocytes: a method for cell selection. Proc Natl Acad Sci USA 75, 2,844–2,848
19. Brewer GJ, Torricelli JR, Evege EK, and Price PJ (1993) Optimized survival of hippocampal neurons in B27-supplemented Neurobasal, a new serum-free medium combination. J Neurosci Res 35, 567–576
20. Mosmann T (1983) Rapid colorimetric assay for cellular growth and survival: application to proliferation and cytotoxicity assays. J Immunol Methods 65, 55–63
21. Manthorpe M, Fagnani R, Skaper SD, and Varon S (1986) An automated colorimetric microassay for neuronotrophic factors. Dev Brain Res 25, 191–198

Chapter 12

Culture of Purified Glial Cell Populations from Optic Nerve

Stephen D. Skaper

Abstract

Glial cells play a key role in nervous system function, providing neurotrophic factor support to neurons as well as taking part in two-way neuron-glia signaling (e.g., neurotransmitter release). White matter-derived glia are important in certain neurodegenerative diseases involving axonal loss, for example in multiple sclerosis. Here we describe procedures for the preparation and culture of mixed nerve cells from postnatal rat optic nerve, followed by protocols which can serve for the purification of individual populations of glia from this tissue, namely O2A progenitors and oligodendrocytes, and astrocytes and astrocyte precursors.

Key words: Optic nerve, O2A progenitors, Oligodendrocytes, Astrocytes, Astrocyte precursors, Immunopanning, Trophic factors, Differentiation, Cell culture

1. Introduction

Recent findings suggest that glial cells, though lacking the excitability usually associated with most neurons, may be more actively involved in brain function than has been previously thought. Collectively, these findings indicate that glial cells can sense, and potentially respond to, a large array of neuronal signals. Because glial cells are intimately associated with most neurons, neurobiologists need to consider the possible significance of active neuronal-glial signaling.

Although a variety of glial preparations have been studied, the basic findings are common and are exemplified by studies of rat optic nerve glia (1–3). Cultures of cells isolated from postnatal optic nerves contain four types of macroglial cells that can be distinguished by their distinct antigenic phenotypes and morphologies: type 1 astrocytes, type 2 astrocytes, oligodendrocytes, and a progenitor cell, termed the O2A progenitor (4–6). The O2A progenitors are developmentally bipotential, giving rise to both oligodendrocytes and type 2 astrocytes in vitro (7, 8). O2A progenitors persist

Stephen D. Skaper (ed.), *Neurotrophic Factors: Methods and Protocols*, Methods in Molecular Biology, vol. 846,
DOI 10.1007/978-1-61779-536-7_12, © Springer Science+Business Media, LLC 2012

in adult animals (9, 10) in large numbers (11). These cell types are not thought to comprise all kinds of central nervous system glia; rather, they may be restricted to the kinds of glia present in white matter. As such, they may be of considerable utility in the study of white matter disease processes affecting the visual system. For example, multiple sclerosis is characterized by optic nerve lesions (12). This chapter details the preparation of mixed optic nerve cells from postnatal or embryonic rat tissue, and serves as the "point of departure" for subsequent methodology dealing with the purification of oligodendrocytes, O2A progenitors, astrocytes, and astrocyte precursor cells of optic nerve origin.

2. Materials

2.1. Equipment and Labware

1. Stereo dissecting microscope (backlighting of stage is preferred) with fiber optic light source.
2. Laminar flow cabinet for dissections.
3. Laminar flow biological safety cabinet (CL2).
4. Humidified, water-jacketed culture incubator at 37°C and 5% CO_2/95% air.
5. Water bath set at 37°C.
6. Dissecting tools (Fine Science Tools).
7. Benchtop centrifuge to accommodate 15- and 50-mL tubes.
8. 15- and 50-mL polypropylene plastic centrifuge tubes (sterile).
9. 12-mL test tubes, sterile.
10. 6-mL flat bottom tubes (Bijou), sterile.
11. 1.5-mL Eppendorf tubes.
12. 10-cm ∅ sterile tissue culture dishes.
13. 6- and 10-cm ∅ sterile Petri culture dishes.
14. #21- and #23-guage hypodermic needles.
15. 5- and 10-mL syringes.
16. Nitex mesh, 20 μm (Tetko HC3-20).
17. 0.22- and 0.45-μm filters (Millipore).
18. Neubauer hemocytometer (Fisher Scientific).
19. Glass coverslips (12 mm ∅ #1, Menzel-Gläser, Menzel GmbH, Braunschweig, Germany) (see Note 1).
20. 24-well tissue culture plates.
21. 96-well poly-D-lysine-treated tissue culture plates (BD Biosciences).

2.2. Reagents

1. Earle's balanced salt solution (EBSS) (Invitrogen).
2. Dulbecco's phosphate buffered saline (D-PBS) (Invitrogen).
3. Minimal essential medium with Earle's salts (MEM) (Invitrogen).
4. L-15 medium with L-glutamine and L-amino acids (Invitrogen).
5. Dulbecco's modified Eagle's medium (DMEM), containing 4.5 g/L glucose, L-glutamine, and pyruvate (Invitrogen).
6. DMEM (American Type Culture Collection, ATCC).
7. RPMI-1640 medium (ATCC).
8. Neurobasal-A medium (Invitrogen).
9. B27 supplement, 50×, with antioxidants (Invitrogen).
10. Fetal calf serum (Invitrogen) (see Note 2).
11. Penicillin/streptomycin, 10,000 U/mL penicillin + 10,000 μg/mL streptomycin (100× stock), sterile, for cell culture (Invitrogen).
12. L-Glutamine (200 mM stock), sterile, for cell culture (Invitrogen).
13. Sodium pyruvate (100 mM stock), sterile, for cell culture (Sigma).
14. Collagenase Type 1-S (Sigma).
15. Papain (Worthington (Lorne)).
16. DNAse, Type I (Sigma).
17. Trypsin, 2.5% solution, sterile, for cell culture (Sigma).
18. Trypsin (0.25%)-EDTA, sterile, for cell culture (Invitrogen).
19. Poly-D-lysine (MW 70,000–150,000), sterile, for cell culture (Sigma).
20. Bovine serum albumin (BSA), essentially fatty acid and globulin free (Sigma).
21. Ovomucoid (trypsin inhibitor from chicken egg white, type III-O, free of ovoinhibitor) (Sigma).
22. L-Cysteine hydrochloride (Sigma).
23. Insulin, from bovine pancreas (Sigma).
24. Transferrin, human, partially iron saturated, tested for cell culture (Sigma).
25. Progesterone (Sigma).
26. Putrescine (Sigma).
27. Sodium selenite (Sigma).
28. T3 (3,3′,5-triiodo-L-thyronine, sodium salt), cell culture tested (Sigma).
29. Thyroxine (T4) (Sigma).
30. Affinity purified goat anti-mouse IgG + IgM (H + L) (Jackson ImmunoResearch).

31. Affinity-purified goat anti-mouse IgM, mu-chain-specific (Jackson ImmunoResearch).
32. A2B5 IgM monoclonal antibody, clone 105 (Millipore).
33. RAN-2 IgM hybridoma TIB-119™ (ATCC).
34. Anti-galactocerebroside, clone mGalC1 antibody (Millipore).
35. Thy 1.1 IgM hybridoma TIB-103™ (name T11D7e2) (ATCC).
36. Anti-neuroepithelial C5 antibody (Prof. Ben Barres, Stanford University).
37. *N*-acetyl cysteine (Sigma).
38. Ciliary neurotrophic factor, human, recombinant (CNTF) (Peprotech).
39. Platelet-derived growth factor, human, recombinant (PDGF-AA) (Peprotech).
40. 8-CPT-2′-O-Me-cAMP (Merck Biosciences).
41. Isobutylmethylxanthine (IBMX) (Sigma).
42. Forskolin (Sigma).
43. Trypan blue stain 0.4%, for cell culture (Invitrogen).
44. Dimethyl sulfoxide, for cell culture (DMSO) (Sigma).

2.3. Stock Solutions

1. *Low ovomucoid solution (10×)*. Dissolve 300 mg BSA and 300 mg ovomucoid in 20 mL D-PBS; adjust pH to 7.4. Pass through 0.45-μm and then 0.22-μm filters attached to a 30-mL syringe. Store in 1 mL aliquots at −20°C.
2. *High ovomucoid solution (10×)*. Dissolve 2 g BSA and 2 g ovomucoid in 20 mL D-PBS; adjust pH to 7.4. Pass through 0.45-μm and then 0.22-μm filters attached to a 30-mL syringe. Store 0.6 mL aliquots at −20°C (see Note 3).
3. *Forskolin (1,000×)*. Add 2.4 mL DMSO to a 10-mg bottle of forskolin (=5 mM). Make 100 μL aliquots and store at −20°C. The final concentration of solvent should not exceed 0.1% in the culture medium.
4. *N-acetyl cysteine (1,000×)*. Dissolve 50 mg *N*-acetyl cysteine in 10 mL of double-distilled water. Sterilize by passing through a 0.22-μm filter. Make 20 μL aliquots and store at −20°C.
5. *IBMX (100 mM)*. Dissolve 22 mg IBMX in 10 mL D-PBS. Sterilize by passing through a 0.22-μm filter. Make 40 μL aliquots and store at −20°C.
6. *Tris pH 9.5 (50 mM)*. Dissolve 12.1 g Trizma base in 200 mL double-distilled water. Adjust pH to 9.5 with HCl.
7. *Sato 100× stock solution*. To 20 mL of Neurobasal-A medium add the following (final concentrations in brackets): 200 mg transferrin [100 μg/mL), 200 mg BSA [100 μg/mL], 5 μL of

stock progesterone solution (stock is 2.5 mg/100 μL ethanol) [60 ng/mL or 0.2 μM], 32 mg putrescine [16 μg/mL], 200 μL of stock sodium selenite (stock is 40 mg in 1 mL of 0.1 N NaOH plus 10 mL Neurobasal-A) [40 ng/mL], 10 μL of stock T3 (stock is 8 mg in 1 mL of 0.1 N NaOH) [30 ng/mL], and 10 μL of stock T4 (stock is 8 mg in 1 mL of 0.1 N NaOH) [40 ng/mL]. Aliquot the above 100× stock solutions (200 μL) and store at −20°C. The ethanol stocks for progesterone, sodium selenite, T3, and T4 should be prepared fresh each time a new batch of 100× Sato medium is made.

2.4. Sato Serum-Free Culture Medium

1. To 20 mL of Neurobasal-A medium add 200 μL Sato 100× stock solution, 400 μL 50× B27 supplement, 200 μL sodium pyruvate 100× stock, 200 μL glutamine 100× stock, 200 μL of 100× penicillin/streptomycin stock (stored at −20°C), 20 μL *N*-acetyl cysteine 1,000× stock.
2. Next add 200 μL of 100× insulin stock (stored at 4°C; stock must be prepared every 4–6 weeks—do not freeze). The 100× insulin stock (0.5 mg/mL) is prepared by adding 2 mg of insulin to 4 mL of tissue culture grade water containing 20 μL of 1 N HCl. Filter the stock through a 0.22-μm filter and store at 4°C.
3. Filter the above medium through a 0.22-μm filter. Before filtering the medium, filter about 5 mL of Neurobasal-A medium to rinse away traces of detergent in the filter. The quality of the filter matters—Millex-GV from Millipore is best. Falcon plastic is preferable, along with Monoject syringes.
4. Recommended trophic factors for mixed optic nerve cultures: Insulin (5 μg/mL) plus CNTF (10 ng/mL). Do not filter the medium once the growth factors are added due to the risk of their sticking to the filter. In addition, forskolin (5 μM) can be included in the medium, as it promotes survival of mixed and purified cultures.

3. Methods

3.1. Preparing an Optic Nerve Cell Suspsension Using Papain

1. Prepare suspensions using papain only when they will be panned or when older nerves are being dissociated. The procedures described here are modified from (12). Otherwise use the simpler trypsin/collagenase protocol (described below—Subheading 3.2). Use sterile technique in all steps (see Notes 4–6).
2. Dissect optic nerves and optic chiasm from postnatal day (P) P6 to P10 rats (if you intend to collect oligodendrocytes use P9–P10 animals to obtain a larger yield). Remove the head and

skull and scoop the brain out—the optic nerves will be lying on the base of the skull. Cut the nerves behind the chiasm and also behind each eye and collect the nerves into a 35-mm Petri dish containing 2 mL of HEPES-buffered MEM (pH 7.3–7.4) (can also use D-PBS containing $CaCl_2$ and $MgCl_2$). Expect a final yield of about 16,000 O2A cells per animal, and base your rat usage on this (i.e., desired final number of cells).

3. Cut each nerve with a fine scissors or scalpel into as many pieces as possible (at least 5–10 per nerve). Transfer the MEM (or D-PBS containing $CaCl_2$ and $MgCl_2$) and nerve pieces to a bijoux or polystyrene 12-mL test tube.
4. About 10 min before use, add 160 U of papain to 5 mL of MEM/HEPES plus DNase stock 1:100 (stock is 0.4%) in a bijou or polystyrene 12-mL test tube. Dissolve the papain by placing this solution in a 37°C water bath for a few minutes. Just before use, add L-cysteine (1 mg) to activate the papain solution, adjust pH to 7.4 (as judged roughly by the color of the phenol red pH indicator in the MEM, with 1 N NaOH). Filter the final papain solution through a 0.22-μM filter. Be sure to rinse the filter first with about 5 mL of MEM solution to remove any traces of detergent in the filter.
5. Using a Pasteur pipette, remove the MEM/HEPES from the vial containing the tissue and add the papain solution. Incubate at 37°C for 75 min for P7 nerves, 90 min for P12 nerves, and 45 min for P0 nerves. Every 15 min gently agitate the tube.
6. Prepare 10 mL of the ovomucoid inhibitor solution: 9 mL of MEM plus 1 mL of the "LOW" ovomucoid 10× stock solution (containing 15 mg of BSA and 15 mg of ovomucoid per mL of MEM) plus 1:100 (100 μL) of the DNase stock solution.
7. Use a Pasteur pipette to gently remove the papain solution. Wash the optic nerve pieces gently by adding 1 mL of inhibitor solution, allow the optic nerve pieces to settle, and then remove the ovomucoid solution with a Pasteur pipette.
8. Serially triturate the nerve pieces in the ovomucoid inhibitor solution, 1 mL at a time, with a 2.5-mL syringe and #21-gauge needle × 2 passes, followed by a #23-gauge needle × up to 8 passes. That is, add 1 mL, pass the pieces through the needle 1 or 2 times, let the pieces settle, collect the suspension above the pieces with a Pasteur pipette into a 10-mL test tube. Add another mL of inhibitor solution, triturate again, and so on. In this way, cells successively removed from the nerve pieces are not repetitively passed through the needle (see Note 7).
9. If the optic nerve pieces fail to completely fall apart, it is likely that enzyme concentration was too low; enzyme incubation was too short; and enzyme lot was too old. Now, cap the tube and gently invert once or twice to mix the cells.

10. Filter the cell suspension through a 20-μm nylon mesh. Nitex mesh from Tetko works well. Sterilize the mesh before use (UV exposure for a few minutes is sufficient), prewet the mesh on the bottom side, then form a cone with the filter on top of a test tube, and pass 1 mL of MEM or inhibitor solution through it before using a Pasteur pipette to transfer the cell suspension through the filter.
11. Transfer a 100-μL sample to a 1.5-mL eppendorf tube for cell counting (see Chapter 3).
12. Collect the cells by centrifuging at 200 × *g* for 10 min.
13. Gently remove the supernatant by aspiration. Using a 1-mL Pipetman, resuspend the cell pellet in another ovomucoid solution containing 5.4 mL MEM plus 0.6 mL "High" ovomucoid 10× stock (contains 60 mg of BSA and 60 mg of ovomucoid, but NO DNase). To resuspend, first add 1 mL of this ovomucoid solution, resuspend with the 1-mL pipetman, then add the remaining 5 mL of the ovomucoid solution, cap tube, and gently invert once or twice to mix (see Note 8).
14. Centrifuge as in step 11, and aspirate supernatant. Use a 200-μL pipetman to gently remove the residual solution, leaving only the cell pellet.
15. Resuspend the cells in Sato serum-free medium at 10^6 cells per mL. Yields are typically about 100,000 cells per pair of P7 optic nerves.
16. Take a 100-μL aliquot for a cell count (see Chapter 3). Spin down the rest of the cells and resuspend in about 7 mL of L15 or other air buffered solution containing insulin (0.5 μg/mL). A 100× insulin stock is prepared by dissolving 2 mg of insulin into 4 mL of H_2O containing 20 μL of 1 N HCl (can be stored at 4°C for 6 weeks). Insulin at high concentration activates insulin-like growth factor-1 receptors on O2A progenitors and oligodendrocytes (13), and is an essential survival factor for O2A cells and oligodendrocytes (14, 15).

3.2. Preparing an Optic Nerve Cell Suspension Using Trypsin and Collagenase

1. Dissect and mince nerves in a 35-mm dish containing EBSS (see Note 9).
2. Prepare enzyme solution containing 4 mL EBSS, 5 mg collagenase, and 200 μL trypsin stock (stock is 2.5% in EBSS or D-PBS and is stored at −70°C). Filter through a 0.22-μm filter. Add DNase 1:100 from stock (stock is 0.4% in EBSS, stored at −20°C).
3. Add nerves to trypsin solution in a bijou (or polystyrene 12-mL test tube) and incubate for 20 min in a water bath at 37°C.
4. After this time add another 200 μL trypsin stock and incubate again for 20 min. (incubate 20 min for P7 nerves and 10–15 min

for P0 nerves). During all incubations agitate gently every 10 min.

5. Remove by aspiration the trypsin-containing supernatant, leaving optic nerves behind.
6. Serial triturate with 2 mL of 10% fetal calf serum in DMEM (#21 needle for two passes, #23 needle for several more, until all nerves are dissociated).
7. Centrifuge at 200 × *g* for 10 min.
8. Resuspend in Sato serum-free medium and culture.

3.3. Adult Optic Nerve Dissociation (Collagenase and Trypsin)

1. Dissect (6–8 adult female rats) nerves into D-PBS, and cut each of the nerves into 15 pieces.
2. Add collagenase (333 U/mL) in 4 mL of D-PBS and pass through a 0.22-μM filter. Incubate nerve pieces with this solution at 37°C for 1 h.
3. Add 4 mL of trypsin (30,000 U/mL) in D-PBS; incubate at 37°C for 20 min.
4. Discard the trypsin solution and add 4 mL of 15,000 U/mL trypsin in D-PBS with 0.27 mM EDTA; incubate at 37°C for 20 min.
5. Add an equal volume of 20% fetal calf serum in D-PBS (with Ca^{2+}, Mg^{2+}) containing 0.08% DNase. Incubate at 37°C for 10 min.
6. Triturate with a 1-mL pipette (2×) before triturating with 21# needle and 23# needle in low ovomucoid.
7. Wash in high ovomucoid as usual.

3.4. Preparation of Embryonic Optic Nerve Cells Using Trypsin

1. Dissociate the nerves with 0.125% trypsin in 4 mL D-PBS containing 50 μL of 4% DNase stock.
2. Incubate at 37°C for 30 min.
3. Triturate in 10% fetal calf serum buffer: (1 mL fetal calf serum, 9 mL D-PBS, 100 μL 4% DNase stock) using #21- and #23-gauge needles as described above in Subheading 3.3.
4. Centrifuge at 200 × *g* for 10 min.
5. Resuspend in Sato serum-free medium and culture as desired.

3.5. Purification Procedure for O2A Progenitors

To purify O2A progenitors, an optic nerve cell suspension is prepared by a papain dissociation procedure (modified from (16)) and passed sequentially over three antibody-coated "panning" (17) dishes (papain must be used and not trypsin, to maximize cell yield and to avoid destroying the RAN-2 surface antigen). First, a RAN-2-coated dish is used which depletes the suspension of macrophages (which stick to the first antibody-coated dish they are placed on), meningeal cells, and type-1 astrocytes. In addition, about 20% of the

O2A progenitors adhere to the RAN-2 dish. The nonadherent cells are then transferred to a galactocerebroside-coated dish, to which only oligodendrocytes adhere. Lastly, the remaining nonadherent cells are transferred to an A2B5 dish. Only O2A progenitors adhere to this dish. The O2A progenitors can be cultured directly on this dish or can be removed with trypsin for culture on coverslips. Provided all three panning steps are used, O2A progenitors can be purified to near-complete homogeneity. If the RAN-2 dish is omitted, a purification of 99.5% is achieved (1 in 200 cells have a flat morphology and are vimentin immunoreactive, identifying them as meningeal or type-1 astrocyte contaminants). If the galactocerebroside dish is omitted, the A2B5 dish will be contaminated by newly formed oligodendrocytes which are still A2B5 immunopositive.

3.5.1. Preparation of Panning Dishes

1. Incubate two 10-cm ∅ Petri dishes overnight at 4°C in 10 mL of 50 mM Tris (pH 9.5) containing 10 μg/mL of affinity-purified goat anti-mouse IgG (H+L), and a third dish with affinity-purified goat anti-mouse IgM, mu-chain-specific. Affinity-purified antibodies must be used, so that irrelevant proteins do not saturate most of the binding sites on the dish (this is also the reason why the primary antibodies cannot be used as a single layer on panning dishes) (see Note 10).
2. Before the papain dissociation, rinse each Petri dish 3 or 4 times with D-PBS.
3. Add the primary antibody in a D-PBS/BSA (0.2%, v/v) solution for at least 1 h. Dish 1: RAN-2 supernatant (1 mL) plus D-PBS/BSA (4 mL). Dish 2: galactocerebroside antibody (5 μg/mL) diluted in 4 mL of PBS/BSA. Dish 3: A2B5 antibody (5 μg/mL) diluted in 4 mL of PBS/BSA. In this case, the third Petri dish can also be coated with goat anti-mouse IgG (H+L), as there will be no irrelevant IgGs to compete with the A2B5 for binding sites.
4. Rinse each dish 3 times with D-PBS.
5. Add enough D-PBS to coat the dish; remove the PBS just before use.

3.5.2. Incubation of the Cell Suspension on the Panning Dishes

1. Transfer the cell suspension to the RAN-2 antibody-coated Petri dish(es) and leave for 30 min (rock gently after 15 min to ensure access of all cells to the panning surface area) (see Note 11).
2. Transfer the nonadherent cells to the galactocerebroside antibody-coated Petri dish for 45 min (agitate gently every 15 min). Only oligodendrocytes will stick to this plate, so if you want pure oligodendrocytes wash and save this plate (see below). This step depletes the suspension of *virtually all* oligodendrocytes.

3. Transfer the nonadherent cells to the A2B5 antibody-coated Petri dish and incubate for 45 min (agitate gently every 15 min) (see Note 12).
4. Wash off nonadherent cells from the A2B5 dish with MEM/HEPES (about 10 mL per wash) or other air-buffered solution about 5 times (follow progress under microscope), being careful not to let the adherent cells dry out. During these washes, periodically rock (agitate) the dish. This last step is critical, as too much rocking will wash some of the O2A cells off the dish, while too little will leave contaminating nonadherent cells.

3.5.3. Removing O2A Cells from the A2B5 Dish with Trypsin

1. Prepare 4 mL of EBSS (preequilibrated as in step 2 below); add 0.2 mL of 2.5% trypsin.
2. Remove the MEM from the A2B5 dish with adherent O2A progenitors and rinse once with EBSS which has been preequilibrated in a 6-cm ∅ dish in the CO_2 incubator for several hours (so that its pH is ok).
3. Remove the EBSS wash and add the 4 mL of EBSS/trypsin (0.125% trypsin final concentration).
4. Incubate in a 37°C incubator for exactly 10 min.
5. Squirt the adherent cells off with a 1-mL pipetman. Be careful not to overdo this action—check under the microscope to make sure that the cells were released.
6. Transfer the 4 mL of trypsin/cell suspension to a 15-mL tube containing 10 mL of 30% fetal calf serum (3 mL of fetal calf serum plus 7 mL of DMEM). This is the most critical step in the entire protocol; if the cells are undertrypsinized (most likely because the batch of trypsin has gone off) they will be removed from the plate but they will not survive past the first 24 h in culture (see Note 13).
7. Take a 100-μL aliquot for a cell count (see Chapter 3).
8. Centrifuge the cell suspension for 15 min at 200 × *g*.
9. Traces of insulin or serum can be washed out by resuspending the cells and washing one last time in L15 or MEM containing 0.5% BSA. It is essential to include this much BSA in any further washes; at this point the purified cells will not survive washing without it.
10. Culture the O2A progenitors on poly-D-lysine-coated glass coverslips (12 mm ∅; at least 5,000 cells/coverslip) in 96-well plates in Sato medium, 100 μL, containing PDGF-AA at 5 ng/mL. See Subheading 3.9.1 for coating of coverslips.
11. Expected yield: 15,000–20,000 O2A cells per P8 rat.

3.6. Adult O2A Panning Procedure

The panning procedure for adult O2As is very similar to that for perinatal O2A progenitors. There are a few differences, as noted below. Apart from these, follow the original panning procedure for perinatal O2A progenitors for the rest of the procedure. The number of dishes and amounts of reagents are adequate for eight adult rats.

1. Prepare adult optic nerve suspension with collagenase and trypsin (instead of papain).
2. Prepare 3–10-cm ∅ Petri plastic panning dishes: 2 RAN-2 dishes and 1 A2B5 dish (instead of RAN-2, galactocerebroside, and A2B5 dishes): Add to each dish 12 mL of the Tris pH 9.5 buffer plus 40 μL of the affinity-purified antibody (anti-mouse IgG + IgM antibody for the RAN-2 dish, and anti-mouse IgM for the A2B5 dish).
3. Incubate the cell suspension on the panning dish as follows: RAN-2 dish, 30 min; A2B5 dish, 45 min.
4. To remove the purified O2A cells from the A2B5 dish with trypsin, use trypsin from Gibco (instead of Sigma).
5. To determine the cell yield, 10 μL (instead of 100 μL) of cell suspension is directly dropped on a culture dish (three separate drops for averaging). Let the cells settle down to the bottom of the dish. Count the total cell number in the drop. Yield = number of cells in the drop × 100 × number of mL.

3.7. Purifying Oligodendrocytes Instead of O2A Progenitors

The simplest way is to let the purified O2A progenitors differentiate into oligodendrocytes by culturing for 2 days in survival factors (CNTF, insulin, forskolin) but not mitogens. To use the purified oligodendrocytes on the galactocerebroside dish, use O1 instead of galactocerebroside antibodies to prepare this panning dish (O1 is an IgM, galactocerebroside is an IgG; it is easier to remove cells from IgM-coated dishes).

3.8. Purification of Astrocytes and Astrocyte Precursor Cells from Rat Optic Nerve

The following protocols are similar to the protocols for purifying oligodendrocyte lineage cells. Only the modifications are described below.

3.8.1. Astrocyte Purification from Perinatal Optic Nerves

1. Dissociate the minced optic nerves with 160 U papain in 5 mL D-PBS with DNase and L-cysteine, and incubate at 37°C for 45 min (P1), 60 min (P4), or 75 min (P8).
2. Triturate the tissue as for purification of O2A progenitors.
3. For panning, coat the first dish with a mouse monoclonal MRC-OX7 anti-Thy1.1 antibody ((5 μg/mL) diluted in 4 mL of PBS/BSA for a 10-cm ∅ dish) to deplete macrophages and fibroblasts.
4. Coat the second dish with the mouse monoclonal anti-A2B5 antibody (for P1). For purification from P4 or P8 optic

nerves, the second dish is also coated with the monoclonal anti-galactocerebroside antibody (to remove oligodendrocytes).

5. Coat the third and last dish with a mouse monoclonal anti-neuroepithelial C5 antibody. The astrocytes will bind to the C5 dish. Astrocytes can then be cultured directly on the C5 dish, or removed with trypsin.

3.8.2. Purification of Astrocyte Precursor Cells from Embryonic Day 17 Optic Nerves

1. Dissociate the nerves with 80 U of papain in 5 mL of D-PBS with DNase and L-cysteine, at 37°C for 30 min.
2. Triturate the tissue as for purification of O2A progenitors.
3. For panning, coat the first dish with Thy1.1 supernatant (1 mL) plus 4 mL of PBS/BSA for a 10-cm ∅ dish.
4. Coat the second dish with a mouse monoclonal anti-neuroepithelial C5 antibody. Astrocyte precursor cells will bind to the C5 dish. The A2B5 antibody dish is omitted because the astrocyte precursor cells are A2B5 positive, as are the O2A progenitors.
5. Astrocyte precursor cells can then be cultured directly on C5 dish, or removed with trypsin.

3.9. Culturing Optic Nerve Cells

3.9.1. Coating Glass Coverslips with Poly-D-Lysine

1. Soak the desired quantity of 12-mm ∅ glass coverslips in 70% ethanol (prepared using MilliQ water) for at least 24 h prior to use. Rinse with clean 70% ethanol and store in 70% ethanol until use.
2. Use sterile forceps to remove coverslips and place these lying flat, but not overlapping each other, in a 10-cm ∅ Petri dish containing autoclaved MilliQ water. Rinse several times with this water to remove ethanol.
3. Add enough poly-D-lysine solution (dilute the 1 mg/mL poly-D-lysine stock, which should be stored at –20°C, to 10 μg/mL in autoclaved MilliQ water) to cover the slips (see Note 14).
4. Incubate the coverslips with poly-D-lysine for a minimum of 30–60 min.
5. Using a sterile forceps, individually pick up each coverslip, rinse 3 times with sterile MilliQ water (6-cm ∅ Petri dish works fine for this), and place a single coverslip in each well of a 24-well tissue culture plate.
6. Remove by aspiration any residual water in each well and allow the coverslips to air dry in a laminar flow hood.

3.9.2. Plating Optic Nerve Cell Suspensions

1. Transfer 20 μL of the cell suspension (resuspended in Sato serum-free medium, as described in preparation of the cell suspension) onto the center of each poly-D-lysine-coated coverslip.
2. Incubate the multiwell plate with coverslips in a 5% CO_2/95% air incubator (37°C) for 15 min to allow the cells to adhere.

3. Add 500 μL of plating medium to each well, and return the plate(s) to the CO_2 incubator.

3.10. Culturing Purified Adult O2A Progenitors

1. *Culture medium*: To 20 mL of DMEM add 200 μL of Sato 100× stock, 400 μL of 50× B27 supplement, 200 μL of 100× penicillin/streptomycin stock, 200 μL of 100× insulin stock.
2. *Biotin* (10 ng/mL final concentration in medium, starting from a 1,000× stock in D-PBS) enhances long-term survival of purified adult O2A cells, and its addition is highly recommended.
3. Addition of *N*-acetyl cycteine is strongly recommended to enhance survival. *N*-acetyl cycteine acts a slow release form of L-cysteine. The DMEM contains cystine but no L-cysteine; this seems to be crucial for the cells to maintain high intracellular glutathione levels.
4. *Peptide factors*: PDGF-AA (10 ng/mL), CNTF (10 ng/mL), 8-CPT-2′-O-Me-cAMP (125 μM) (or 10 μM forskolin and 0.1 mM IBMX). 8-CPT-2′-O-Me-cAMP is a cAMP agonist, while forskolin activates adenylylcyclase and IBMX inhibits cyclic AMP phosphodiesterase. These agents act to elevate intracellular levels of cyclic AMP, which promotes survival of O2A progenitors.
5. To plate cells on coverslips, preplate cells (3,000–5,000 cells in 20–30 μL) on each poly-D-lysine-coated coverslip and let the cells settle for 10 min before adding 500 μL of medium.
6. To plate cells on dish, dilute cells to about 30,000/1 mL, plate 100 μL on a poly-D-lysine-coated 6-cm ∅ dish and immediately spread the medium drop with cells evenly on the dish using a glass Pasteur pipette that had been bent at a 90° angle by heating. Allow the cells to settle in the 37°C incubator for 10–15 min before adding 2.5 mL of medium.

4. Notes

1. The source of glass is very important for the coverslips: German suppliers are highly recommended, in particular the one listed here.
2. Heat inactivation of fetal calf serum is recommended to destroy heat-labile complement. Thaw the bottle of serum in advance, using a 37°C water bath. Next, immerse the serum bottle in the water bath after reequilibrating to 56°C and leave for 30 min. Swirl the bottle occasionally to ensure proper mixing. Allow the serum to cool to room temperature, aliquot into 50-mL tubes and store at –20°C.

3. If you encounter difficulty dissolving the BSA and ovomucoid for the "HIGH" ovomucoid stock solution, it's ok also to prepare a 5× stock and freeze down aliquots of 1.2 mL.
4. Fresh papain should be obtained every month.
5. Never freeze the papain stock solution.
6. Never prepare papain in L15 medium since it inhibits the papain.
7. The quality of the syringe plastic matters, as plastic stabilizers can reach out and cause toxicity, especially when using serum-free medium.
8. Never shake tubes containing a cell suspension.
9. Hank's balanced salt solution (lacking calcium and magnesium) can be used in place of EBSS.
10. Rabbit antisera and IgG ascites can generally not be used for panning procedures, as irrelevant IgGs contributed by these would compete with the A2B5 for binding sites. This is a common mistake in immunopanning.
11. If optic nerves from more than 25 animals are used, transfer the cells to another RAN-2 panning dish for 20 min. The RAN-2 dishes eliminate meningeal and astrocyte contamination of the final, A2B5 dish. However, it does not deplete the suspension of all astrocytes, as many astrocytes in the optic chiasm (but not nerve) are RAN-2 negative.
12. If you want to recover live O2A cells from the A2B5 dish, the A2B5 concentration must be titered so that it is low enough to easily remove the cells with trypsin, but not so low that cells don't adhere strongly enough to remain once nonadherent cells are washed from the dish.
13. O2A progenitors from the A5B5 dish can be cultured directly on this dish, if there is no need to culture on particular surfaces (e.g., coverslips) where detachment and cell counting is called for.
14. Only use poly-D-lysine which is cell culture grade.

References

1. Barres BA, Chun LLY, and Corey DP (1988) Ion channel expression by white matter glia: I. Type 2 astrocytes and oligodendrocytes. *Glia* 1, 10–30
2. Barres B.A, Koroshetz WJ, Swartz KJ, Chun LLY, and Corey DP (1990) Ion channel expression by white matter glia: the 02A glial progenitor cell. *Neuron* 4, 507–524
3. Barres BA, Koroshetz WJ, Chun LLY, and Corey DP (1990) Ion channel expression by white matter glia: the type-l astrocyte. *Neuron* 5, 527–544
4. Raff MC, Abney ER, Cohen J, Lindsay R, and Noble M (1983) Two types of astrocytes in cultures of developing rat white matter: differences in morphology, surface gangliosides, and growth characteristics. *J Neurosci* 3, 1289–1 300
5. Miller RH, and Raff MC (1984) Fibrous and protoplasmic astrocytes are biochemically and developmentally distinct. *J Neurosci* 4, 585–592

6. Miller RH, ffrench-Constant C, and Raff MC (1989) The macroglial cells of the rat optic nerve. *Annu Rev Neurosci* 12, 517–534
7. Raff MC, Miller RH, and Noble M (1983) A glial progenitor cell that develops in vitro into an astrocyte or an oligodendrocyte depending on culture medium. *Nature* 303, 390–396
8. Raff MC, Williams BP, and Miller RH (1984) The *in vitro* differentiation of a bipotential glial progenitor cell. *EMBO J* 3, 1857–1864
9. ffrench-Constant C, and Raff MC (1986) The oligodendrocyte-type 2 astrocyte cell lineage is specialized for myelination. *Nature* 323, 335–338
10. Wolswijk G, and Noble M (1989) Identification of an adult-specific glial progenitor cell. *Development* 105, 387–400
11. Barres BA (1991) New roles for glia. *J Neurosci* 11, 3685–3694
12. Green AJ, McQuaid S, Hauser SL, Allen IV, and Lyness R (2010) Ocular pathology in multiple sclerosis: retinal atrophy and inflammation irrespective of disease duration. *Brain* 133, 1591–1601
13. Zhao M, Wan ZL, Whittaker L, Xu B, Phillips NB, Katsoyannis PG, et al (2009) Design of an insulin analog with enhanced receptor binding selectivity: rationale, structure, and therapeutic implications. *J Biol Chem* 284, 32178–32187
14. Pang Y, Zheng B, Fan LW, Rhodes PG, and Cai Z (2007) IGF-1 protects oligodendrocyte progenitors against TNF α-induced damage by activation of PI3K/Akt and interruption of the mitochondrial apoptotic pathway. *Glia* 55, 1099–1107
15. Wilczak N, Chesik D, Hoekstra D, and De Keyser J (2008) IGF binding protein alterations on periplaque oligodendrocytes in multiple sclerosis: implications for remyelination. *Neurochem Int* 52, 1431–1435
16. Huettner JE and Baughman RW (1986) Primary culture of identified neurons from the visual cortex of postnatal rats. *J Neurosci* 6, 3044–3060
17. Williams DL, Lo JL, and Amborski JF (1986) Enrichment of T lymphocytes from bovine peripheral blood mononuclear cells using immunoaffinity depletion technique ("panning"). *Vet Immunol Immunopathol* 11, 199–204

Chapter 13

Isolation and Culture of Rat Cone Photoreceptor Cells

Stephen D. Skaper

Abstract

In retinal diseases characterized by photoreceptor degeneration, the main cause of clinically significant vision loss is cone, rather than rod, loss. Photoreceptor apoptosis can be affected by the availability and/or delivery of neurotrophic proteins, and cultures of photoreceptors are valuable for studying these processes. In the present study, a technique was designed to purify cones to make it possible to screen for neuroprotective molecules. The present chapter describes a technique for preparing cultures of purified rat retina cone photoreceptors based upon panning with peanut agglutinin lectin, which selectively binds to cones. In addition, we describe a protocol for the purification and culture of retinal pigmented epithelial cells from postnatal rat.

Key words: Cone photoreceptors, Retina, Pigmented epithelial cells, Rat, Cell culture, Neurotrophic factors

1. Introduction

Photoreceptors of the retina are highly specialized neurons that transduce light stimuli to membrane potential changes signaled to second-order neurons. These cells are essential for normal vision, but they degenerate in a number of conditions, including genetic diseases such as retinitis pigmentosa (1), environmental insults such as light damage (2), and as a result of normal aging (3, 4). Although most of the mutations responsible for retinitis pigmentosa in humans and animal models affect rod photoreceptor–specific genes, rod apoptosis is followed by secondary cone degeneration. Photoreceptor rescue or neuroprotection are topics of great current interest that are necessary for formulating therapeutic approaches (5, 6).

Nonmammalian vertebrates have the remarkable ability to replace neurons lost through damage. Fish, and to a limited extent birds, replace lost neurons by the dedifferentiation of Müller glia

Stephen D. Skaper (ed.), *Neurotrophic Factors: Methods and Protocols*, Methods in Molecular Biology, vol. 846,
DOI 10.1007/978-1-61779-536-7_13, © Springer Science+Business Media, LLC 2012

to a progenitor state followed by the replication of these neuronal progenitor cells (7). Recent studies have investigated whether regeneration can be stimulated in the mouse and rat retina. Several groups have reported that at least some types of neurons can be regenerated in the mammalian retina in vivo or in vitro and that the regeneration of neurons can be stimulated using growth factors, transcription factors, or subtoxic levels of excitatory amino acids. For example, basic fibroblast growth factor supports survival of cultured rat (8) and porcine (9) photoreceptors and delays photoreceptor degeneration in inherited retinal dystrophy in rats (10). Ciliary neurotrophic factor alone (11, 12) or in combination with brain-derived neurotrophic factor (13) promotes photoreceptor development and survival. An analog of ciliary neurotrophic factor (Axokine) delayed photoreceptor loss in a feline model of hereditary retinal degeneration (14).

Retinal pigmented epithelial (RPE) cells are melanin-containing cells that constitute a monolayer between the neural retina and the choroid. RPE cells play a key role in maintaining the normal function of retina and can express several neurotrophic factors such as platelet-derived growth factor, epidermal growth factor, vascular endothelial growth factor, and pigment-derived epithelial factor (15). In addition, RPE cells secrete glial cell line–derived neurotrophic factor and brain-derived neurotrophic factor (BDNF) (16). Delivery of pigment epithelium–derived factor is protective in experimental animal models of retinal degeneration involving photoreceptors (17–19). The identification of other cone survival factors and screening for factors supporting their survival requires the availability of purified cultures of mammalian cones. The following chapter describes a detailed protocol for the culture of cone photoreceptors from rat retina based on a peanut agglutinin (PNA) lectin–panning procedure that allows their selective isolation. The utility of the technique benefits from the fact that PNA lectin binds to cone photoreceptors from various species in both normal and pathologic conditions. In addition, the preparation of RPE cells from rat is described.

2. Materials

2.1. Equipment and Labware

1. Stereo dissecting microscope (backlighting of stage is preferred) with fiber optic light source.
2. Laminar flow cabinet for dissections.
3. Laminar flow biological safety cabinet (CL2)
4. Humidified, water-jacketed culture incubator at 37°C and 5% CO_2/95% air.
5. Water bath set at 37°C.

6. Dissecting tools (Fine Science Tools).
7. Bench centrifuge to accommodate 15- and 50-mL tubes.
8. 15- and 50-mL polypropylene plastic centrifuge tubes (sterile).
9. 10-cm ∅ sterile tissue culture dishes (any supplier)
10. 3.5-cm ∅ and 10-cm ∅ sterile petri plastic culture dishes (any supplier).
11. Glass coverslips (12 mm ∅ #1, Menzel-Gläser, Menzel GmbH, Braunschweig, Germany) (see Note 1)
12. 96-well poly-D-lysine-treated tissue culture plates (BD Biosciences).
13. 0.22-μm filters (Millipore).
14. Neubauer hemocytometer (Fisher Scientific).

2.2. Reagents

1. Dulbecco's phosphate-buffered saline, pH 7.4, sterile (DPBS) (Invitrogen).
2. Water, sterile, tissue culture grade.
3. L-15 medium, with L-glutamine and L-amino acids (Invitrogen).
4. Neurobasal-A medium (Invitrogen).
5. B27 supplement, 50×, with antioxidants (Invitrogen).
6. Fetal calf serum (FCS) (Invitrogen) (see Note 2).
7. Penicillin/streptomycin (100× stock), sterile, for cell culture (Invitrogen).
8. Gentamicin (50 mg/mL stock), sterile, for cell culture (Sigma).
9. Papain (Worthington (Lorne)).
10. DNase, type I (Sigma).
11. Trypsin, 2.5% solution, sterile, for cell culture (Sigma).
12. Collagenase (Sigma).
13. Poly-D-lysine (MW 30,000–70,000), sterile, for cell culture (Sigma).
14. Metrizamide (Sigma).
15. Merosin, human (Millipore).
16. Goat antirabbit IgG directed against PNA lectin (Sigma).
17. PNA lectin (affinity-purified) (Sigma).
18. Affinity-purified goat antimouse IgG+IgM (H+L) (Jackson ImmunoResearch).
19. CD90/Thy1 antibody (MRC OX-7) (Abcam).
20. Bovine serum albumin, essentially fatty acid and globulin free, (BSA) (Sigma).
21. Ovomucoid (trypsin inhibitor from chicken egg white, type III-O, free of ovoinhibitor) (Sigma).

22. L-Cysteine.
23. Insulin, from bovine pancreas (Sigma).
24. Transferrin, human.
25. Progesterone.
26. Putrescine.
27. Sodium selenite.
28. T3 (3,3′,5-triiodo-L-thyronine, sodium salt), cell culture–tested (Sigma).
29. Sodium pyruvate (100 mM stock), sterile, for cell culture (Sigma).
30. L-Glutamine (200 mM stock), sterile, for cell culture (Invitrogen).
31. BDNF, human, recombinant (Peprotech).
32. CNTF, human, recombinant (CNTF) (Peprotech).
33. Forskolin.
34. *N*-acetyl cysteine.
35. Phenol red (0.5%), sterile, for cell culture (Sigma).

2.3. Stock Solutions

1. *10× EBSS stock solution without Ca2+, Mg2+, and bicarbonate.* 1.16 M NaCl (68 g/L), 54 mM KCl (4 g/L), 10 mM $NaH_2PO_4{\cdot}H_2O$ (1.4 g/L), 10 mM D-glucose (10 g/L), and 0.05% phenol red (10 mL/L of 0.5% solution). Filter-sterilize and store at 4°C.
2. *D-Glucose, 30% (w/v).* Dissolve 15 g of D-glucose in 50 mL double-distilled water.
3. *$NaHCO_3$, 1 M.* Dissolve 4.2 g of $NaHCO_3$ in 50 mL double-distilled water.
4. *$MgSO_4$, 100 mM.* Dissolve 1.23 g of $MgSO_4{\cdot}7H_2O$ in 50 mL double-distilled water.
5. *EDTA, 0.5 M.* Dissolve 7.3 g of EDTA in 50 mL double-distilled water.
6. *EBSS-papain solution with Mg2+, EDTA, and bicarbonate.* Combine 20 mL of 10× EBSS solution, 2 mL of 100 mM $MgSO_4$, 2.4 mL of 30% D-glucose, 0.4 mL of 0.5 M EDTA, and 5.2 mL of 1 M $NaHCO_3$. Bring to a final volume of 200 mL with double-distilled water.
7. *Borate buffer 0.15 M, pH 8.4.* Dissolve 28.6 g of sodium borate ($Na_2B_4O_7{\cdot}10H_2O$) in 500 mL double-distilled water (pH will be ~9.2). Adjust pH to 8.4 with 5 N HCl. Filter-sterilize and store at 4°C (up to 6 months).
8. *Low ovomucoid solution.* Mix together 9 mL DPBS, 1 mL 10× low ovomucoid stock (15 mg each BSA and ovomucoid/mL DPBS), and 200 μL 0.4% DNase stock (made by dissolving

4 mg/mL of DNase in L-15 medium; filter-sterilize, aliquot, and store at −20°C). Check pH and adjust to 7.4 if necessary. Filter-sterilize the final solution.

9. *High ovomucoid solution.* Mix together 4.8 mL DPBS and 1.2 mL 6× high ovomucoid stock (50 mg each BSA and ovomucoid/mL DPBS). Filter-sterilize.
10. *Forskolin (1,000×).* Add 2.4 mL dimethyl sulfoxide to a 10-mg bottle of forskolin (=5 mM). Make 100 μL aliquots and store at −20°C. The final concentration of solvent should not exceed 0.1% in the culture medium.
11. *N-acetyl cysteine (1,000×).* Dissolve 60 mg *N*-acetyl cysteine in 1 mL Neurobasal-A medium. Filter-sterilize, make 20 μL aliquots, and store at −20°C.
12. *Metrizamide (prepare day before experiment).* Prepare a 20% (1 g per 5 mL) metrizamide solution in L-15 medium. Prepare a 40% (2 g in 5 mL) metrizamide solution in DPBS. Filter each of the above solutions using a 0.22-μm filter attached to a 5-mL syringe. Prepare 5 mL of each solution (will keep up to 1 month at 4°C).
13. *Sato 100× stock solution.* To 20 mL of Neurobasal-A medium, add the following: 200 mg transferrin (100 μg/mL), 200 mg BSA (100 μg/mL), 5 μL of stock progesterone solution (stock is 2.5 mg/100 μL ethanol) (final = 60 ng/mL or 0.2 μM), 32 mg putrescine (16 μg/mL), 200 μL of stock sodium selenite (stock is 40 mg in 1 mL of 0.1 N NaOH plus 10 mL Neurobasal-A) (final = 40 ng/mL), and 10 μL of stock T3 (stock is 8 mg in 1 mL of 0.1 N NaOH) (final = 30 ng/mL). Aliquot the above 100× stock solutions (200 μL) and store at −20°C. The ethanol stocks for progesterone, sodium selenite, and T3 should be prepared fresh each time a new batch of 100× Sato medium is made.

2.4. High-Insulin Sato Culture Medium

1. To 20 mL of Neurobasal-A medium, add 200 μL Sato 100× stock solution, 100 μL 50× B27 supplement, 200 μL sodium pyruvate 100× stock, 200 μL glutamine 100× stock, 200 μL of 100× penicillin/streptomycin stock (stored at −20°C), and 20 μL *N*-acetyl cysteine 1,000× stock.
2. Next, add 200 μL of 100× insulin stock (stored at 4°C; stock must be prepared every 4–6 weeks—do not freeze). The 100× insulin stock (0.5 mg/mL) is prepared by adding 2 mg of insulin to 4 mL of tissue culture grade water containing 20 μL of 1 N HCl. Filter the stock through a 0.22-μm filter and store at 4°C.
3. Filter the above medium through a 0.22-μm filter. Before filtering the medium, filter about 5 mL of Neurobasal-A medium to rinse away traces of detergent in the filter. The quality of the

filter matters—Millex-GV from Millipore—is best. Falcon plastic is preferable, along with Monoject syringes.

4. Before using the medium, add neurotrophic factors (BNDF, 20 μL of a 5-μg/mL stock; CNTF, 20 μL of a 1-μg/mL stock) and forskolin (5 μM; prepare 1,000× stock in dimethyl sulfoxide so that final concentration of solvent does not exceed 0.1%) to promote cell survival.

2.5. Coating of Culture Dishes or Coverslips (Prepare Day Before Experiment)

Merosin only:

1. Prepare the merosin coating solution as follows: To 5 mL of Neurobasal-A medium, add 10 μL merosin and 50 μL of 100× penicillin/streptomycin stock solution.
2. If using coverslips, first wash 3 times with sterile water.
3. For coating, use 0.4 mL merosin per cm^2 surface area (e.g., 0.5 mL merosin solution per well of a 24-well plate and 0.1 mL per well of a 96-well plate).
4. Coverslips, if used, can be sterilized by dipping in 70% ethanol and briefly passing through a flame (see Note 3).
5. Incubate with merosin solution overnight at 37°C.

Poly-D-lysine only:

1. Dissolve a 5-mg bottle of poly-D-lysine in 5 mL borate buffer. Add 200 μL of this poly-D-lysine stock (1 mg/mL) to 20 mL tissue culture grade sterile water (see Notes 4 and 5).
2. For coating, use 0.2 mL poly-D-lysine per cm^2 surface area.
3. Incubate with poly-D-lysine at room temperature (20–22°C) for 60 min.
4. Remove poly-D-lysine solution and rinse wells 3 times with sterile water.

Merosin ± Poly-D-lysine:

1. Coat first with poly-D-lysine, then add merosin, and incubate overnight.

2.6. Coat Panning Dishes with Second Antibody and PNA (Prepare Day Before Experiment)

1. Prepare 50 mM Tris by dissolving 0.79 g of Tris in 100 mL double-distilled water and adjusting pH to 9.5 with 5 N NaOH.
2. To a 10-cm ∅ petri dish, add 10 mL of 50 mM Tris, pH 9.5 containing 10 μg PNA/mL (PNA dish).
3. To a second 10-cm ∅ petri dish, add 10 mL Tris to OX-7 dish at 10 μg/mL (Thy-1 dish).

4. To a third 10-cm ∅ petri dish, add 30 μL of affinity-purified goat antimouse IgG + IgM (2.4 mg/mL).
5. Incubate all dishes on a flat surface (to ensure uniform coating) at 4°C overnight.

3. Methods

3.1. Cone Photoreceptors

3.1.1. Complete Preparation of Antibody Dishes (Day of Experiment)

1. Rinse PNA dish 4 times with DPBS (mark side to pour) and rinse very well. Store in DPBS until ready to use (in the cold).
2. Add 1 mL OX-7 + 0.2 mL 4% BSA (40 mg/mL DPBS) + 3 mL DPBS to a 10-cm ∅ OX-7 dish.
3. Incubate the OX-7 dish at room temperature for 2 h on a flat surface.
4. Rinse the OX-7 dish 3 times with DPBS and keep in DPBS until needed (in the cold).

3.1.2. Equilibrate EBSS and Sterilize Mesh Filter

1. Aliquot 4 mL EBSS (lacking calcium and magnesium) into one universal 10-mL tube (bijou) and 6 mL EBSS into another tube.
2. Leave them to equilibrate in the 37°C 5%CO_2/95% air incubator by leaving caps on loose. These solutions should change color.
3. Expose the nitex nylon mesh (70-μm pore size) filter to UV light under the vertical flow culture hood 20 min per side.

3.1.3. Dissociation of Retina

1. Dissect out postnatal day P17 mouse or rat retinas. Store minced retinas in 35-mm petri dish in DPBS.
2. Equilibrate EBSS-papain (1 mM EDTA) buffer with 5%CO_2/95% O_2 for 15 min in a 37°C water bath.
3. Dissolve 100 U papain in 5 mL preequilibrated papain buffer. Add 30 μL of a 2.4-mg/mL L-cysteine solution made up in DPBS. Sterilize with a 0.22-μm filter. Remove the DPBS and add retinas to papain by pouring. Add 200 μL of 0.4% DNase and place in the 37°C bath.
4. Digest tissue for 90 min; gently agitate the tissue every 15 min.
5. Add DNase when needed (cells get sticky and do not settle). This usually requires 400 μL of 0.4% DNase added over the 90-min period.
6. While waiting, make the following solutions: Panning buffer containing 15 mL DPBS, 15 μL 100× insulin stock (see high-insulin Sato medium), and 75 μL of 4% BSA.

7. Remove the supernatant (step 5 tissue digestion) by aspiration.
8. Wash the tissue with 4 mL of low ovomucoid solution and allow the cells to settle. Remove the supernatant and discard.
9. Add 2 mL of low ovomucoid solution and triturate gently 5 times with a 1-mL pipetman.
10. Let cells settle, then transfer the supernatant to a new centrifuge tube.
11. Again, add 2 mL of low ovomucoid solution and triturate as before. In subsequent triturations, use slightly more force until all cells are in suspension.
12. Centrifuge cells at 200 × *g* for 15 min at room temperature; discard the supernatant.
13. Resuspend the cell pellet in high ovomucoid solution and centrifuge at 200 × *g* for 15 min.
14. Remove the supernatant and resuspend the cell pellet in 1 mL of panning buffer. Bring the volume to 10 mL.
15. Close the centrifuge tube tightly and incubate in a 37°C water bath for 90 min to let cells recover.
16. Wet the nitex mesh with leftover panning buffer and build a filter funnel using forceps.
17. Filter the cells with a 1-mL pipette and bring the volume up to 15 mL.

3.1.4. Panning

1. Remove the DPBS from the OX-7 dish.
2. Add the cell suspension to the dish, keeping this on a flat surface.
3. Wait 30 min, rocking vigorously after 15 min (but avoid splashing the medium out of the dish). Check for cell adhesion under the microscope.
4. Remove the DPBS from the PNA dish. First, shake the dish hard, then transfer the supernatant from the OX-7 dish by pouring. Use a pipette to remove the remaining cells.
5. Incubate the OX-7 dish for 30 min, shaking gently every 15 min.
6. During this time, make the following: FCS buffer 10 mL Neurobasal-A medium plus 3 mL heat-inactivated FCS (see Note 2) and trypsin solution (4 mL preequilibrated EBSS plus 200 μL 2.5% trypsin). Filter-sterilize both solutions.
7. Discard the supernatant from the PNA dish and wash gently 5–7 times with DPBS to remove all nonadhering cells. Do this by adding 10 mL of DPBS, swirling, and dumping the solution. Check under the microscope to ensure that all nonadherent cells have been removed.
8. Add 6 mL of preequilibrated EBSS to the dish and discard quickly.

9. Add 4 mL of trypsin solution (from step 6) and incubate 10 min in the 37°C incubator.
10. Squirt cells off the dish quickly with a 1-mL pipette (working around the dish once), then add 2 mL of FCS buffer (from step 6), and transfer the solution to a 15-mL centrifuge tube.
11. Add 5 mL of FCS buffer to the plate and again pipette once around. Add the cells recovered from this step to the 15-mL centrifuge tube in step 10.
12. Invert the centrifuge tube and remove 100 μL to count cells.
13. Centrifuge the tube with cells at 200 × *g* for 15 min and resuspend the cell pellet in high-insulin Sato medium to give a density of 10^6 cells/mL.
14. Plate the cells according to the experimental design. If the final surface coating is merosin, first rinse 3 times with Neurobasal-A medium before cell plating.

3.2. Culture of RPE Cells

3.2.1. RPE Dissociation

1. Cut off anterior portion of eye (make sure to remove all of the iris). Remove the lens, vitreous, cornea, and retina.
2. Remove the RPE by pulling the RPE/choroid out of the eye socket with forceps and store in Ca^{2+}/Mg^{2+}—free DPBS. Remove the optic nerve.
3. Equilibrate the EBSS-papain (1 mM EDTA) solution with 5% CO_2/95% O_2 for 15 min in a 37°C incubator.
4. Dissolve 100 U of papain and 333 U/mL of collagenase in 5 mL preequilibrated papain buffer. Next, add 0.8 mg of L-cysteine to this solution.
5. Sterilize the solution in point 4 using a 0.22-μm filter. Remove the DPBS and add retinas to papain by pouring in. Next, add 200 μL 0.4% DNase.
6. Digest the tissue in 37°C bath for 90 min with gentle agitation of the tissue every 15 min.
7. Remove the supernatant by aspiration, being careful not to draw up the digested tissue.
8. Wash the tissue with 4 mL of the low ovomucoid solution and let cells settle.
9. Remove supernatant and discard.
10. Add 2 mL of the low ovomucoid solution. Triturate gently with a long (9-in.) sterile Pasteur glass pipette using 5 up-and-down strokes.
11. Let cells settle, remove the supernatant, and transfer the cells to a new centrifuge tube.
12. Add again 2 mL of low ovomucoid solution and repeat the trituration step.

13. In subsequent trituration steps, use slightly more force until the RPE is released from the choroid and cells are in suspension.
14. Choroid does not dissociate and so should be removed and discarded.
15. Centrifuge the cells at 200 × *g* for 15 min at 25°C.
16. Discard the supernatant and resuspend the cells in high ovomucoid solution.
17. Centrifuge at 200 × *g* for 15 min.
18. Remove the high ovomucoid solution but leaving 2 mL to resuspend the cells.

3.2.2. Purification of RPE Cells

1. Using a 15-mL polystyrene centrifuge tube, layer metrizamide (6 mL) with 40% on the bottom.
2. Carefully add the 2 mL of high ovomucoid solution containing cells (see Subheading 3.1, step 18) to the top of the metrizamide and centrifuge at 500 × *g* for 15 min at 20–22°C.
3. Prepare a 4% BSA solution (0.4 g BSA in 10 mL L-15) and filter-sterilize. This will be used as a cushion for the 40% metrizamide layer (step 5 below).
4. Using a hypodermic syringe and needle, carefully remove about 1 mL of the cell suspension between the 40% layer of metrizamide and the high ovomucoid solution.
5. Resuspend the 1 mL of metrizamide cell suspension in 6 mL of L-15.
6. To a fresh 15-mL centrifuge tube, to which has been added 1 mL of 4% BSA solution, now place the cell suspension from step 5.
7. Centrifuge at 200 × *g* for 15 min (this step may be difficult because the metrizamide is very dense).
8. Remove the supernatant and resuspend cells in high-insulin Sato culture medium.
9. Count cells (see Chap. 3) and plate at a density of 50,000–100,000 cells cm^2.

4. Notes

1. The source of glass is very important for the coverslips: German suppliers are highly recommended, in particular the one listed here.
2. Heat inactivation of FCS is recommended to destroy heat-labile complement. Thaw the bottle of serum in advance, using a 37°C water bath. Next, immerse the serum bottle in the water bath

after reequilibrating to 56°C and leave for 30 min. Swirl the bottle occasionally to ensure proper mixing. Allow the serum to cool to room temperature, aliquot into 50-mL tubes, and store at −20°C.

3. When flame-sterilizing cover glasses, it is recommended to "wave" the cover glass through the air immediately after flaming. Because of their fragile nature, heat buildup may cause frequent shattering of the glass—hence the need to dissipate the heat.
4. Poly-D-lysine is very hygroscopic and difficult to weigh. It is advisable to purchase the material as a 5-mg quantity and then dissolve the entire amount in borate buffer. The solutions can be stored at 4°C for up to 6 months.
5. Working solutions of poly amino acids should be prepared in sterile tissue culture plastic polypropylene tubes. *Take care*: The poly amino acid will adhere to glass surfaces and also to polystyrene plastic.

References

1. Sancho-Pelluz J, Arango-Gonzalez B, Kustermann S, Romero FJ, van Veen T, Zrenner E, et al (2008) Photoreceptor cell death mechanisms in inherited retinal degeneration. Mol Neurobiol 38, 253–269
2. LaVail MM, Yasumura D, Matthes MT, Lau-Villacorta C, Unoki K., Sung C-H, et al (1998) Protection of mouse photoreceptors by survival factors in retinal degenerations. Invest Ophthalmol Vis Sci 39, 592–602
3. Gao H, Hollyfield JG (1992) Aging of the human retina: differential loss of neurons and retinal pigment epithelial cells. Invest Ophthalmol Vis Sci 33, 1–17
4. Curcio CA, Allen KA, Kalina RE (1993) Aging of the human photoreceptor mosaic: evidence for selective vulnerability of rods in central retina. Invest Ophthalmol Vis Sci 34, 3278–3296
5. Karl MO, Reh TH (2010) Regenerative medicine for retinal diseases: activating endogenous repair mechanisms Early changes in synaptic connectivity following progressive photoreceptor degeneration in RCS rats. Trends Mol Med 16, 193–202
6. Wenzel A, Grimm C, Samardzija M, Remé CE (2005) Molecular mechanisms of light-induced photoreceptor apoptosis and neuroprotection for retinal degeneration. Prog Retin Eye Res 24, 275–306
7. Cuenca N, Pinilla I, Sauvé Y, Lund R (2005) Early changes in synaptic connectivity following progressive photoreceptor degeneration in RCS rats. Eur J Neurosci 22, 1057–1072
8. Fontaine V, Kinkl N, Sahel J, Dreyfus H, Hicks D (1998) Survival of purified rat photoreceptors in vitro is stimulated directly by fibroblast growth factor-2. J Neurosci 18, 9662–9672
9. Traverso V, Kinkl N, Grimm L, Sahel J, Hicks D (2003) Basic fibroblast and epidermal growth factors stimulate survival in adult porcine photoreceptor cell cultures. Invest Ophthalmol Vis Sci 44, 4550–4558
10. Faktorovich EG, Steinberg RH, Yasumura D, Matthes MT, LaVail MM (1990) Photoreceptor degeneration in inherited retinal dystrophy delayed by basic fibroblast growth factor. Nature 347, 83–86
11. Fuhrmann S, Kirsch M, Hofmann HD (1995) Ciliary neurotrophic factor promotes chick photoreceptor development in vitro. Development 121, 2695–2706
12. Kassen SC, Thummel R, Campochiaro LA, Harding MJ, Bennett NA, Hyde DR (2009) CNTF induces photoreceptor neuroprotection and Müller glial cell proliferation through two different signaling pathways in the adult zebrafish retina. Exp Eye Res 2009 88, 1051–1064
13. Caffé AR, Söderpalm AK, Holmqvist I, van Veen T (2001) A combination of CNTF and BDNF rescues rd photoreceptors but changes rod differentiation in the presence of RPE in retinal explants. Invest Ophthalmol Vis Sci 42, 275–282

14. Chong NH, Alexander RA, Waters L, Barnett KC, Bird AC, Luthert PJ (1999) Repeated injections of a ciliary neurotrophic factor analogue leading to long-term photoreceptor survival in hereditary retinal degeneration. Invest Ophthalmol Vis Sci 40, 1298–1305
15. Cepeda IL, Flores J, Cornfeldt ML, O'Kusky JR, Doudet DJ (2007) Human retinal pigment epithelial cell implants ameliorate motor deficits in two rat models of Parkinson disease. J Neuropathol Exp Neurol 66, 576–584
16. Ming M, Li X, Fan X, Yang D, Li L, Sheng Chen S, et al (2009) Retinal pigment epithelial cells secrete neurotrophic factors and synthesize dopamine: possible contribution to therapeutic effects of RPE cell transplantation in Parkinson's disease. J Trans Med 2009, 7:53 doi:10.1186/1479-5876-7–53
17. Murakami Y, Ikeda Y, Yonemitsu Y, Onimaru M, Nakagawa K, Kohno R, et al (2008) Inhibition of nuclear translocation of apoptosis-inducing factor is an essential mechanism of the neuroprotective activity of pigment epithelium-derived factor in a rat model of retinal degeneration. Am J Pathol 173, 1326–1338
18. Miyazaki M, Ikeda Y, Yonemitsu Y, Goto Y, Kohno R, Murakami Y, et al (2008) Synergistic neuroprotective effect via simian lentiviral vector-mediated simultaneous gene transfer of human pigment epithelium-derived factor and human fibroblast growth factor-2 in rodent models of retinitis pigmentosa. J Gene Med 10, 1273–1281
19. Miyazaki M, Ikeda Y, Yonemitsu Y, Goto Y, Sakamoto T, Tabata T, et al (2003) Simian lentiviral vector-mediated retinal gene transfer of pigment epithelium-derived factor protects retinal degeneration and electrical defect in Royal College of Surgeons rats. Gene Ther 10 1503–1511

Chapter 14

Culture of Rat Retina Pigmented Epithelial Cells

Stephen D. Skaper

Abstract

Retinal pigment epithelium cells play a key role in maintaining the normal function of retina and can express several neurotrophic factors, which support the neurosensory retina and may also provide trophic signals to the host dopaminergic neurons. The following chapter describes a protocol for the purification and culture of retinal pigmented epithelial cells from postnatal rat.

Key words: Pigmented epithelium, Retina, Cell culture, Neurotrophic factors

1. Introduction

Retinal pigmented epithelial (RPE) cells are melanin-containing cells that constitute a monolayer between the neural retina and the choroid. In RPE cells, tyrosine is catalyzed by tyrosinase to L-dopa that is polymerized to form melanin (1). L-dopa in RPE cells is converted into dopamine in the terminals of nigrostriatal dopaminergic neurons and provides dopamine to the nigrostriatal system directly after RPE cells are transplanted (2, 3). RPE cells play a key role in maintaining the normal function of retina and can express several neurotrophic factors such as platelet-derived growth factor, epidermal growth factor, vascular endothelial growth factor, and pigment-derived epithelial factor (PEDF) (4), which support the neurosensory retina and may also provide trophic signals to the host dopaminergic neurons. In addition, RPE cells secrete glial cell line–derived neurotrophic factor and brain-derived neurotrophic factor (BDNF) (3), which may contribute to the beneficial effects of RPE cells transplantation in an animal model of Parkinson's disease. In this context, RPE-conditioned medium protects cultured primary dopaminergic neurons against rotenone- and 6-hydroxydopamine-induced cell loss (3).

Stephen D. Skaper (ed.), *Neurotrophic Factors: Methods and Protocols*, Methods in Molecular Biology, vol. 846, DOI 10.1007/978-1-61779-536-7_14, © Springer Science+Business Media, LLC 2012

PEDF is a 50-kDa secreted glycoprotein and a noninhibitory member of the serine protease inhibitor gene family and was originally discovered in conditioned media from RPE cells but since has been identified throughout the central nervous system (5). PEDF has neuroprotective properties for various types of neurons including cerebellar granule neurons (6), retinal neurons (5), hippocampal neurons (7), spinal cord motor neurons (8), striatal (enkephalinergic) neurons (9), and mesencephalic (dopaminergic) neurons (9, 10). PEDF upregulation induced by memantine, an *N*-methyl-D-aspartate receptor antagonist, is involved in increased proliferation of hippocampal progenitor cells (11).These observations raise the possibility that application of PEDF may be helpful in designing new therapeutic strategies for neurodegenerative diseases such as amyotrophic lateral sclerosis, Parkinson's disease, Huntington's disease, Alzheimer's disease, and brain ischemia (12). The following chapter provides a protocol for the culture of RPE cells from rat.

2. Materials

2.1. Equipment and Labware

1. Stereo dissecting microscope (backlighting of stage is preferred) with fiber optic light source.
2. Laminar flow cabinet for dissections.
3. Laminar flow biological safety cabinet (CL2).
4. Humidified, water-jacketed culture incubator at 37°C and 5% CO_2/95% air.
5. Water bath set at 37°C.
6. Dissecting tools (Fine Science Tools).
7. Bench centrifuge to accommodate 15- and 50-mL tubes.
8. 10-cm ∅ sterile tissue culture dishes.
9. Glass coverslips (12 mm ∅ #1, Menzel-Gläser, Menzel GmbH, Braunschweig, Germany) (see Note 1).
10. 96-well poly-D-lysine-treated tissue culture plates (BD Biosciences).
11. 15- and 50-mL polypropylene plastic centrifuge tubes (sterile).
12. 0.22-μm filters (Millipore).
13. Neubauer hemocytometer (Fisher Scientific).

2.2. Reagents

1. Dulbecco's phosphate-buffered saline, pH 7.4, sterile (DPBS) (Invitrogen).
2. Water, sterile, tissue culture grade.
3. L-15 medium, with L-glutamine and L-amino acids (Invitrogen).
4. Neurobasal-A medium (Invitrogen).

5. B27 supplement, 50×, with antioxidants (Invitrogen).
6. Fetal calf serum (Invitrogen) (see Note 2).
7. L-Cysteine.
8. Insulin, from bovine pancreas (Sigma).
9. Transferrin, human.
10. Progesterone.
11. Putrescine.
12. Sodium selenite.
13. T3 (3,3′,5-triiodo-L-thyronine, sodium salt), cell culture tested (Sigma).
14. Penicillin/streptomycin (100× stock), sterile, for cell culture (Invitrogen).
15. Gentamicin (50 mg/mL stock), sterile, for cell culture (Sigma).
16. Papain (Worthington (Lorne)).
17. Collagenase (Sigma).
18. DNase (type I) (Sigma).
19. Sodium pyruvate (100 mM stock), sterile, for cell culture (Sigma).
20. L-Glutamine (200 mM stock), sterile, for cell culture (Invitrogen).
21. Bovine serum albumin, essentially fatty acid and globulin free, (BSA) (Sigma).
22. Ovomucoid (trypsin inhibitor from chicken egg white, type III-O, free of ovoinhibitor) (Sigma).
23. Poly-D-lysine (MW 30,000–70,000), sterile, for cell culture (Sigma).
24. Metrizamide (Sigma).
25. Merosin, human (Millipore).
26. BDNF, human, recombinant (Peprotech).
27. Ciliary neurotrophic factor, human, recombinant (Peprotech).
28. Forskolin.
29. *N*-acetyl cysteine.
30. Phenol red (0.5%), sterile, for cell culture (Sigma).

2.3. Stock Solutions

1. *10× EBSS stock solution without* $Ca2^+$, $Mg2^+$, *and bicarbonate.* 1.16 M NaCl (68 g/L), 54 mM KCl (4 g/L), 10 mM $NaH_2PO_4{\cdot}H_20$ (1.4 g/L), 10 mM D-glucose (10 g/L), and 0.05% phenol red (10 mL/L of 0.5% solution). Filter-sterilize and store at 4°C.

2. D-*Glucose, 30% (w/v)*. Dissolve 15 g of D-glucose in 50 mL double-distilled water.
3. *$NaHCO_3$, 1 M*. Dissolve 4.2 g of $NaHCO_3$ in 50 mL double-distilled water.
4. *$MgSO_4$, 100 mM*. Dissolve 1.23 g of $MgSO_4 \cdot 7H_2O$ in 50 mL double-distilled water.
5. *EDTA, 0.5 M*. Dissolve 7.3 g of EDTA in 50 mL double-distilled water.
6. *EBSS-papain solution with $Mg2^+$, EDTA, and bicarbonate*. Combine 20 mL of 10× EBSS solution, 2 mL of 100 mM $MgSO_4$, 2.4 mL of 30% D-glucose, 0.4 mL of 0.5 M EDTA, and 5.2 mL of 1 M $NaHCO_3$. Bring to a final volume of 200 mL with double-distilled water.
7. *Borate buffer 0.15 M, pH 8.4*. Dissolve 28.6 g of sodium borate ($Na_2B_4O_7 \cdot 10H_2O$) in 500 mL double-distilled water (pH will be ~9.2). Adjust pH to 8.4 with 5 N HCl. Filter-sterilize and store at 4°C (up to 6 months).
8. *Low ovomucoid solution*. Mix together 9 mL DPBS, 1 mL 10× low ovomucoid stock (15 mg each BSA and ovomucoid/mL DPBS), and 200 μL 0.4% DNase stock (made by dissolving 4 mg/mL of DNase in L-15 medium; filter-sterilize, aliquot, and store at −20°C). Check pH and adjust to 7.4 if necessary. Filter-sterilize the final solution.
9. *High ovomucoid solution*. Mix together 4.8 mL DPBS and 1.2 mL 6× high ovomucoid stock (50 mg each BSA and ovomucoid/mL DPBS). Filter-sterilize.
10. *Forskolin (1,000×)*. Add 2.4 mL dimethyl sulfoxide to a 10-mg bottle of forskolin (=5 mM). Make 100 μL aliquots and store −20°C. The final concentration of solvent should not exceed 0.1% in the culture medium.
11. *N-acetyl cysteine (1,000×)*. Dissolve 60 mg *N*-acetyl cysteine in 1 mL Neurobasal-A medium. Filter-sterilize, make 20 μL aliquots, and store at −20°C.
12. *Metrizamide (prepare day before experiment)*. Prepare a 20% (1 g/5 mL) metrizamide solution in L-15 medium. Prepare a 40% (2 g in 5 mL) metrizamide solution in DPBS. Filter each of the above solutions using a 0.22-μm filter attached to a 5-mL syringe. Prepare 5 mL of each solution (will keep up to 1 month at 4°C).
13. *Sato 100× stock solution*. To 20 mL of Neurobasal-A medium, add the following: 200 mg transferrin (100 μg/mL), 200 mg BSA (100 μg/mL), 5 μL of stock progesterone solution (stock is 2.5 mg/100 μL ethanol) (final = 60 ng/mL or 0.2 μM), 32 mg putrescine (16 μg/mL), 200 μL of stock sodium selenite

(stock is 40 mg in 1 mL of 0.1 N NaOH plus 10 mL Neurobasal-A) (final = 40 ng/mL), and 10 μL of stock T3 (stock is 8 mg in 1 mL of 0.1 N NaOH) (final = 30 ng/mL). Aliquot the above 100× stock solutions (200 μL) and store at –20°C. The ethanol stocks for progesterone, sodium selenite, and T3 should be prepared fresh each time a new batch of 100× Sato medium is made.

2.4. High-Insulin Sato Culture Medium

1. To 20 mL of Neurobasal-A medium, add 200 μL Sato 100× stock solution, 100 μL 50× B27 supplement, 200 μL sodium pyruvate 100× stock, 200 μL glutamine 100× stock, 200 μL of 100× penicillin/streptomycin stock (stored at –20°C), and 20 μL *N*-acetyl cysteine 1,000× stock.
2. Next, add 200 μL of 100× insulin stock (stored at 4°C; stock must be prepared every 4–6 weeks—do not freeze). The 100× insulin stock (0.5 mg/mL) is prepared by adding 2 mg of insulin to 4 mL of tissue culture grade water containing 20 μL of 1 N HCl. Filter the stock through a 0.22-μm filter and store at 4°C.
3. Filter the above medium through a 0.22-μm filter. Before filtering the medium, filter about 5 mL of Neurobasal-A medium to rinse away traces of detergent in the filter. The quality of the filter matters—Millex-GV from Millipore—is best. Falcon plastic is preferable, along with Monoject syringes.
4. Before using the medium, add neurotrophic factors (BNDF, 20 μL of a 5 μg/mL stock; ciliary neurotrophic factor, 20 μL of a 1 μg/mL stock) and forskolin (5 μM; prepare 1,000× stock in dimethyl sulfoxide so that final concentration of solvent does not exceed 0.1%) to promote cell survival.

2.5. Coating of Culture Dishes or Coverslips (Prepare Day Before Experiment)

1. Prepare the merosin coating solution as follows: To 5 mL of Neurobasal-A medium, add 10 μL of merosin and 50 μL of 100× penicillin/streptomycin stock solution.
2. For coating, use 0.4 mL merosin per cm^2 surface area.
3. Coverslips, if used, can be sterilized by dipping in 70% ethanol and briefly passing through a flame (see Note 3).
4. Incubate with merosin solution overnight at 37°C.
5. Poly-D-lysine coating solution: Dissolve a 5-mg bottle of poly-D-lysine in 5 mL borate buffer. Add 200 μL of this poly-D-lysine stock (1 mg/mL) to 20 mL of tissue culture grade sterile water (see Notes 4 and 5).
6. For coating, use 0.2 mL poly-D-lysine per cm^2 surface area.
7. Incubate with poly-D-lysine at room temperature (20–22°C) for 60 min.

3. Methods

3.1. Dissociation of RPE

1. Cut off anterior portion of eye (make sure to remove all of the iris). Remove the lens, vitreous, cornea, and retina.
2. Remove the RPE by pulling the RPE/choroid out of the eye socket with forceps and store in Ca^{2+}/Mg^{2+}—free DPBS. Remove the optic nerve.
3. Equilibrate the EBSS-papain (1 mM EDTA) solution with 5% CO_2/95% O_2 for 15 min in a 37°C incubator.
4. Dissolve 100 U of papain and 333 U/mL of collagenase in 5 mL preequilibrated papain buffer. Next, add 0.8 mg of L-cysteine to this solution.
5. Sterilize the solution in point 4 using a 0.22-μm filter. Remove the DPBS and add retinas to papain by pouring in. Next, add 200 μL 0.4% DNase.
6. Digest the tissue in 37°C bath for 90 min with gentle agitation of the tissue every 15 min.
7. Remove the supernatant by aspiration, being careful not to draw up the digested tissue.
8. Wash the tissue with 4 mL of the low ovomucoid solution and let cells settle.
9. Remove supernatant and discard.
10. Add 2 mL of the low ovomucoid solution. Triturate gently with a long (9-in.) sterile Pasteur glass pipette using five up-and-down strokes.
11. Let cells settle, remove the supernatant, and transfer the cells to a new centrifuge tube.
12. Add again 2 mL of low ovomucoid solution and repeat the trituration step.
13. In subsequent trituration steps, use slightly more force until the RPE is released from the choroid and cells are in suspension.
14. Choroid does not dissociate and so should be removed and discarded.
15. Centrifuge the cells at 200 × *g* for 15 min at 25°C.
16. Discard the supernatant and resuspend the cells in high ovomucoid solution.
17. Centrifuge at 200 × *g* for 15 min.
18. Remove the high ovomucoid solution but leaving 2 mL to resuspend the cells.

3.2. Purification of RPE

1. Using a 15-mL polystyrene centrifuge tube, layer metrizamide (6 mL) with 40% on the bottom.

2. Carefully add the 2 mL of high ovomucoid solution containing cells (see Subheading 3.1, step 18) to the top of the metrizamide and centrifuge at 500 × *g* for 15 min at 20–22°C.
3. Prepare a 4% BSA solution (0.4 g BSA in 10 mL L-15) and filter-sterilize. This will be used as a cushion for the 40% metrizamide layer (step 5 below).
4. Using a hypodermic syringe and needle, carefully remove about 1 mL of the cell suspension between the 40% layer of metrizamide and the high ovomucoid solution.
5. Resuspend the 1 mL of metrizamide cell suspension in 6 mL of L-15.
6. To a fresh 15-mL centrifuge tube, to which has been added 1 mL of 4% BSA solution, now place the cell suspension from step 5.
7. Centrifuge at 200 × *g* for 15 min (this step may be difficult because the metrizamide is very dense).
8. Remove the supernatant and resuspend cells in high-insulin Sato culture medium.
9. Count cells (see Chap. 3) and plate at a density of 50,000–100,000 cells cm^2.

4. Notes

1. The source of glass is very important for the coverslips: German suppliers are highly recommended, in particular the one listed here.
2. Heat inactivation of fetal calf serum is recommended to destroy heat-labile complement. Thaw the bottle of serum in advance, using a 37°C water bath. Next, immerse the serum bottle in the water bath after reequilibrating to 56°C and leave for 30 min. Swirl the bottle occasionally to ensure proper mixing. Allow the serum to cool to room temperature, aliquot into 50-mL tubes and store at −20°C.
3. When flame-sterilizing cover glasses, it is recommended to “wave” the cover glass through the air immediately after flaming. Because of their fragile nature, heat buildup may cause frequent shattering of the glass—hence the need to dissipate the heat.
4. Poly-D-lysine is very hygroscopic and difficult to weigh. It is advisable to purchase the material as a 5-mg quantity and then dissolve the entire amount in borate buffer. The solutions can be stored at 4°C for up to 6 months.

5. Working solutions of poly amino acids should be prepared in sterile tissue culture plastic polypropylene tubes. *Take care*: The poly amino acid will adhere to glass surfaces, and also to polystyrene plastic.

References

1. Aroca P, Urabe K, Kobayashi T, Tsukamoto K, and Hearing VJ (1993) Melanin biosynthesis patterns following hormonal stimulation. J Biol Chem 268, 25650–25655
2. Watts RL, Raiser CD, Stover NP, Cornfeldt ML, Schweikert AW, Allen RC, et al (2003) Stereotaxic intrastriatal implantation of human retinal pigment epithelial (hRPE) cells attached to gelatin microcarriers: a potential new cell therapy for Parkinson's disease. J Neural Transm Suppl 65, 215–227
3. Ming M, Li X, Fan X, Yang D, Li L, Sheng Chen S, et al (2009) Retinal pigment epithelial cells secrete neurotrophic factors and synthesize dopamine: possible contribution to therapeutic effects of RPE cell transplantation in Parkinson's disease. J Trans Med 2009, 7:53 doi:10.1186/1479-5876-7-53
4. Cepeda IL, Flores J, Cornfeldt ML, O'Kusky JR, and Doudet DJ (2007) Human retinal pigment epithelial cell implants ameliorate motor deficits in two rat models of Parkinson disease. J Neuropathol Exp Neurol 66, 576–584
5. Tombran-Tink J, Barnstable CJ (2003) PEDF: a multifaceted neurotrophic factor. Nat Rev Neurosci 4, 628–636
6. Yabe T, Wilson D, and Schwartz JP (2001) NFkappaB activation is required for the neuroprotective effects of pigment epithelium-derived factor (PEDF) on cerebellar granule neurons. J Biol Chem 276, 43313–43319
7. DeCoster MA, Schabelman E, Tombran-Tink J, and Bazan NG (1999) Neuroprotection by pigment epithelial-derived factor against glutamate toxicity in developing primary hippocampal neurons. J Neurosci Res 56, 604–610
8. Bilak MM, Corse AM, Bilak SR, Lehar M, Tombran-Tink J, and Kuncl RW (1999) Pigment epithelium-derived factor (PEDF) protects motor neurons from chronic glutamate-mediated neurodegeneration. J Neuropathol Exp Neurol 58, 719–728
9. McKay BS, Goodman B, Falk T, and Sherman SJ (2006) Retinal pigment epithelial cell transplantation could provide trophic support in Parkinson's disease: results from an in vitro model system. Exp Neurol 201, 234–243
10. Falk T, Zhang S, and Sherman SJ (2009) Pigment epithelium derived factor (PEDF) is neuroprotective in two in vitro models of Parkinson's disease. Neurosci Lett 458, 49–52
11. Namba T, Yabe T, Gonda Y, Ichikawa N, Sanagi T, Arikawa-Hirasawa E, et al (2010) Pigment epithelium-derived factor up-regulation induced by memantine, an N-methyl-D-aspartate receptor antagonist, is involved in increased proliferation of hippocampal progenitor cells. Neuroscience 167, 372–383
12. Yabe T, Sanagi T, and Yamada H (2010) The neuroprotective role of PEDF: implication for the therapy of neurological disorders. Curr Mol Med 10, 259–266

Chapter 15

Mammalian Growth Cone Turning Assays Identify Distinct Cell Signalling Mechanisms That Underlie Axon Growth, Guidance and Regeneration

Andrew J. Murray, Andrew G. Peace, Steven J. Tucker, and Derryck A. Shewan

Abstract

The cell signalling mechanisms underlying mammalian central nervous system axon growth and guidance change during development, such that axons that establish appropriate connectivity in the embryo fail to regenerate after injury to the adult nervous system. The growth cone turning assay has been used in *Xenopus* neurons to elucidate mechanisms of axon guidance during development. Here, we describe how we have adapted this assay for rat dorsal root ganglion neurons to study the influence of extracellular secreted factors causing growth cone attraction and repulsion. Additionally, we describe how this method can be combined with small interfering RNA and cDNA transfections to manipulate protein expression in growth cones, and fluorescence resonance energy transfer to monitor the activity of signalling pathways in live neurons. This assay provides the unique ability to manipulate and visualise the internal status of growth cone signalling whilst challenged with extracellular chemotropic signalling molecules, and can be used to develop strategies to promote axon regeneration in the mature mammalian central nervous system.

Key words: Axon guidance, Regeneration, Growth cone, Turning assay, Fluorescence resonance energy transfer, Cell signalling, cAMP

1. Introduction

The growth cone is the motile tip of a growing axon responsible for integrating signals from the extracellular environment and controlling the rate and direction of axonal growth. The growth cone senses its molecular environment and makes directional growth decisions according to both the spatial expression of extracellular repulsive and attractive guidance cues, and the internal status of growth cone signalling molecules that ultimately mediate

Stephen D. Skaper (ed.), *Neurotrophic Factors: Methods and Protocols*, Methods in Molecular Biology, vol. 846, DOI 10.1007/978-1-61779-536-7_15,

cytoskeletal regulation. When axons are severed in the mature nervous system, the distal tip of the severed axon quickly forms a growth cone and attempts to mount a regenerative response. However, a variety of growth-inhibiting molecules present in the mature nervous system blocks axon regrowth (1).

Since the development of a reliable method for characterising growth cone turning to gradients of secreted factors (2), numerous insights have been gained into the intracellular and extracellular mechanisms of growth cone guidance. These include descriptions of how manipulating signalling pathways within growth cones alters the attractive and repulsive responses to extracellular factors (3, 4). These studies into growth cone turning during development have been mainly carried out in spinal and retinal neurons in embryos of the South African clawed-toed frog *Xenopus laevis.* However, the study of responses to extracellular factors by developing or regenerating mammalian neurons requires a number of adaptations because, curiously, mammalian (or even zebrafish) neurons temporarily halt neurite outgrowth when exposed to temperatures higher than ambient room temperature. Here, we describe a growth cone turning assay that can be used to study the responses of mature mammalian neurons to gradients of extracellular secreted factors. We have previously used this assay to examine cyclic nucleotide-mediated signalling pathways and their potential to initiate growth cone attraction (5, 6) and switch repulsion to attraction when stimulated in the presence of secreted molecules such as myelin associated glycoprotein and netrin-1 (7). We describe how this assay can be used to manipulate growth cone intracellular signalling pathways with pharmacological agents and, additionally, demonstrate how the growth cone turning assay can be evolved to combine chemotropic observations with efficient transfection of mammalian dorsal root ganglion (DRG) neurons with small interfering RNA or cDNA, further enhancing the repertoire of manipulations available through protein knockdown, knockout or over-expression. Furthermore, we use fluorescence resonance energy transfer (FRET) to visualise signalling pathways activated in growth cones in real time, gaining unique insights into the mechanisms that control axon growth and guidance. We have used this method to study the responses of rat DRG growth cones to extracellular guidance molecules as well as dissect intracellular signalling pathways involved in DRG growth cone turning, but we envisage that this could be easily employed to study turning responses in a variety of mammalian neurons.

2. Materials

2.1. Dorsal Root Ganglion Neuron Culture

1. Bottenstein and Sato's fluid (BSF). Ham's F12 culture medium (Invitrogen) supplemented with 16 μg/mL putrescine, 100 μg/mL transferrin, 0.3% bovine serum albumin (BSA), 100 U/mL penicillin/streptomycin, 0.06 μg/mL progesterone, 0.16 μg/mL sodium selenite, 20 μg/mL insulin (all Sigma-Aldrich) and 0.1 μg/mL nerve growth factor (NGF) (Serotec) (see Note 1).
2. Low-walled, glass bottom culture dishes (MatTek).
3. Poly-L-lysine (70,000–150,000 mol. weight) (PLL) (Sigma-Aldrich) and appropriate growth substratum, for example laminin (Invitrogen).
4. 15% BSA dissolved in Ham's F12 and stored in 2 mL aliquots in 15-mL centrifuge tubes (Sterilin) at −20°C.
5. 37°C cell culture incubator with 5% CO_2/95% air and class I tissue culture hood.
6. Haemocytometer for counting cell numbers.
7. Trypsin inhibitor/DNAse. Hank's balanced salt solution (HBSS; Invitrogen) containing 0.002% trypsin inhibitor, 0.025% DNAse and 1% BSA (all Sigma-Aldrich).

2.2. DRG Neuron Transfection

1. Amaxa nucleofector machine (Lonza AG).
2. Nucleofector solution (Lonza AG).
3. Nucleofection cuvettes and Pasteur pipettes (Lonza AG).
4. Appropriate concentration of siRNA or transfection quality DNA.

2.3. Growth Cone Turning Assays

1. Glass pipettes (Intracell) pulled on a Sutter Pipette Puller to have a 1 μm bore width.
2. Picospritzer III with a TSS10 dual pulse stimulator (both Intracell) connected to a compressed air supply and micropipette holder held by a micromanipulator.
3. Inverted phase contrast microscope with heated stage and CO_2 supply and camera, connected to a computer equipped with live imaging software. If CO_2 is not available, HEPES buffer can be used as an alternative.
4. ImageJ software for analysis of growth cone turning, available freely from http://rsbweb.nih.gov/ij/.

2.4. FRET Analysis

1. cDNA constructs designed to encode particular signalling proteins with appropriate FRET-paired fluorophores.
2. Leica AF6000LX imaging system or equivalent FRET-capture imaging system.
3. Leica application software or equivalent for analysis of FRET time-lapse experiments.

3. Methods

DRG neurons represent an important model for nervous system regeneration. They are unique among peripheral neurons in that they project one axonal branch to the periphery and another into the central nervous system during development, but when injured in the adult, only the peripheral branch can regenerate whilst the central branch is inhibited at the boundary with the spinal cord (8). The procedure described here is based on the protocol developed for *Xenopus* neurons (2). A major advantage of the *Xenopus* system is that neurons develop at room temperature (15–20°C). Mammalian neurons, however, require a temperature of 37°C and a CO_2-buffered atmosphere.

One major problem with examining the growth of mammalian neurons is the poor growth rate of axons when exposed to an atmosphere at 37°C. This is likely due to the heated microscope stage causing rapid evaporation of the culture medium and a lack of 5% CO_2 in the atmosphere. The lack of CO_2, in the short term, can be overcome with the addition of 15 μM HEPES to the culture medium to buffer the pH under normal atmospheric conditions (9). To overcome the exposure of the culture medium to the heated atmosphere, we overlay the medium with a thin layer of oil, preventing evaporation and gas exchange. Figure 1a shows that the growth rate of DRG neurites is comparable in a sealed culture dish at 37°C and culture medium overlayed with oil. Typical responses of DRG growth cones to a gradient of NGF are shown in Fig. 1b.

3.1. Preparing Growth Substrata

1. Prepare 10 μg/mL PLL in sterile distilled H_2O and gently pipette 500 μL of this solution onto each central glass well of MatTek low-walled 50-mm petri dishes under aseptic conditions and leave overnight.
2. The next day, remove the PLL and allow the dishes to dry completely in a tissue culture hood (see Note 2).
3. Prepare a solution of 2 μg/mL L-laminin in F12 medium and gently pipette 500 μL into the central well of each MatTek dish.
4. Place dishes in a 37°C incubator for at least 2 h, but ideally overnight. Other extracellular matrix substrates, such as fibronectin, may also be applied in this way.

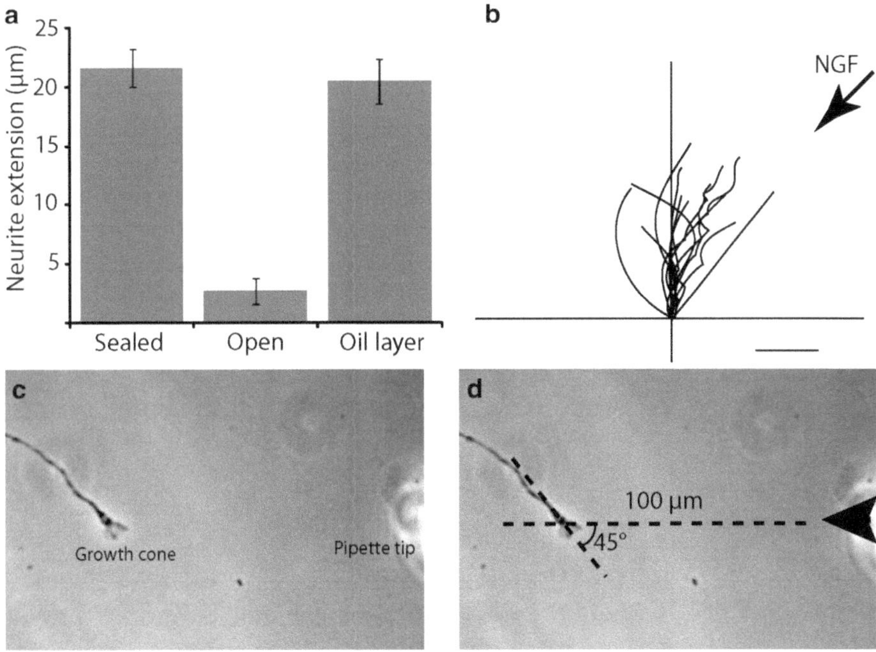

Fig. 1. (**a**) Growth rates of DRG neuron neurites in a sealed culture dish, exposed to a heated (37°C) atmosphere and 'sealed' with a layer of oil. Note that the oil layer allows similar levels of neurite outgrowth to that seen in a sealed tissue culture dish. (**b**) Typical responses of DRG neuron growth cones to a gradient of NGF emitted from the micropipette. The bias of neurites growing in the direction of the gradient indicates an attractive response. (**c**) Position of the axon/growth cone (*left*) compared to the micropipette (*right*). (**d**) Image from (**c**) showing the angle of the growth cone and distance from the micropipette.

3.2. Preparation of Dissociated DRG Neurons for Growth Cone Turning Assays

Here, we describe the culture of dissociated adult DRG neurons, as we believe they represent a good model for studying axon regeneration. However, we have also used this assay with DRG neurons at other developmental stages to study intrinsic cell signalling changes in developing neurons (5, 7), and the procedure can be used to study other mammalian neuronal types (10):

1. Dissect the spinal column from a sacrificed rat and remove as much connective tissue and muscle as possible, then use cuticle nippers to remove dorsal vertebral plates to expose the spinal cord from the dorsal side.
2. Locate the DRGs lateral to the spinal cord on either side at each vertebral segment, carefully cut the dorsal roots and place the DRGs in warm F12 culture medium in a 35-mm petri dish.
3. Trim the roots from the DRGs and incubate the ganglia in F12 medium containing 0.125% collagenase for 1 h at 37°C in a 5% CO_2/95% air incubator.
4. Transfer the ganglia/collagenase to a 15-mL centrifuge tube containing 10 mL F12 medium to wash out the collagenase. Remove all F12 solution leaving the ganglia behind and add

2 mL of warm BSF and gently triturate the ganglia using a sterile glass Pasteur pipette followed by a Gilson P1000 pipette (see Note 3). Then add a further 2 mL BSF medium.

5. Allow the larger debris to settle to the bottom of the tube and carefully transfer 1 mL of the solution to 15-mL centrifuge tubes each containing 2 mL 15% BSA, slowly pipetting the solution down the side of the tube so that the cell suspension sits on top of the BSA cushion (see Note 4). Centrifuge the tubes at 100 × *g* for 5 min. Myelin and fibrous debris will stay at the media/BSA interface, and neurons will pellet at the bottom of the tube.
6. Gently remove the tubes from the centrifuge and carefully remove the F12/BSA without disturbing the pellet of neurons at the bottom of each tube. Gently re-suspend the pellet in nucleofector solution (100 μL per transfection; pool the pellets from all tubes used) (see Note 5).
7. Add the appropriate concentration of DNA plasmid or siRNA duplex, transfer to a transfection cuvette and pulse in the nucleofector machine set to programme G13. Immediately after transfection, take a small sample (around 5 μL) to count on a haemocytometer and add the rest to 900 μL of BSF to a 1.5-mL Eppendorf tube (see Note 6).
8. Count the neurons on the haemocytometer and dilute the dissociated neurons to 2,500 per mL in BSF. Remove low-walled culture dishes from the CO_2 incubator and remove the laminin solution from each well (see Note 7). Add 500 μL of the diluted cell suspension to the central well of each MatTek dish, being careful not to spill the suspension outside of the coverslip. Carefully return the dishes to the incubator and leave overnight.
9. The following morning, gently add 4.5 mL of BSF medium to each culture dish and check for signs of neurite outgrowth (see Note 8).

3.3. Set-up of Picospritzer and Equipment for Growth Cone Turning Assays

1. Using a Sutter Pipette Puller, pull 10-cm borosilicate glass pipettes with filaments, O.D. 1.0 mm, I.D. 0.5 mm (Intracell), so that they have a bore width of 1 μm (see Note 9).
2. Connect a compressed air supply to a Picospritzer III under the control of a TSS10 dual pulse stimulator (or equivalent) to deliver air pulses at a frequency of 2 Hz for 20 ms duration, at a pressure of 3 psi and a delay of 0.1 ms.
3. Using Eppendorf Microloader pipette tips and a P20 Gilson (or equivalent) pipette, fill the tip of the glass micropipette with the reagent to be assayed (4 or 5 μL will suffice), at a 1,000-fold greater concentration to that required at the growth cone (2), and connect to the Picospritzer pipette holder attached to a micromanipulator.

4. Preheat the microscope chamber to 37°C and equilibrate to 5% CO_2. If CO_2 is not available, add 15 mM HEPES buffer to the neuronal culture medium (see Note 10).
5. Remove the culture dishes from the incubator and gently overlay the medium with a thin (2–3 mL) layer of vegetable oil that has been pre-warmed to 37°C. Make sure that the oil covers the entire surface and does not mix with the medium (see Note 11). Vegetable oil does not seem to adversely affect neuronal cultures, but other types of oil, for example mineral oil, may be effective.
6. If using any pharmacological reagents to manipulate the intracellular signalling pathways in the neurons, bath apply to the culture medium at least 30 min before beginning an assay (although this depends on the agent being used).
7. Using phase optics, identify a growth cone to be analysed and position it so it is visible on the computer screen. Position the micropipette so that it is 100 μm from the growth cone at an angle of 45° (see Fig. 1; Note 12).
8. Position the micropipette so that it is on or just above the surface of the culture dish.
9. Take an image of the growth cone/pipette.
10. Switch on the Picospritzer/stimulator to begin the assay. Leave running for 30–60 min, depending on the typical rate of neurite outgrowth.
11. After this time, take a second image of the growth cone position. Elevate the micropipette and reposition the culture dish for the next growth cone (see Note 13).

3.4. Analysis of Growth Cone Turning

1. Using ImageJ (NIH), measure the start angle of growth cone with respect to the micropipette by drawing a line through the centre of the growth cone from the most proximal segment of the neurite (this angle should be ~45° to the horizontal position of the micropipette, from either above or below the horizontal; see Fig. 1c, d).
2. Repeat step 1 with the image from the end of the assay (30 or 60 min); measure the angle between the final position of the growth cone, its original position and the original direction of growth at time zero. A final position of the growth cone on the pipette side of the original 45° direction of growth represents growth cone attraction to the gradient, whilst a final position on the other side of the original 45° direction of growth represents growth cone repulsion.
3. Measure the distance that the growth cone has advanced during the assay. If a growth cone has not advanced at least 10 μm, we consider it unresponsive because filopodia can

advance such distances without growth cone advance, and therefore we do not include such growth cones in our data.

3.5. FRET Combined with an Applied Gradient

It is assumed that growth cone turning is achieved by biased expression of intracellular signalling molecules leading to unilateral assembly or disassembly of the cytoskeleton. We can now directly measure protein activity across a growth cone when challenged with a molecular gradient by combining the turning assay with FRET. When a gradient is applied, changes in the distribution of FRET across the growth cone indicate whether the extracellular gradient activates or inhibits intracellular signalling proteins, revealing an asymmetrical distribution of activity.

1. Plate DRG neurons transfected with appropriate FRET cDNA constructs onto MatTek dishes coated as described in Subheading 3.1, but use 5–10 μg/mL laminin to enhance numbers of adherent transfected cells.
2. Analyse cells 24–48 h post-transfection using a FRET-capturing system (e.g. see Subheading 2.4) using an inverted-stage microscope set-up as described in Subheading 2.3.
3. Ensure FRET system and filters are properly configured for the FRET probes being used and that appropriate correction images are taken. These include untransfected, donor only and acceptor only controls to correct for cross-talk between fluorescent channels. Using FRET analysis software, these can be used to generate correction factors in line with the algorithm used by Wouters et al. (11).
4. Care should also be taken to avoid bleaching of the fluorophores during time-lapse imaging. Control experiments should be conducted to optimise exposure times, fluorescent light intensity and frequency of image capture.
5. Select a growth cone that is flat, spread and contains lamellipodia and filopodial protrusions. Care must also be taken to choose a growth cone emitting suitable levels of fluorescence. Rotate the dish so that the growth cone is facing the side of the stage with the micromanipulator, at a 45° angle.
6. Having completed set-up and bleaching control measurements, time-lapse experiments should be conducted by capturing consecutive images in three separate channels: the donor channel (donor fluorophore excitation wavelength and donor fluorophore emission capture), the FRET channel (donor fluorophore excitation wavelength and acceptor fluorophore emission capture) and the emission channel (acceptor fluorophore excitation wavelength and acceptor fluorophore emission capture). These channels are combined by the software to create a pseudocolour sensitised FRET emission channel (FRET SE),

which quantifies the degree of FRET across the specimen being examined. In the case of using an applied gradient, a phase contrast image should also be taken to visualise the tip of the micropipette.

7. Load the micropipette with the desired protein, agonist or pharmacological reagent as described in Subheading 3.2.
8. Take an image (three FRET channels + one differential interference contrast/phase), indicating the start of the time lapse (t0).
9. Lower the micropipette into position under differential interference contrast/phase optics, placing the tip 100 μm from the growth cone.
10. Repeat step 8 at the end of the time course. In our experimental set-up, 5 min (t5) is sufficient to detect a change in the distribution of the FRET sensor, although different time lapses may be suitable depending on protein analysed or gradient used. However, growth cones change morphology and orientation over longer time points, and selecting similar regions of interest may become more challenging.
11. Raise the micropipette and select a new growth cone.

3.6. Analysis of FRET Distribution

1. Regions of interest (ROIs) are outlined within the growth cone for both the start and end point images. One ROI is drawn outlining 50% of the surface area of the growth cone on the side facing the pipette and the other is drawn outlining 50% of the growth cone on the side further from the pipette, with the junction at which they meet bisecting the central axis of the growth cone in line with the neurite shaft immediately proximal to the growth cone.
2. A ratiometric value for the distribution of FRET across the growth cone is obtained for each image using the formula: NEAR SIDE FRET SE value/FAR SIDE FRET SE value. A value <1 suggests the protein is less active on the near side of the growth cone, whilst a value of >1 suggests the protein is more active on the near half of the growth, i.e. the side exposed to the gradient.
3. The procedure should be repeated for at least 15 growth cones, with values for t0 and t5 plotted on a cumulative distribution graph (see Fig. 2). Statistical comparison can be performed comparing t0 and t5 values; a significant mean increase at t5 indicates a near side bias of signalling protein activity caused by the gradient, whereas a significant mean decrease at t5 indicates a far side bias. Over longer time periods, this bias in distribution of signalling molecule activity can be related to growth cone attraction or repulsion.

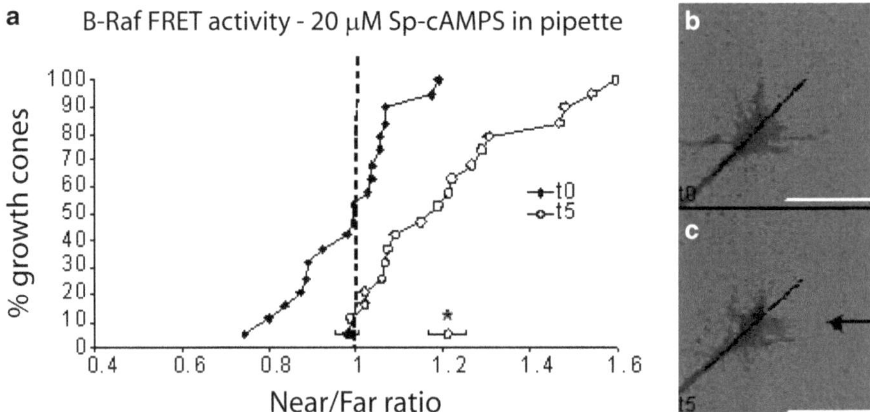

Fig. 2. Detection of asymmetric signalling activity within growth cones exposed to a gradient of the cAMP agonist, Sp-cAMPS. (**a**) Near/Far ratio of FRET emission (abscissa) indicating the distribution of B-Raf FRET activity across embryonic rat DRG growth cones plotted against the percentage of growth cones that display that ratio or less (ordinate axis) at the beginning of the experiment (t0, *dark circles*) and 5 min after application of the gradient (t5, *light circles*), with each point representing one growth cone. The mean ratio for each group is indicated by corresponding symbols above the abscissa. High values represent higher B-Raf activity on the side closest to the pipette, and values less than 1 represent higher B-Raf activity on the side further from the pipette. Sp-cAMPS clearly instigates higher B-Raf activity on the side of the growth cone towards the gradient. (**b**, **c**) Grayscale pictures of pseudocoloured FRET images at t0 (**b**) and t5 (**c**), with darker colours indicating low FRET and lighter colours indicating high FRET. *Arrow* in (**c**) indicates the direction of the applied gradient. Scale bars represent 10 μm. *$P<0.05$, student's *t* test.

4. Notes

1. BSF without insulin and NGF is stable at 4°C for several weeks. After supplementing medium with NGF/insulin, BSF can be stored for approximately 2 weeks at 4°C.
2. The PLL solution should be completely dried from the well before applying the growth substrate, as liquid PLL is toxic to cells.
3. Be very careful not to damage the neurons during the trituration step. Generally, a few passes using the glass Pasteur pipette and Gilson P1000 pipette are sufficient.
4. 15% BSA can be prepared in advance and stored at −20°C. It must be thawed in a 37°C water bath and mixed carefully before use so that the BSA is evenly distributed.
5. If you are planning to carry out growth cone turning assays on untransfected neurons, re-suspend the neuron pellet in 100 μL trypsin inhibitor/DNAse, before quickly adding 900 μL of BSF medium and moving on to step 8.
6. The optimum concentration of siRNA or plasmid DNA needs to be established for each study. The transfection stage should be completed as quickly as possible, ideally in less than 2 min,

as the nucleofector solution is toxic with prolonged exposure. Following transfection, care should be taken not to agitate the culture as the neurons are particularly fragile at this stage.

7. A thin film should be visible over the glass coverslip if the laminin has adhered to the substrate. Be careful to not let the coverslip dry out between removing the laminin solution and adding the cell suspension.
8. Mature DRG neurons that have been transfected with either siRNA or plasmid DNA using the Amaxa nucleofector machine typically require 48 h to produce pronounced neurite outgrowth. This time period also allows for significant knockdown of target proteins using siRNA (5) and expression of fluorescently tagged proteins used for FRET analysis (7).
9. An accurate bore width is very important as slight deviations in the diameter of the glass pipette can dramatically alter the gradient. A stage micrometre can be used to check bore width.
10. Without CO_2, but with HEPES buffered medium, we routinely see good neurite outgrowth for at least 6 h.
11. We have tested various types of oil for their ability to 'seal' the culture medium and prevent evaporation and gas exchange whilst still allowing the micropipette to pass through without becoming blocked.
12. It is important to carry out turning assays with the pipette angled both above and below the growth cone at the start of the assay. Growth cones have been reported to turn in a clockwise direction, even in the complete absence of extracellular cues, due to the interactions of myosin Va and myosin Vb with actin (12). Therefore, carrying out equal numbers of assays from above and below the pipette avoids any bias due to this natural turning.
13. Due to the small bore width of the pipette, it is easily blocked by proteins and chemicals, and therefore we routinely check that the pipette is still ejecting at the beginning and end of each assay by aligning the pipette with small items of unattached debris that can be seen hovering above the coverslip surface. Solutions containing BSA are particularly prone to blockage.

References

1. Filbin MT (2003) Myelin-associated inhibitors of axonal regeneration in the adult mammalian CNS. Nat Rev Neurosci 4, 703–713
2. Lohof AM, Quillan M, Dan Y, Poo MM (1992) Asymmetric modulation of cAMP activity induces growth cone turning. J Neurosci 12, 1253–1261
3. Song H, Poo MM (1999) Signal transduction underlying growth cone responses to diffusible factors. Curr Opin Neurobiol 9, 355–363
4. Shewan D, Dwivedy A, Anderson R, Holt CE (2002) Age related changes underlie switch in netrin-1 responsiveness as growth cones advance along the visual pathway. Nat Neurosci 5, 955–962
5. Murray AJ, Shewan DA (2008) Epac mediates cAMP-dependent axon growth, guidance and regeneration. Mol Cell Neurosci 38, 578–588

6. Murray AJ, Peace AG, Shewan DA (2009) cGMP promotes neurite outgrowth and growth cone turning and improves axon regeneration on spinal cord tissue in combination with cAMP. Brain Res 1294, 12–21
7. Murray AJ, Tucker SJ, Shewan DA (2009) cAMP-dependent axon guidance is distinctly regulated by Epac and protein kinase A. J Neurosci 29, 15434–15444
8. Golding J, Shewan D, Cohen J (1997) Maturation of the mammalian dorsal root entry zone—from entry to no entry. Trends Neurosci 20, 303–308
9. Shipman C Jr (1969) Evaluation of 4-(2-hydroxyethyl)-1-piperazineethanesulfonic acid (HEPES) as a tissue culture buffer. Proc Soc Exp Biol Med 130, 305–310
10. Erskine L, Reijntjes S, Pratt T, Denti L, Schwarz Q, Vieira JM, Alakakone B, Shewan D, Ruhrberg C (2011) VEGF signaling through neuropilin 1 guides commissural axon crossing at the optic chiasm. Neuron 70, 951–965
11. Wouters FS, Verveer PJ, Bastiaens PIH (2001) Imaging biochemistry inside cells. Trends Cell Biol 11, 203–211
12. Tamada A, Kawase S, Murakami F, Kamiguchi H (2010). Autonomous right screw rotation of growth cone filopodia drives neurite turning. J Cell Biol 188, 429–441

Chapter 16

Culture of Dissociated Sensory Neurons from Dorsal Root Ganglia of Postnatal and Adult Rats

Davina E. Owen and Julie Egerton

Abstract

The development of new therapeutics for management of pain is likely to become much more mechanism based, and therefore, we need a more thorough understanding of the different pain development pathways. The afferent fibers of sensory neurons, with their cell bodies in the dorsal root ganglia (DRG), are thought to be key in pain mechanisms. DRG neurons can be prepared from embryonic, postnatal, or adult tissue. Embryonic preparations have the advantage of higher cell yields and greater proportion of neurons, but they are dependent on neurotrophins for the first week of culture. In contrast, dissociated postnatal and adult DRG sensory neurons offer the possibility to study mature neurons that may better resemble the in vivo characteristics of these cells. Here, we describe the dissociation of DRG sensory neurons from postnatal and adult rats. DRG are dissected and dissociated using a prolonged trypsin/collagenase treatment, followed by mechanical separation of the neurons. We have routinely prepared them for electrophysiological studies by the methods outlined in this chapter and describe some of the pitfalls that we have encountered, with hints of how to overcome them.

Key words: Dorsal root ganglia, Dissociated cell culture, Growth factors, Rat, Sensory neurons, Adult, Postnatal

1. Introduction

Dorsal root ganglion (DRG) neurons extend processes that are located both in the peripheral (PNS) and central nervous systems (CNS), grow in both environments during development, and subsequently are used for the study of aspects of both systems. Sensory neurons that originate in the DRG are unique in that, instead of the typical axon-dendrite structure of other neurons, they are unipolar neurons with a single axon stem bifurcating into a peripheral branch that goes to the distal tissues and a central branch that goes to spinal cord. Many components of the PNS and CNS may be affected by pathologies leading to neuropathic or

Stephen D. Skaper (ed.), *Neurotrophic Factors: Methods and Protocols*, Methods in Molecular Biology, vol. 846,
DOI 10.1007/978-1-61779-536-7_16, © Springer Science+Business Media, LLC 2012

inflammatory pain, but the sensory neurons are often affected early and more severely. They are a key component of the nociceptive pathway, and as such, we need to have a better understanding of their role, in their response to and generation of inflammatory mediators, and the control of their excitability in the development of chronic pain.

The use of dissociated neurons, in contrast to acute preparations or slice cultures, means that a defined mechanism can be more easily studied. The isolated neuron's environment can easily be manipulated in vitro to try and reflect the in vivo environment or even the disease state. DRG neurons are very accessible and easy to harvest in embryonic, young, and adult animals, and they survive well in culture. Typically, DRG neurons are cultured from rats, mice, or chicken embryos, and in most cases, one can obtain many neurons from a single animal. For example, a complete preparation from an adult rat yields ~40 DRGs or ~100,000 neurons. The yield from an average age pup is ~500,000 neurons.

Postnatal DRG cultures can be used as a screening tool as the yields are considerably greater and they grow well on multiwell plate formats, which makes them open to fluorometric imaging plate reader/calcium imaging (1), neurite outgrowth, cell health assays, and cocultures. Unlike embryonic DRG neurons, postnatal and adult DRG neurons do not need nerve growth factor (NGF) or serum for survival (2), although the presence of NGF does affect their phenotype. For example, NGF regulates the capsaicin sensitivity of adult DRG neurons (3). The protocol in this chapter, which involves dissociation of rat DRGs using a prolonged trypsin/collagenase treatment, followed by mechanical separation of the neurons, has been modified from Lindsay (2) and adapted for optimum electrophysiological experimentation.

2. Materials

2.1. Postnatal DRG Neurons

1. Hank's balanced salt solution (HBSS) (calcium-/magnesium-free, Invitrogen).
2. 0.05% Trypsin-EDTA (Invitrogen) (see Note 1).
3. 0.1% Collagenase Type 1A-S (Sigma). Dilute 1 mL of 0.5% stock diluted in 4 mL HBSS. The 0.5% stock solution is prepared by dissolving 50 mg of collagenase in sterile HBSS. This can be divided into 1 mL aliquots and stored at –20°C for at least 6 months.
4. Human β-NGF (R&D Systems). Reconstitute a 100-μg vial of β-NGF in 2 mL of phosphate buffered saline (pH 7.4) (PBS) + 0.1% bovine serum albumin (BSA) to yield a 0.05 μg/mL

stock. Filter-sterilize, aliquot into 50-μL portions, and store at −20°C for at least 6 months.

5. N2 supplement (100×, Invitrogen).
6. Penicillin/streptomycin solution ("Pen/Strep"), 100× (10,000 U/mL penicillin and 10,000 μg/mL streptomycin) (Invitrogen).
7. 7.5% BSA solution (Invitrogen).
8. Mouse laminin, 1 mg/mL (Sigma). Prepare a working solution of 5 μg/mL by diluting 50 μL of the supplier's stock solution in 10 mL of PBS.
9. Poly-L-lysine hydrobromide (Sigma P2636): Dilute in sterile distilled water to give a working solution of 100 μg/mL.
10. *Dissociation medium*: Prepare Dulbecco's modified Eagle's medium (DMEM; high glucose, with sodium pyruvate, Invitrogen) containing 10% (v/v) heat-inactivated fetal calf serum (FCS) and 50 U/mL penicillin + 50 μg/mL streptomycin.
11. Preparation of substrate for plating should be started the day before. Coat tissue cell culture dishes/plates/coverslips with 100 μg/mL poly-L-lysine, leave overnight at room temperature, and wash off well with sterile H_2O. Next, coat the cell culture dishes/plates/coverslips with 5 μg/mL laminin, and leave at 37°C (for at least 50 min before plating cells). Remove all the laminin just before plating DRG cells (see Note 2).
12. *Growth medium*: Supplement 50 mL of DMEM to contain 50 ng/mL β-NGF (add 50 μL from a 50 μg/mL stock), N2 supplements (500 μL from 100× stock), 0.05% BSA (334 μL from 7.5% BSA solution), 50 U/mL penicillin, and 50 μg/mL streptomycin (5 mL of 100× stock)—note that this medium is serum-free.

2.2. Adult DRG Neurons

1. Nutrient Mixture F-12 (with phenol red; Invitrogen).
2. Ultroser G (Biosepra); reconstitute with 20 mL distilled water, filter-sterilize, and add fresh to culture medium to 4% w/v.
3. Poly-DL-ornithine hydrobromide (Sigma P8638): Prepare a 25 mg/mL stock solution in distilled water and filter-sterilize. Next, prepare a working solution of 500 μg/mL poly-DL-ornithine by adding 200 μL of the 25 mg/mL stock to 10 mL of sterile water.
4. Collagenase (Type IV, Sigma): Prepare a 6.25 mg/mL stock solution (0.625% w/v) in F-12 medium (no phenol red) by dissolving 50 mg in 8 mL medium and filter-sterilize. Further dilute the collagenase to 0.625 mg/mL (0.0625% w/v) by adding 200 μL of the 6.25 mg/mL stock to 1.8 mL of F-12 medium.

5. BSA (essentially fatty acid–free, Sigma): Prepare a 15% (w/v) solution by dissolving 1.5 g in 10 mL of F-12 medium and filter-sterilize. It is best to make this on the day that DRG cells are prepared for culture.
6. Mouse nerve growth factor (mNGF), 7.5S murine (Promega): Prepare a 100 μg/mL stock solution in F-12 medium, filter-sterilize, and store as 30 μL aliquots at –20°C. Next, dilute 5 μL of the mNGF stock solution in 5 mL of F-12 medium to give 100 ng/mL mNGF.
7. Growth medium: To 500 mL of nutrient mixture F-12, add 5 mL of "Pen/Strep" stock (100 U/mL) penicillin and 100 μg/mL streptomycin, 5 mL of 100× N2 supplement, and 20 mL of 100% Ultroser G (4% w/v final).
8. The plating substratum should be started before preparation of the dissociated cells. Coat tissue cell culture dishes/plates/coverslips with 500 μg/mL poly-DL-ornithine and leave for at least 30 min at room temperature under the culture cabinet. Wash twice with sterile distilled water and air-dry uncovered in the cabinet. Next, coat the cell culture dishes/plates/coverslips with freshly prepared laminin solution (5 μg/mL) and incubate at 37°C for at least 3.5 h. Aspirate off the laminin solution under a culture cabinet and allow to dry uncovered in the cabinet for at least 10 min prior to seeding with DRG cells.

2.3. Apparatus

1. Dissecting microscope preferably in a laminar flow cabinet.
2. Laminar flow hood.
3. Dissecting instruments (Fine Science Tools).
4. Humidified tissue culture incubator at 37°C and 5% CO_2.
5. Water bath set at 37°C.
6. Bench centrifuge to spin 15-mL falcon tubes.

3. Methods

3.1. Postnatal DRG Neurons

3.1.1. Dissection

1. Sterilize dissection instruments by treating with 70% v/v ethanol and allow to dry in the dissection cabinet.
2. Collect the DRG from 7- to 9-day-old rat pups (Sprague Dawley). Cull the rat pups following Schedule 1 (cervical dislocation followed by decapitation) according to Home Office protocols.
3. Remove the tail and make an incision through the skin following the line of the spinal column from the head to the tail. Peel the skin away and trim away the adipose tissue from the shoulders.

4. Using a small pair of sharp pointed scissors (Stevens tenotomy, straight/sharp scissors) make an incision through the whole length of the dorsal column only. This is achieved by putting the lower blade of the scissors into the spinal column and cutting through the dorsal spinal vertebrae from the head to the tail to reveal the spinal cord.
5. The vertebrae should be trimmed on one side to just above the level of the row of DRG (see Note 3). Remove the spinal cord by scooping it out from underneath starting at the neck. Remove vertebrae from the other side of the column again to just above the level of the DRG which can now be seen.
6. Remove the DRGs one at a time using forceps (Inox 5) which are placed behind the cell body easing it out of the pocket it sits in. Use the forceps or small scissors in the other hand to cut off the roots as close to the cell body as possible. Transfer the DRGs to a 60-mm ∅ petri dish containing cold HBSS (on ice). Continue in this way until all the DRGs have been collected from each pup. It should be possible to collect 30–35 DRGs (maybe 40 with more experience and care when exposing the DRGs) from each pup.
7. Clean the DRG in the cold HBSS to remove meninges and excess roots that may still be attached. Transfer the clean ganglia to another 60-mm petri dish containing 5 mL of trypsin-EDTA. The trypsin-EDTA should be only partially defrosted to keep the ganglia as cold as possible.

3.1.2. Dissociation

1. Incubate the ganglia in the trypsin-EDTA for 15–20 min at 37°C.
2. Add 5 mL DMEM (10% FCS) to the dish with ganglia and transfer to a 15-mL tube. Leave the DRG to settle to the bottom or centrifuge at 200 × g for 3 min.
3. Carefully remove the supernatant using a 5-mL Stripette (do not use vacuum aspiration).
4. Add 0.1% collagenase solution (1 mL of 0.5% stock plus 4 mL HBSS) to the ganglia and incubate for 90 min at 37°C.
5. Add 5 mL DMEM (10% FCS) and centrifuge at 200 × g for 3 min.
6. Carefully remove the supernatant using a 5-mL Stripette (do not use vacuum aspiration).
7. Add 1 mL DMEM (10% FCS) and mechanically dissociate the ganglia by triturating no more than 6–10 times using a 1-mL pipette tip.
8. Dilute the cell suspension (point 7) to 10 mL with DMEM (10% FCS) and plate onto a 100-mm ∅ tissue culture dish for 1 h at 37°C. The nonneuronal cells and large-diameter DRG neurons

will stick to the dish, whereas the small-/medium-diameter neurons will float in the medium. The longer the incubation, the more likely that the small/medium diameter neurons will stick to the dish, in which case, wash the dish 1–2 times with the medium in the dish (see Note 4).

9. Collect the medium (containing unattached neurons) and transfer to a 15-mL tube. Centrifuge at 200 × *g* for 5 min.
10. Remove the medium using a Stripette and resuspend the cell pellet with 1 mL serum-free DMEM (50 ng/mL β-NGF, N2 supplements, and 0.05% BSA) using a 1-mL pipette tip to break up cell pellet (see Notes 5 and 6).
11. Count cells with a hemocytometer (see Note 7).
12. Plate cells directly into growth medium at a density appropriate to the experimental application. Examples of some plating densities for different applications are given below as a guide:

96-well plates	30,000 cells per well in 100 μL growth medium
384-well plates	20,000 cells per well in 50 μL growth medium
Electrophysiology	200,000 cells per 22-mm coverslip/35-mm dish in 2 mL medium, or 100,000 cells (in 0.5 mL) spotted onto the coverslip only and incubate for 1 h to allow neurons to attach before flooding the culture dish with 2 mL medium
Immunocytochemistry	100,000 cells per coverslip/35-mm dish in 2 mL medium
Single cell imaging	70,000–100,000 cells in 50–100 μL spotted onto a coverslip, incubate for 1 h to allow neurons to attach before flooding the culture dish with 2 mL medium

13. Contaminating glia that may still be present in the cultures can be reduced by adding the antimitotic cytosine β-D-arabinoside to a final concentration of 1 μM after 24 h in culture. This is conveniently done by preparing a stock solution of 1 mM cytosine β-D-arabinoside in water, filter-sterilizing, and storing as aliquots at −20°C. This can be added 1:1,000 to the cultures.

3.2. Adult DRG Neurons

3.2.1. Dissection

1. Sterilize dissection instruments by treating with 70% v/v ethanol, and allow to dry in the dissection cabinet.
2. Weigh the rats and then sacrifice by exposure to a rising concentration of CO_2. Confirm the death of each animal by exposing the heart and severing one of the main cardiac blood vessels. Perform exsanguination by placing each rat on its side

on absorbent bench coat (this minimizes the amount of blood in the spinal column) (see Note 8).

3. In the dissection cabinet, remove the head of the rat to gain access to the spinal column. Make an incision through the skin following the line of the spinal column from the head to the tail and peel back the skin. Use a pair of scissors to remove any excess tissue surrounding the dorsal column.
4. Using a pair of pointed bone cutters (Littauer bone cutters, straight), cut through each side of the dorsal vertebrae, remove each vertebrae away to expose the spinal cord, and move in the head to tail direction taking care not to damage the spinal cord. Carefully pick up and lift out the spinal cord using a pair of small standard forceps (standard pattern) while cutting the DRG nerve branches close to the spinal cord (this will minimize the loss of any DRG).
5. Remove the DRG one at a time using forceps (Inox 5) which are placed behind the cell body, easing it out of the pocket it sits in. Use the forceps or small scissors in the other hand to cut off the roots as close to the cell body as possible. Transfer the DRG to a 60-mm petri dish containing cold HBSS (on ice). Continue in this way until all the DRGs have been collected from each rat. It should be possible to collect 30–35 DRGs from each rat. 40 DRGs per rat may even be achievable once the operator is skilled and careful when exposing the cord.
6. Trim roots and nonneuronal tissue from each set of ganglia with fine forceps and fine spring-loaded scissors (treated with 70% ethanol) by removing the roots and any attached non-DRG tissue under the dissecting microscope. Place the trimmed DRG into new 35-mm dishes containing 1.8 mL fresh HBSS.

3.2.2. Dissociation

1. Add 200 μL of 6.25 mg/mL collagenase (0.0625% final concentration) to the medium containing the DRG and swirl gently to mix. Incubate the ganglia for 2–3 h at 37°C, 5% CO_2.
2. Using a 5-mL Stripette, carefully transfer the ganglia to a 15-mL conical-bottomed centrifuge tube and wash twice with 10 mL prewarmed growth medium to remove the collagenase.
3. Remove the medium by aspiration and resuspend the ganglia in 2 mL growth medium. Triturate the DRG approximately 20 times through a flame-polished Pasteur pipette (smallest hole possible) to obtain a single cell suspension.
4. Slowly drip the cell suspension onto a cushion of 2 mL 15% BSA in growth medium in a 15-mL centrifuge tube and centrifuge at 200 × *g* at room temperature for 10 min to remove nonneuronal cells and debris.
5. Carefully remove the supernatant by aspiration until approximately 50 μL remains covering the cell pellet. Resuspend the cell pellet

in 1 mL of prewarmed growth medium containing mNGF (1 μL mNGF for every 1 mL medium) using a 1-mL pipette. Transfer the cell dissociate to a sterile 5-mL tube ready for seeding.

6. Count cells with a hemocytometer (see Note 7).
7. Plate cells directly into growth medium at a density appropriate to the experimental application. It is our experience that the number of neuronal cells from adult DRG preparations is much lower than those achieved from postnatal rats. Only a small number of wells of a 96-well plate would be possible, and so, we would conclude that adult DRGs are not suitable for higher-throughput formats. For electrophysiology, we have used 12 mm ∅ coverslips and plated cells at 2,000 cells per coverslip.
8. Contaminating glia that may still be present in the cultures can be reduced by adding the antimitotic cytosine β-D-arabinoside to a final concentration of 1 μM after 24 h in culture.

4. Notes

1. We purchase 0.5% (10×) trypsin-EDTA and dilute with calcium- and magnesium-free PBS. It is crucial that the trypsin-EDTA is kept at <4°C before use and in the absence of calcium and magnesium; otherwise, the trypsin will become inactive, and the dissociation, incomplete. This reduces the quality of the final cell preparation.
2. We have utilized both commercially available poly-L-lysine–coated plates and coverslips and ones that we have made. It seems the cells are slightly better for some applications on the homemade substrates, but this should be determined for your application.
3. The DRG cannot be seen at this time, so it is wise to cut a little higher until greater confidence is achieved. More vertebrae can always be trimmed off once the spinal cord has been removed and the DRG can be seen.
4. This plating step can be omitted if it is not necessary to remove the contaminating glial cells from your preparation. The quality of the neurons may well be improved by eliminating this step, as the preparation time is reduced and the neuronal cells do not have an increased exposure to FCS. The presence of glial cells in the preparation may well produce trophic factors to support the DRG neurons and make the in vitro preparation more comparable to in vivo. The nonneuronal cells can easily be left in the preparation for the purposes of electrophysiology

or single cell injection, for example, where the operator should be able to distinguish morphologically between neuronal and nonneuronal cell types.

5. If the DMEM stated in the methods is not used and a different recipe DMEM substituted, this may affect the expression/function of receptors and ion channels. In our experience, the removal of pyruvate from the medium causes loss of calcium responses to capsaicin. It is possible to culture both postnatal and adult DRG neurons in Neurobasal medium plus B27 supplement (Invitrogen). A comparison may be necessary to establish if cellular functionality is affected.
6. The addition of growth factors to the growth medium should be investigated according to the functionality required. For example, brain-derived growth factor (20 ng/mL) addition to postnatal DRG neurons is required for purinergic P2X 2/3 receptor–responsive cells in place of NGF.
7. Automated cell counters such as Cedex may not be suitable for counting DRG neurons as the small neurons may be too small for such devices to detect. Also, most automated devices take large amounts of sample to complete a count, which is not usually practical with DRG preparations. New-generation automated cell counters which use much smaller sample volumes and have higher resolution for cell detection could be compared with manual hemocytometer counts to check their accuracy.
8. In practice, 2–4 adult rats can be dissected per session per person due to time constraints; cell viability and responsiveness decrease if dissection time is prolonged.

References

1. Wood JN, Winter J, James IF, Rang HP, Yeats J, and Bevan S (1988) Capsaicin-induced ion fluxes in dorsal root ganglion cells in culture. J Neurosci 8, 3208–3220
2. Lindsay RM (1988) Nerve growth factors (NGF, BDNF) enhance axonal regeneration but are not required for survival of adult sensory neurones. J Neurosci 8, 2394–2405
3. Winter J, Forbes CA, Sternberg J, and Lindsay RM (1988) Nerve growth factor (NGF) regulates adult rat cultured dorsal root ganglion neuron responses to capsaicin. Neuron 1, 973–981

Chapter 17

Culture and Proliferation of Highly Purified Adult Schwann Cells from Rat, Dog, and Man

Kirsten Haastert-Talini

Abstract

This chapter presents fast and easy protocols to obtain highly purified cultures of proliferating adult rat, canine, and human Schwann cells. Cell preparation from predegenerated adult sciatic nerves combined with the use of melanocyte growth medium supplemented with forskolin, fibroblast growth factor-2, pituitary extract, and heregulin as selective, serum-free culture medium and two methods for a consecutive cell-enrichment step are described. Our protocols result in approximately 90% pure Schwann cell cultures (or higher). The average time to obtain highly purified in vitro cultures of adult Schwann cells is 21 days.

Key words: Adult schwann cells, Rat, Human, Dog, Enrichment, In vitro

1. Introduction

The culture of adult Schwann cells from laboratory animals and humans is of high relevance for therapeutic approaches, e.g., transplantation of these cells to promote central or peripheral nervous system regeneration. Furthermore, such cultures could be of interest to investigate cell–cell interactions between neurons and Schwann cells or stem cells and Schwann cells in cell replacement approaches. The culture of highly purified Schwann cells could further be of interest with regard to developmental myelination or demyelination and remyelination in peripheral neuropathies.

To study the role of Schwann cells in peripheral neuropathies in vitro or the effects of adult Schwann cell transplantation in regenerative medicine in vivo, it is needed to culture highly purified adult Schwann cell populations. The number of contaminating fibroblasts which might induce bias of any study result should be reduced as much as possible. Numerous publications exist on

Stephen D. Skaper (ed.), *Neurotrophic Factors: Methods and Protocols*, Methods in Molecular Biology, vol. 846,
DOI 10.1007/978-1-61779-536-7_17,

the purification and in vitro culture of adult Schwann cells from different species (1–15). We established techniques for isolation and high enrichment of rat (16, 17), dog (18, 19), and human (17, 20) adult Schwann cells. The use of fresh tissue results in poor primary cell yields; therefore, we recommend predegenerated tissue as source for adult Schwann cells. Predegeneration of peripheral nerve tissue allows Wallerian degeneration to take place. In response to nerve injury, Schwann cells switch from a myelinating to a non-myelinating phenotype, and together with macrophages, they remove necrotic tissue and myelin debris. Furthermore, Schwann cells proliferate to form bands of Büngner which help the regrowing axons to elongate their growth cones in the direction of the denervated targets (21–23). Thus, predegeneration specifies the induction of Schwann cell dedifferentiation and proliferation prior to tissue dissociation; predegeneration in vitro additionally promotes fibroblast outgrowth from the cultured tissue (16, 18). Our methods are based on cell harvest from predegenerated peripheral nerve tissue, 20-h dissociation of tissue, culture of cells in serum-free melanocyte growth medium (MGM) supplemented with Schwann cell mitogens, and either an easy and fast washing step ("cold jet") or an immunopurification step to reduce contaminating fibroblasts from mixed peripheral nerve cell cultures. The medium composition was basically determined in adult rat Schwann cell cultures (16) and later adopted for culturing adult Schwann cells from canine and human origins. We use MGM instead of Dulbecco's modified Eagle's medium (DMEM) because in our hands, the use of DMEM was inductive to fibroblast overgrowth. The same holds true for the use of 5% or 10% fetal calf serum (FCS) in the medium; in our hands, only serum-free culture conditions successfully suppressed overgrowth of Schwann cells by fibroblasts. The mitogens used in our culture medium are commonly used in other protocols as well. We used to produce the fibroblast growth factor-2 (FGF-2) in our lab, but the commercially available FGF-2^{18kD} exerts the same beneficial effects. We found the use of mitotic toxins like arabinoside-C not useful in enrichment of adult rat and human Schwann cells because it inhibited growth of both fibroblasts and Schwann cells. However, for cell cultures from canine origin, it was more selective for the suppression of fibroblast growth. Final removal of contaminating fibroblasts from rat and human cultures is based on different attachment properties of cultured adult rat and human peripheral nerve cells. In these cultures, fibroblasts grow mainly flat on the cell culture surfaces, and the adult Schwann cells predominantly grow on top of the fibroblasts. The "cold jet" technique was modified from an enrichment method to obtain Schwann cell–free cultures of dorsal root ganglion neurons (24) and includes a washing step. A technique we also use in neonatal rat Schwann cell enrichment is the removal of fibroblast via magnetic bead–based cell separation (25). While this technique resulted in a reduction of the cell yield for adult rat and human Schwann cell cultures, it was more effective than "cold jet" in the purification of adult canine Schwann cells.

2. Materials

2.1. In Vitro Predegeneration

1. DMEM, high glucose without L-glutamine (Gibco BRL).
2. Penicillin/streptomycin (Gibco BRL).
3. Amphotericin (Gibco BRL).
4. Collagen G (Biochrom).
5. Melanocyte Growth Medium, MGM (plus manufacturer's supplements) (PromoCell).
6. Scalpels (No. 21, Fine Science Tools).
7. Dumont forceps (Dumont No. 5, Fine Science Tools).
8. Vannas microscissors (No. 15003-08, Fine Science Tools).
9. Stereo microscope (Stemi SV 6, Carl Zeiss AG).
10. FCS (gold, Gibco BRL).

2.2. Dissociation of Predegenerated Tissue and Initial Seeding

1. Distilled water (aqua ad iniectabila, Ampuwa® Fresenius Kabi, Germany).
2. Poly-L-ornithine hydrobromide (PORN), 30–70 kDa; (Sigma-Aldrich) stock solution: boric acid/NaOH at 0.15 M each, adjusted to pH 8.4 by using HCl: 0.1 N, store at +4°C for weeks (at −20°C for months).
3. Laminin (Becton Dickinson), store in aliquots at −80°C.
4. Poly-L-lysine hydrobromide (PLL, 70–150 kDa) (Sigma-Aldrich; 0.5 mg/mL in distilled water, store at +4°C for weeks).
5. Collagenase (Type IV, 160 U/mg, Gibco BRL).
6. Dispase (Roche Diagnostics).
7. DNase I (Roche Diagnostics).
8. Hank's Balanced Salt Solution, (without Mg and Ca (HBSS), Gibco, Invitrogen).
9. Trypan blue solution (Gibco BRL).
10. Forskolin (FK, 7-deacetyl-7-[*O*-(*N*-methylpiperazino)-γ-butyryl]-, dihydrochloride, Calbiochem).
11. FGF-2^{18kD} (PeproTech).
12. Bovine pituitary extract (BPE-26, PromoCell).
13. Recombinant human neuregulin 1-β1/heregulin-β1 epidermal growth factor (EGF) domain (rhuHRG, [rHRG-$\beta 1_{176-246}$]), *E. coli* (R&D Systems).
14. Bovine serum albumin (BSA, Fraction V, Sigma-Aldrich).

2.3. Removal of Fibroblasts/ Enrichment of Adult Schwann Cell Cultures

1. Cytosine arabinoside (Sigma-Aldrich).
2. Dynabeads® sheep anti-rat IgG (Dynal Biotech, Invitrogen).
3. Rat anti-canine Thy1-antibody (anti-Cdw 90 canine, Serotec).
4. Dynal® MPC-S (example for magnet, Dynal Biotech, Invitrogen).
5. Trypsin/EDTA (0.5 g/L trypsin, 0.2 g/L EDTA; Gibco BRL).
6. Phosphate-buffered salt solution (PBS, pH 7.1).
7. Roto-Shake Genie® (Scientific Industries Inc.).

2.4. Analysis of Schwann Cell Purity by Immunocytochemistry

1. 4% Paraformaldehyde in PBS.
2. Blocking buffer: 5% BSA in PBS or 3% goat serum (Gibco, Invitrogen) in PBS.
3. Primary antibody for *rat Schwann cells*: mouse anti-rat $p75^{\text{low-affinity nerve growth factor receptor}}$ ($p75^{LNGFR}$) antibodies (from a hybridoma cell line in our lab; dilute 1:5 in blocking buffer, antibody concentration 35 μg/mL (16)).
4. Primary antibody for *human Schwann cells*: mouse anti-human $p75^{LNGFR}$ antibodies (Accurate Chemical & Scientific Corporation, USA; dilute 1:20 in blocking buffer (20)).
5. Primary antibody for *rat and human Schwann cells*: rabbit anti-cowS100calcium-bindingproteinantibodies(DakoCytomation; dilute 1:200 in blocking buffer/0.1% Triton-X-100 (Sigma-Aldrich) (16, 20)).
6. Primary antibody for *canine Schwann cells*: rabbit anti-human $p75^{LNGFR}$ (dilute 1:200 in blocking buffer, p75 pAb Promega).
7. Primary antibody for *rat fibroblasts*: mouse anti-rat Thy1-antibodies (from a hybridoma cell line in our lab; dilute 1:3 in blocking buffer, antibody concentration 20 μg/mL (16)).
8. Primary antibody for *canine fibroblasts*: rat anti-canine Thy1-antibody (anti-Cdw 90 canine, dilute 1:1,000 in blocking buffer), Serotec.
9. Secondary antibodies: Cy3-conjugated (dilute 1:400 in blocking buffer; Jackson ImmunoResearch) or goat anti-rabbit Alexa 555 (dilute 1:500 in blocking buffer, Invitrogen) to visualize primary antibody binding in fluorescence microscopy.
10. 4′,6′-Diamidino-2-phenylindole (DAPI, dilute 1:1,000 in PBS; Sigma-Aldrich) for nuclear counterstaining.

3. Methods

3.1. In Vitro Predegeneration

1. Harvest fresh peripheral nerve tissue and transfer it into DMEM + 1% Pen/Strep (see Note 1).
2. Depending on the nerve anatomy, the procedure to remove the epineurium with the help of Vannas microscissors and

Dumont (No. 5) forceps is slightly different. The epineurium of rat nerves can be stripped off as whole, while isolated nerve fascicles have to be pulled out of the connective tissue from canine and human peripheral nerve tissue (see Note 2). To facilitate this step, the nerve tissue could be cut in segments of 1 cm in length.

3. Collect the epineurium-free nerve tissue in DMEM + 1% Pen/Strep (e.g., 10 mL in a 15-mL tube).
4. Transfer the preparation to a sterile work bench and wash the tissue with DMEM.
5. For in vitro predegeneration, place groups of four to five 1-cm segments into the wells of an uncoated 6-well plate (see Note 3).
6. Cover the tissue segments with 1.5–2.0 mL of 10% FCS in culture medium: MGM (+ manufacturer supplements) + 2 μM forskolin + 5 μg/mL BPE-26 + 10 ng/mL FGF-2^{18kD} + 1% Pen/Strep (see Note 4).
7. Over the next 7–10 days, exchange the predegeneration medium every second day (see Note 5).

3.2. Dissociation of Predegenerated Tissue and Initial Cell Seeding

For initial seeding of adult Schwann cells, the culture surfaces have to be coated to enhance cell adherence.

3.2.1. Cell Culture Surface for Adult Rat and Human Schwann Cell Cultures

1. Twenty-four hours before initial seeding, remove the PORN stock solution from the +4°C storage and dilute 1:10 with ice-cold sterile distilled water (final concentration 1 mg PORN/mL). Keep the PORN solution on ice and remove an aliquot of laminin from −80°C and thaw it on ice.
2. Mix the ice-cold PORN solution with the laminin to give a final concentration of 6-μg laminin/mL PORN.
3. For coating, cover the cell culture wells thoroughly (e.g., 6-well plate: 1.5 mL/well) with ice-cold PORN-laminin solution and incubate for 24 h at room temperature (see Note 6).
4. Wash the wells with either distilled water or MGM and keep them covered with either solution to avoid drying.

3.2.2. Cell Culture Surface for Adult Canine Schwann Cell Cultures

Use PLL to coat plates for canine Schwann cell cultures.

1. Two hours before initial seeding, remove the prepared PLL solution from the +4°C storage and thoroughly cover the cell culture wells with it.
2. Incubate for 1 h, then wash the wells with either distilled water or MGM and keep them covered with either solution to avoid drying.

3.2.3. Dissociation of Predegenerated Nerve Tissue

1. Prepare the solution for enzymatic dissociation as follows: DMEM + 10% FCS + 1% Pen/Strep + 0.125% collagenase + 1.25 U/mL dispase (1.5–2 mL/predegeneration well on 6-well plate).
2. Remove the medium from predegeneration dishes and use a sterile scalpel to cut the nerve tissue into short segments of 2–4 mm in length.
3. Add dissociation solution to each predegeneration well and incubate the tissue for 2 h in a 37°C and 5% CO_2 incubator.
4. Transfer the solutions and tissue residue into an adequate number of 15-mL tubes (one tube per predegeneration well). Use an equal volume of HBSS to wash the wells and to reduce the enzymatic activity.
5. Mechanically separate the tissue residue into single cells by gentle pipetting on and off the suspension through fire-polished glass pipettes until the solution becomes optically homogeneous.
6. Centrifuge the single-cell solution at 235 × *g* for 5 min at +21°C.
7. Remove the supernatant and suspend the pellet in DMEM. Repeat the centrifugation (see Note 7).

3.2.4. Initial Seeding

1. Dilute the cell pellet in culture medium (see Note 8). To enhance the seeding efficiency, add 1% BSA to the culture medium for the first 24 h (see Note 9).
2. It is important to realize that seeding efficiency, initial survival rate, and proliferation rate of primary adult Schwann cells depend crucially on optimal cell densities in culture (26). Therefore, count the number of viable cells prior to seeding, e.g., after 1:2 dilution of a 10-μL cell suspension aliquot in trypan blue (will stain dead/dying cells blue).
3. Plate primary adult peripheral nerve cells at a density of 10^6 living (trypan blue-negative) cells/well on a 6-well plate (35 mm^2).
4. The next day, change the medium to culture medium without BSA and culture the primary peripheral nerve cells for 2–5 days at +37°C and 5% CO_2 (see Note 10).

3.3. Enrichment of Adult Schwann Cell Cultures/Removal of Fibroblasts

The fact that the adult Schwann cells grow mainly on top of the more flattened and more securely adherent fibroblasts allows separation of the two cell types by an easy washing step.

3.3.1. Enrichment of Adult Rat and Human Schwann Cells

1. Remove the medium and slowly add 1 mL of ice-cold PBS and immediately aspirate for 2–3 times in order to flush the well (6-well plate) and loosen the Schwann cells.
2. Apply a stream of 1-mL ice-cold culture medium, using a blue 1-mL Gilson tip and pipette on and off for several times, detaching the Schwann cells from the underlying fibroblast

layer. Especially those parts of the wells with high cell densities have to be rinsed. To reduce the risk of loosening also fibroblasts, monitor the detachment using a phase-contrast microscope.

3. Transfer the suspension of floating cells, mainly Schwann cells, to a 15-mL tube (see Note 11) and centrifuge (235 × *g* for 5 min at +21°C).
4. Resuspend the cell pellet in culture medium, determine the viable cell number with trypan blue staining (see above), and seed the enriched adult Schwann cells into freshly prepared PORN-laminin-coated wells.

3.3.2. Removal of Fibroblasts from Adult Canine Schwann Cell Cultures

For enrichment of adult canine Schwann cells, "cold jet" does not work efficiently; however, these cells can be immunopurified.

1. Wash magnetic Dynabeads® sheep anti-rat IgG in 0.1% BSA in PBS (10-μL beads in 500-μL 0.1% BSA). Initially, the Dynabeads have to be fluffed up and then mixed with the washing solution inside an eppendorf cup (1.5 mL).
2. Bring the magnet close to the wall of the cup; this draws the magnetic Dynabeads toward the magnet, allowing the supernatant to be easily removed. Repeat the procedure 2 more times.
3. To couple the Dynabeads with rat anti-canine Thy1-antibody (dilute 1:300 in MGM), add 10 μL of the washed Dynabeads to 300 μL of the antibody solution in another eppendorf cup.
4. Securely seal the vial with parafilm, put it into an ice-filled 50-mL tube, and rotate it around its horizontal axis 45 min at +4°C (e.g., place Roto-Shake Genie® in +4°C room). This step facilitates equal coating of the beads with the antibodies.
5. Wash the antibody-coupled beads 3 times with 0.1% BSA as before.

In parallel, all cells are enzymatically detached from the wells.

6. Remove the medium and incubate the cells with trypsin/EDTA for 1 min at +37°C (within the CO_2 incubator), followed by further incubation at room temperature and observation under phase-contrast microscopy.
7. As soon as all cells are detached, add culture medium/collagenase IV (dilute 1:1), pipette the cells on and off, and finally, transfer them into a 15-mL tube for centrifugation later on.
8. Again, wash each well with 500-μL culture medium; add these washes to the cell suspension.
9. Centrifuge at 235 × *g* for 5 min at +21°C.
10. Resuspend the cell pellet in culture medium/DNase I (dilute 10:1) and centrifuge again.
11. Resuspend the final cell pellet in culture medium (1×10^5 cells/300 μL).

Then, incubate the cell suspensions with anti-Thy1-coupled magnetic beads to allow the beads to bind to the Thy1 (CD90) surface antigens on fibroblasts (10 μL beads + 300 μL cell suspension).

12. Carefully mix beads and cell suspension within an eppendorf cup, securely seal the vial with parafilm, put into an ice-filled 50-mL tube, and rotate it around its horizontal axis for 45 min at +4°C.
13. Bring the magnet close to the wall of the cup; this draws the fibroblasts, which are now covered by magnetic Dynabeads, toward the magnet, and the supernatant, including the enriched Schwann cells, can easily be collected.
14. Determine the number of viable cells by trypan blue staining and seed the enriched adult Schwann cells into freshly prepared PORN-laminin-coated wells.
15. After the cells have grown to 90% confluency (about 10 days), perform a second immunopurification, as above.

3.3.3. Conditions for Propagation of Adult Schwann Cell In Vitro

After final purification of adult Schwann cells, the cultures can be split whenever they reach confluency. For this, the cells are detached with the use of trypsin/EDTA and reseeded on freshly prepared PORN-laminin-coated culture dishes. Appropriate cell densities for seeding enriched adult Schwann cells (purity > 95%) are as follows:

- Adult rat Schwann cells: 5×10^4 cells/cm$^{2.}$
- Adult canine and human Schwann cells: 2×10^5/cm^2.

We do not include a proliferation step in our protocol as we found proliferation rates to be enhanced in adult human Schwann cells by permanently supplementing fresh heregulin to each medium exchange (20). Elevated proliferation rates were induced in adult rat Schwann cell cultures just by passaging the cells with the “cold jet” technique (16). Adult canine Schwann cells show a proliferation rate under the given conditions, which does not require any further culture supplement.

3.4. Analysis of Schwann Cell Purity by Immunocytochemistry

Although phase-contrast microscopy may give an impression of cell culture purity with the flatter morphology of fibroblasts and a clearer nucleus-soma relation in Schwann cells, it is necessary to check for Schwann cell purity of cultures before and after enrichment. Immunocytochemistry should be used to rigorously determine culture purity. For this purpose, sister cultures should be grown on, e.g., 24-well plates and analyzed (see Note 12). Adult Schwann cells can be identified by staining either for intracellular calcium-binding protein S100 or for cell surface p75^{LNGFR} (16–20). Nuclear counter-staining with DAPI will visualize S100- or p75^{LNGFR}-immunonegative

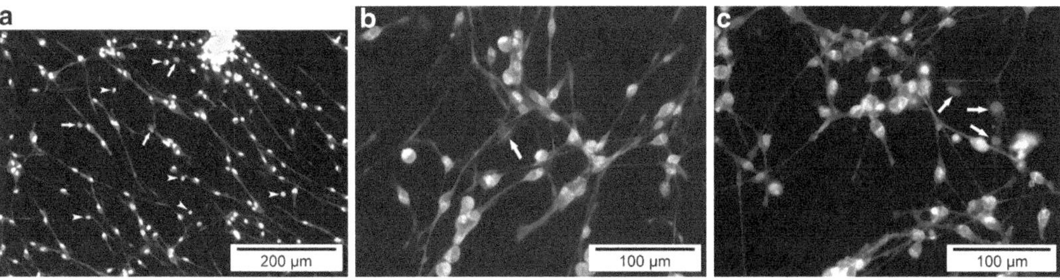

Fig. 1. Highly purified human Schwann cell culture characterized by immunocytochemistry against the intracellular calcium-binding protein S100 (**a**, **b**) or the cell surface molecule p75^{LNGFR} (**c**). Nuclear counterstain with DAPI visualizes S100- or p75^{LNGFR}-immunonegative cells, e.g., fibroblasts, which additionally display darker and bigger nuclei in DAPI staining (indicated by *white arrows*). *Arrow heads* in (**a**) indicate bright nuclei of dying cells.

cells, e.g., fibroblasts. Figure 1 illustrates a highly purified human Schwann cell culture, characterized by immunoreactivity against S100 (see Fig. 1a, b) and p75^{LNGFR} (see Fig. 1c). Fibroblasts eventually display a weak staining but could additionally be identified by their multipolar, flattened morphology with darker and larger nuclei in DAPI-staining, while adult Schwann cells show a typical bipolar or tripolar morphology and smaller nuclei.

1. Remove the culture medium and fix the cells with 4% paraformaldehyde (15 min at room temperature).
2. Remove the fixative and incubate the cells with blocking buffer (30 min at room temperature).
3. Add a single type of primary antibody (see Note 13) and incubate overnight in the dark at +4°C.
4. Wash the cells with blocking buffer.
5. Incubate with secondary antibody for 2 h at room temperature (in the dark!).
6. Counterstain cell nuclei with DAPI (15 min at room temperature, in the dark!).
7. Wash the cells with PBS and keep them covered with PBS for analysis under the inverted fluorescence microscope.

4. Notes

1. Due to logistical circumstances, human nerve tissue (residues of sural nerve transplants harvested during reconstructive surgery in vivo, donor age (13–60 years), male or female (20)) or dog nerve tissue (donor animals euthanized because of infaust prognosis with no peripheral neuropathies (18, 19)) could be stored at +4°C for a maximum of 20 h.

2. This work has to be done under the stereo microscope because it is of high importance that exclusively nerve fascicles are harvested, those appear dense and never hollow-like, e.g., blood vessels.
3. For predegeneration of canine nerve samples, it is eventually of advantage to precoat the cell culture surface with 2 mL collagen G (1:1 in MGM + manufacturer supplements). Therefore, the mixture is allowed to solidify for 1 h at +37°C. Attachment of the epineurium-free nerve fascicles to the collagen coating is facilitated by incubation of the tissue in dry wells for 15 min at +37°C.
4. For canine tissue, also 1% amphotericin should be added to the predegeneration culture medium.
5. The medium has to be removed very carefully, and before adding new medium, the "dry" plates could eventually be incubated for 15 min at +37°C to enhance tissue adherence on the culture surfaces.
6. To avoid evaporation and drying of the coating, the well plate has to be covered and sealed with parafilm (eventually an additional wrap with aluminum foil could also help).
7. In case there are cell clusters or tissue clumps remaining, gentle pipetting could also be performed with culture medium + collagenase, (0.125%) again followed by centrifugation and another optional pipetting with a 10:1 mixture of culture medium and DNase before final centrifugation. However, it is most important to treat the cells with absolute care. One must avoid triturating already separated cells; therefore, all single-cell solution supernatants need to be collected and diluted to reduce enzymatic activity within the solution. Furthermore, unnecessary centrifugation should be avoided.
8. The right volume of culture medium to be added to the final cell pellet has to be carefully calculated (e.g., 0.5 mL/tissue dissociated in 1 well on a 6-well plate), as one needs to avoid additional centrifugation prior to initial seeding of the cells.
9. For culturing cells of human origin, 10 nM of fresh rhuHRG (heregulin) has to be added to the culture medium to increase the proliferation rate whenever the medium is exchanged.
10. For culturing cells of canine origin, 1 mM cytosine arabinoside should be added from day 2 to 6 in order to reduce the number of fibroblasts.
11. If necessary, "cold jet" can be repeated once or twice with culture medium.
12. In a 24-well plate, a volume of 150 μL of buffer or antibody in buffer is sufficient for incubation. During incubation, the wells

should be covered to avoid evaporation and the plates should at best be wrapped in aluminum foil for darkness.

13. It is advised to use only single staining against Schwann cells or fibroblasts.

Acknowledgments

In the first place, I wish to thank all persons who helped to establish and to advance the described techniques: Dr. Christina Mauritz, Dipl. Biol.; Maike Wesemann; Sukhada Chaturvedi, PhD; Dr. Peer Seef; Dr. Ruth Schmitte; and Silvana Taubeler-Gerling. Many thanks also to Prof. Dr. Claudia Grothe for creating a supportive working atmosphere. Special thanks go to our clinical colleagues and their patients for providing us with human and canine nerve tissue samples: Prof. Dr. Cordula Matthies; Prof. Dr. Götz Penkert; Dr. Veronika Stein, PhD; Dr. Henning Schenk, PhD; Dr. Cornelia Flieshardt; and Prof. Dr. Andrea Tipold.

References

1. Ansselin AD, Corbeil SD, Davey DF (1995) Culture of Schwann cells from adult animals. In Vitro Cell Dev Biol Anim 31, 253–254
2. Calderon-Martinez D, Garavito Z, Spinel C, Hurtado H (2002) Schwann cell-enriched cultures from adult human peripheral nerve: a technique combining short enzymatic dissociation and treatment with cytosine arabinoside (Ara-C). J Neurosci Methods 114, 1–8
3. Casella GT, Bunge RP, Wood PM (1996) Improved method for harvesting human Schwann cells from mature peripheral nerve and expansion in vitro. Glia 17, 327–338
4. Guenard V, Kleitman N, Morrissey TK, Bunge RP, Aebischer P (1992) Syngeneic Schwann cells derived from adult nerves seeded in semipermeable guidance channels enhance peripheral nerve regeneration. J Neurosci 12, 3310–3320
5. Verdu E, Rodriguez FJ, Gudino-Cabrera G, Nieto-Sampedro M, Navarro X (2000) Expansion of adult Schwann cells from mouse predegenerated peripheral nerves. J Neurosci Methods 99, 111–117
6. Morrissey TK, Kleitman N, Bunge RP (1991) Isolation and functional characterization of Schwann cells derived from adult peripheral nerve. J Neurosci 11, 2433–2442
7. Levi AD (1996) Characterization of the technique involved in isolating Schwann cells from adult human peripheral nerve. J Neurosci Methods 68, 21–26
8. Keilhoff G, Fansa H, Schneider W, Wolf G (1999) In vivo predegeneration of peripheral nerves: an effective technique to obtain activated Schwann cells for nerve conduits. J Neurosci Methods 89, 17–24
9. Keilhoff G, Fansa H, Smalla KH, Schneider W, Wolf G (2000) Neuroma: a donor-age independent source of human Schwann cells for tissue engineered nerve grafts. Neuroreport 11, 3805–3809
10. Peulve P, Laquerriere A, Paresy M, Hemet J, Tadie M (1994) Establishment of adult rat Schwann cell cultures: effect of b-FGF, alpha-MSH, NGF, PDGF, and TGF-beta on cell cycle. Exp Cell Res 214, 543–550
11. Komiyama T, Nakao Y, Toyama Y, Asou H, Vacanti CA, Vacanti MP (2003) A novel technique to isolate adult Schwann cells for an artificial nerve conduit. J Neurosci Methods 122, 195–200
12. Scarpini E, Kreider BQ, Lisak RP, Pleasure DE (1988) Establishment of Schwann cell cultures from adult rat peripheral nerves. Exp Neurol 102, 167–176

13. Vroemen M, Weidner N (2003) Purification of Schwann cells by selection of p75 low affinity nerve growth factor receptor expressing cells from adult peripheral nerve. J Neurosci Methods 124, 135–143
14. Pauls J, Nolte C, Forterre F, Brunnberg L (2004) [Cultivation and expansion of canine Schwann cells using reexplantation]. Berl Munch Tierarztl Wochenschr 117, 341–352
15. Techangamsuwan S, Imbschweiler I, Kreutzer R, Kreutzer M, Baumgartner W, Wewetzer K (2008) Similar behaviour and primate-like properties of adult canine Schwann cells and olfactory ensheathing cells in long-term culture. Brain Res 1240, 31–38
16. Mauritz C, Grothe C, Haastert K (2004) Comparative study of cell culture and purification methods to obtain highly enriched cultures of proliferating adult rat Schwann cells. J Neurosci Res 77, 453–461
17. Haastert K, Mauritz C, Chaturvedi S, Grothe C (2007) Human and rat adult Schwann cell cultures: fast and efficient enrichment and highly effective non-viral transfection protocol. Nat Protoc 2, 99–104
18. Haastert K, Seef P, Stein VM, Tipold A, Grothe C (2009) A new cell culture protocol for enrichment and genetic modification of adult canine Schwann cells suitable for peripheral nerve tissue engineering. Res Vet Sci 87, 140–142
19. Schmitte R, Tipold A, Stein VM, Schenk H, Flieshardt C, Grothe C, et al (2010) Genetically modified canine Schwann cells-In vitro and in vivo evaluation of their suitability for peripheral nerve tissue engineering. J Neurosci Methods 186, 202–208
20. Haastert K, Mauritz C, Matthies C, Grothe C (2006) Autologous adult human Schwann cells genetically modified to provide alternative cellular transplants in peripheral nerve regeneration. J Neurosurg 104, 778–786
21. Bunge RP (1993) Expanding roles for the Schwann cell: ensheathment, myelination, trophism and regeneration. Curr Opin Neurobiol 3, 805–809
22. Bunge RP (1994) The role of the Schwann cell in trophic support and regeneration. J Neurol 242(1 Suppl 1), S19–S21
23. Stoll G, Jander S, Myers RR (2002) Degeneration and regeneration of the peripheral nervous system: from Augustus Waller's observations to neuroinflammation. J Peripher Nerv Syst 7, 13–27
24. Jirsova K, Sodaar P, Mandys V, Bar PR (1997) Cold jet: a method to obtain pure Schwann cell cultures without the need for cytotoxic, apoptosis-inducing drug treatment. J Neurosci Methods 78, 133–137
25. Haastert K, Grosskreutz J, Jaeckel M, Laderer C, Bufler J, Grothe C, et al (2005) Rat embryonic motoneurons in long-term co-culture with Schwann cells-a system to investigate motoneuron diseases on a cellular level in vitro. J Neurosci Methods 142, 275–284
26. Casella GT, Wieser R, Bunge RP, Margitich IS, Katz J, Olson L, et al (2000) Density dependent regulation of human Schwann cell proliferation. Glia 30, 165–177

Chapter 18

Use of PC12 Cells and Rat Superior Cervical Ganglion Sympathetic Neurons as Models for Neuroprotective Assays Relevant to Parkinson's Disease

Cristina Malagelada Grau and Lloyd A. Greene

Abstract

Cellular models composed of primary neuronal cultures or neuron-like cell lines are commonly used to study neuron cell death and to test the neuroprotective properties of specific compounds. Cellular models are easily accessible, permitting dissection and modulation of signaling pathways involved in neuron death. For example, drug or shRNA delivery is more straightforward since there is no blood–brain barrier to cross. However, since these models have their limitations, any important findings should ultimately be verified with animal models and human samples. Here, we describe two cellular models that can be used as a highly informative and easy to use starting point for testing potential neuroprotective drugs for Parkinson's disease: PC12 cells and sympathetic neuronal cell cultures. We describe in detail the protocols needed to apply these models to study neuroprotection in the context of Parkinson's disease.

Key words: PC12, Parkinson's disease, 6-Hydroxydopamine, 1-Methyl-4-phenylpyridinium, Nerve growth factor

1. Introduction

Tumor-derived cell lines have been extensively used as models to study pathological conditions. The features of such proliferating cells—homogenous and available in large numbers and easy to grow and to transfect or to infect with viral particles—make them highly suitable to screen drugs, manipulate genes, and explore signaling pathways. For neurodegeneration studies, one of the most used and cited culture systems has been the PC12 cell line which was derived from a transplantable rat pheochromocytoma (1). These cells have the machinery to synthesize, release, take up, and store catecholamines, the major species of which is dopamine. A notable feature of PC12 cells is that they respond to nerve growth factor (NGF). In response to NGF, PC12 cells are converted from proliferating

Stephen D. Skaper (ed.), *Neurotrophic Factors: Methods and Protocols*, Methods in Molecular Biology, vol. 846, DOI 10.1007/978-1-61779-536-7_18,

chromaffin-like cells to nondividing sympathetic-neuron-like cells that extend axons and become electrically excitable (2, 3). As will be described here, generating and using PC12 cells for neurodegeneration experiments is a relatively simple process.

With respect to their use for studying neurodegeneration, PC12 cells are sensitive to mitochondrial toxins that mimic Parkinson's disease (PD), such as 6-hydroxydopamine (4), 1-methyl-4-phenylpyridinium (MPP^+) (5), rotenone (6, 7), or paraquat (8). These toxins induce cellular degeneration within 24–48 h after exposure. The cells are also reported to undergo degeneration and death in response to overexpression of mutant forms of synuclein that are associated with familial PD (9).

These properties make PC12 an uncomplicated and convenient in vitro model to study causes and possible treatments for PD. However, one must keep in mind that, as with all models, PC12 cells have their limitations. Any result obtained with this tumor cell line does not assure that the same finding will hold in in vivo models or in the human disease. Therefore, any hypotheses arising from PC12 cell studies should be tested also in primary neuronal cell cultures and in vivo models and, ultimately, wherever possible, in PD itself (10).

One type of primary culture used to study PD is composed of neurons derived from dissociated rodent superior cervical sympathetic ganglia. Like dopaminergic neurons in the substantia nigra, noradrenergic sympathetic neurons of PD patients develop Lewy bodies and undergo degeneration (11, 12). From an experimental point of view, sympathetic neurons have the advantage that they are relatively easy to obtain and culture (13), are a homogeneous population, and also succumb to PD mimetics such as 6-hydroxydopamine (6-OHDA) and MPP^+ (5, 14, 15). They are also useful to assess the role of specific genes by knockdown or by overexpression via transfection or viral infection (13, 16). In addition, they are the primary neuronal counterparts of NGF-treated PC12 cells and as such represent the next step in progressing from experiments in a cell line to cultures of primary neurons.

In this chapter, we will first guide the reader through the process of culturing PC12 cells and sympathetic neurons. We will then describe how to use these cultures to monitor cell viability when testing compounds and treatments for their capacity to protect against PD mimetics.

2. Materials

All work should be carried out in a cell culture hood using sterile technique. It is recommended to disinfect surfaces with 70% ethanol before and after the proceedings.

2.1. Coating Plastic Culture Ware Substrates

1. Rat tail collagen (Roche). Stock solution in 0.2% acetic acid in sterile double distilled or deionized water. Working dilution 1:15 in sterile double distilled or deionized water.

2.2. Coating Glass Coverslip Substrates

1. Poly-D-lysine (Millipore) molecular weight >300 kDa. Dilute stock solution (1 mg/mL) in sterile double distilled water to the final working concentration of 50 μg/mL.
2. Rat tail collagen (Roche). Stock solution in 0.2% acetic acid in sterile double distilled or deionized water. Working dilution 1:15 in sterile double distilled or deionized water.

2.3. PC12 Cells

1. *Complete medium*: RPMI 1640 cell culture medium containing 10% heat inactivated horse serum (Sigma) (see Note 1), 5% fetal bovine serum, and penicillin/streptomycin (50 units/50 μg of each per mL).
2. *Differentiation medium*: RPMI 1640 cell culture medium containing 1% heat-inactivated horse serum and penicillin/streptomycin (50 units/50 μg of each per mL) and 50 ng/mL recombinant human or murine NGF (commercially available from suppliers such as Alomone). This should be added from 1,000× stock just before use.
3. *Freezing medium*: complete medium with 10% dimethyl sulfoxide (DMSO).

2.4. Sympathetic Neuron Cultures

1. Differentiation medium.
2. Uridine/fluorodeoxyuridine: a 500-μM stock solution in sterile double distilled or deionized water.
3. Trypsin (Invitrogen): 0.25% (w/v) in phosphate buffered saline (PBS) (sterile and without EDTA).
4. RPMI 1640 cell culture medium.
5. One tube with 15 mL RPMI 1640 cell culture medium + NGF (50 ng/mL) without serum and other additives.
6. One tube with 25–30 mL of RPMI1640 cell culture medium without serum and other additives.
7. One bucket of ice.
8. Two insulin needles.
9. One polystyrene support for dissection (5 × 5 × 1 cm). Wrapping the support in aluminum foil makes sterilization with 70% ethanol much more easier.
10. 70% Ethanol.
11. Two micro dissecting tweezers (Roboz).
12. One dissecting tweezers (Roboz).
13. One dissecting scissors (Roboz).
14. One regular scissors (Roboz).

2.5. PD Toxin Treatments

1. 10-mM 6-OHDA (Tocris) stock in sterile double distilled or deionized water
2. 100-mM MPP^+ (Sigma) stock in sterile double distilled or deionized water

2.6. Viability Assay

1. *10× nuclei buffer* (100 mL): cetyldimethyl-ethanolammonium bromide (5 g), NaCl (0.165 g), glacial acetic acid (2.8 mL), 10% Triton X-100 (50 mL), 1 M $MgCl_2$ (2 mL), 10× PBS (10 mL), and H_2O (35.2 mL). The working dilution is 1× in distilled water (17).
2. Neubauer hemacytometer chamber with coverslip.

3. Methods

We will start by describing how to grow PC12 cells and how to treat them with NGF. We will then discuss how to prepare rat sympathetic neuron cultures. We also provide instructions on how to perform treatments of the cultures with PD mimetic toxins that induce cell death, and finally, we will suggest several options to assess cell viability.

3.1. Growing PC12 Cells (see Note 2)

3.1.1. Coating Plasticware

To coat the plates, pipette the necessary amount of collagen (1:15) to uniformly cover the surface. For a 10-cm culture dish, 1 mL is sufficient, and this volume can be accordingly adjusted for culture dishes/wells of different sizes. Alternatively, one can add a larger volume and then remove the excess with a sterile pipette. Allow the plates to dry uncovered under a tissue culture hood for at least an hour. After drying, plates can be stored at room temperature, wrapped in aluminum foil or in the initial plastic bags, for 2–3 weeks (see Notes 3 and 4).

3.1.2. Thawing and Plating PC12 Cells

1. If PC12 cells are received or stored in a frozen state (–80°C or in liquid nitrogen), they will be in complete medium and 10% DMSO. Thaw the vial at 37°C in a water bath. This step should be done quickly to diminish the toxicity of DMSO.
2. Prepare a sterile falcon tube with 10 mL of warm (37°C) complete medium.
3. Add the thawed content of the cryotube to the 10 mL of complete medium.
4. Mix well and centrifuge the cells for 5 min at 240×*g* at room temperature in a table top centrifuge.
5. Remove the supernatant (containing DMSO) and add 10 mL of fresh complete medium to resuspend the cell pellet.
6. Mix well by trituration in a pipette and plate the contents in a 10-cm collagen-coated tissue culture plate. This step also serves

to break up cell clumps. Shake the plate well, back and forth and left to right, to spread the cells uniformly. Avoid shaking circularly or otherwise the cells will tend to concentrate at the center of the plate.

7. Once the cells are plated, renew the culture medium every 2–3 days. Approximately 2/3 of the medium should be exchanged. Maintain a volume of 5–8 mL of medium.

3.2. Subculturing PC12 Cells

It is recommended to subculture the cells when 90–95% confluence is reached. One plate can be split in two or three plates.

1. Before detaching the cells from the plasticware, aspirate some of the medium from the plate to reach a volume of 3–5 mL. This avoids potential splashing of the medium out of the plate during cell detachment.
2. Incline the plate at about 30°, and with a plastic sterile pipette, eject a few milliliters of medium over the cell monolayer. Cells will detach with the pressure exerted by the medium.
3. After applying medium in this way to the entire surface, triturate the cells several times with the pipette to break up cell clumps. Split the resulting cell suspension into two or three plates (see Notes 5 and 6).
4. Bring the total volume per plate to 5–8 mL of complete medium.

3.3. Freezing PC12 Cells

1. Detach cells from the plates.
2. Centrifuge the cell suspension at 242 × *g* for 5 min.
3. Resuspend the pellet from one plate in 2 mL of complete medium containing 10% DMSO.
4. Pipette the suspended cells into cryotubes and place them in an isopropanol-freezing container.
5. Place the container with the cells in a −80°C freezer. The isopropanol container will freeze the cryotubes about 1°C/min (see Note 7). The cryotubes can be stored at −80°C for up to several months and transferred to liquid nitrogen for longer periods of storage.

3.4. Treating PC12 Cells with NGF

PC12 cells dramatically change their phenotype when they are exposed to NGF. They exit the cell cycle, project long neurite-like processes, and take on many properties of differentiated sympathetic neurons, including synthesis, storage, and release of catecholamines (principally dopamine).

1. Detach PC12 cells from the plate as described in Subheading 3.2.
2. Dilute the cell suspension with differentiating medium (RPMI1640, 1% horse serum, and 50 ng/mL NGF). Typically,

one confluent culture of PC12 cells can be used to seed 5–20 cultures of the same size for NGF treatment. Note that the NGF stock (50 μg/mL) should be stored at 4°C and should be freshly added to RPMI 1640/1% horse serum medium as needed.

3. Plate the cells on the appropriate collagen-coated plasticware.
4. Exchange ¾ of the culture medium every 2–3 days with differentiating medium plus fresh NGF to a final dilution of 50 ng/mL.
5. The cell should begin to extend neurites within 24 h and will continue to do so over the next 7–10 days.

3.5. Cultivating Sympathetic Neurons from Rat Superior Cervical Ganglia

From one rat litter of 12–14 pups, a 24- or 48-well plate can be obtained. The cell density can be adjusted as desired by counting the cells at the end of the dissection. The steps below describe the preparation of cultures from one litter of newborn rat pups. A video of the dissection and culture procedure can be found at (13).

1. Soak all the dissection tools in 70% ethanol for at least 10 min.
2. Spray or wipe down with 70% ethanol all the surfaces of the dissecting hood, the microscope, and the fiber optic lights. Also, disinfect with 70% ethanol the small polystyrene foam support wrapped in aluminum foil. Put a sterile piece of gauze over the support.
3. Prepare two 10-cm plates with 5 mL of RPMI or PBS. These will be used to rinse the tools during the dissection to wash off any adhering tissue.
4. Place the lids of the 10-cm plates face up in the hood and pipette onto them one or two drops of RPMI for each pup that will be dissected. After dissection, the ganglia will be transferred to these drops for cleaning under the dissecting microscope.
5. Put the tubes with RPMI on ice.
6. Keep the pups in a clean box. Dissect one pup at a time.
7. Spray the pup for dissection with 70% ethanol and immediately decapitate it using a pair of sharpened small dissecting scissors. Remove the head closer to the shoulders than to the jaw so as to avoid cutting through the ganglia.
8. Place the head on the polystyrene support facing up.
9. With the fine scissors, cut the skin of the neck. Pin the skin flaps with the insulin needles to the polystyrene support. This will hold the head in place and prevent the skin from folding over the field of dissection.
10. Pipette RPMI onto the exposed tissue to remove blood cells and debris and to prevent drying.

11. Under the dissecting microscope, with two pairs of fine-tip tweezers, remove the salivary glands and connective tissue.
12. Remove the sternocleidomastoid muscle on the side of dissection to have a better view of the carotid artery. The carotid branches into the external and the internal carotids. The sympathetic ganglion is right beneath the branch point of the artery. With one set of tweezers, pull the lower end of the carotid up, and with the other, remove the translucent tear-shaped ganglion attached to the vessel, that is, the sympathetic ganglion. Close to the external carotid, there is another rounder and smaller translucent ganglion. This is the nodose ganglion. Be careful not to confuse the two ganglia.
13. Put the ganglion in one of the RPMI1640 drops on the culture plate lid.
14. Repeat the dissection to remove the second ganglion on the other side of the neck.
15. Under the dissecting microscope, clean the ganglia by teasing away any remains of the carotid and surrounding connective tissue. This will diminish contamination of the cultures by nonneuronal cells.
16. Transfer the cleaned ganglia to the 15-mL tube with RPMI plus NGF on ice. Collect all ganglia in this way until the dissection of all the pups is finished.
17. Centrifuge the ganglia at 168 × *g* for 3 min. Take care because the ganglia tend to attach to the wall of the tube.
18. Remove the supernatant carefully and add 1 mL of 0.25% trypsin in PBS (sterile). Incubate at 37°C for 30 min.
19. To neutralize the trypsin after the 30-min incubation, add 10 mL of RPMI medium with serum. Centrifuge at 168 × *g* for 3–5 min, remove the supernatant, and add 2 mL of RPMI with 1% horse serum and 50 ng/mL NGF.
20. Triturate the ganglia for 40–50 strokes in a glass Pasteur pipette with a fire-polished tip. A second Pasteur pipette with a narrower fire-polished tip can be used for the final 15 strokes.
21. Plate the cells (normally two ganglia per well in a 24-well plate) at the desired density, adding more RPMI with 1% horse serum and NGF to bring to the appropriate volume for plating.
22. Twenty-four hours later, add uridine/fluorodeoxyuridine to the wells to reach a final concentration of the antimitotic of 10 μM (see Note 8).
23. Change the medium of the cultures every 2–3 days. Add 10-μM uridine/fluorodeoxyuridine to the fresh medium for the first two medium changes.
24. Use the cultures between 7 and 15 days after plating.

3.6. Toxin Treatments

Both NGF-differentiated PC12 cells and sympathetic neurons will be ready to use after at least 7 days following plating.

1. Prepare stock solutions of the toxins. Use stock concentrations to minimize the volume of solution added to the cultures.
2. Change the medium right before any treatment.
3. If a specific compound is tested in conjunction with the toxins, pretreatment should be done at least 30 min to 1 h before adding the toxin. This is to permit entry into the cells. Always include controls of sister cultures treated with the vehicles used for delivery of toxin or tested agent. If the vehicle is DMSO, the final amount should be no more than 0.1% of the final volume in the culture well.
4. To treat neuronal PC12 cells with (see Notes 9–12):
   - 6-OHDA: a stock of 10 mM is recommended to treat the cells at a final concentration of 50–100 μM.
   - MPP^+: a stock of 100 mM is recommended to treat the cells at a final concentration of 500 μM–1 mM.
5. To treat rat sympathetic neurons, prepare stocks as above, but the final concentrations of the toxins should be much lower: 5–10 μM for 6-OHDA and 50–100 μM for MPP^+ (see Notes 9–12).
6. After treatments, wait for the desired time to assess the cultures. If viability is monitored, 24 and/or 48 h are suggested.
7. For analyzing the samples by Western immunoblotting, choose shorter incubation times.

3.7. Viability Assessment

To quantify viable cells, there are several methods to choose from. However, 6-OHDA and MPP^+ with their capability to inhibit mitochondrial complexes can interfere with tetrazolium salt assays, leading to an overestimation of cell death. We routinely use a method with a detergent solution that dissolves the cell plasma membrane and leaves the nuclear membrane intact (17). This permits counting of the numbers of surviving cells in the cultures. With a Neubauer hemacytometer chamber, healthy round nuclei can be counted under a phase microscope and distinguished from the smaller, darker dead or degenerating nuclei.

The procedure is as follows:

1. Aspirate to fully remove the medium.
2. Add 0.5 mL of detergent solution to the wells (in a 48-well plate, add accordingly for other-size dishes or wells) and mix well (see Note 13).
3. Place approximately 10 μL of the nuclear suspension under the coverslip of the Neubauer chamber.
4. Count at least 100 nuclei (see Note 14).

5. Cell survival can be expressed as the percentage of viable cells in experimental cultures compared with sister control cultures.

For transfection experiments, cell viability (after treatment) can be assessed by counting living green fluorescent protein-positive cells in strips under a fluorescence microscope. If the transfection rate is low (as in superior cervical ganglion cultures), counting all the labeled cells in the culture is recommended.

4. Notes

1. If horse serum needs to be inactivated, thaw the bottle and put it in a water bath at 56°C for 30 min.
2. RPMI medium requires an atmosphere enriched in CO_2 to maintain the proper pH. Therefore, an incubator should be used that maintains at 7.5–8% CO_2 atmosphere.
3. Coating glass substrates (such as coverslips or glass slide cultures) for immunofluorescence. To perform confocal microscopy on either neuronal PC12 cells or superior cervical ganglion neurons, it is necessary to culture them on glass. This requires coating the surface with a substrate that permits cell attachment. To coat the glass, first incubate overnight in a solution containing 50 μg/mL of poly-D-lysine. Following this, wash with sterile distilled water, coat with collagen, and let dry as described above.
4. Do not irradiate collagen plates with UV. This damages collagen polymers and decreases cell-binding capability.
5. It is easier to carry out initial experiments (e.g., concentration-response curves) with neuronal PC12 cells, and afterward, to confirm the main results with sympathetic neurons. It is also useful to use 48-well plates to maximize the number of replicate cultures. Routinely, PC12 cells are grown in 10-cm dishes. When using differentiated PC12 cells, the type of plates used for experiments can be adapted to optimize use. For instance, if the experiment is testing viability, it is better to use 24- or 48-well plates to permit multiple (at least 4) replicates. On the other hand, if the object is to collect protein extracts, it is better to use 6-well plates or even 10-cm dishes.
6. PC12 cells tend to detach from the plasticware easily, and even more so, if the plate has been coated with collagen more than 3 weeks earlier. If such detachment starts and the cells are not dying, try to replate them on a fresh collagen coated plate. This problem is common in neuronal PC12 cells after transfection: the older the coating, the more they will detach after transfection.

7. If you do not have an isopropanol container, you can place the cryotubes on ice for 10 min and then freeze them at –80°C.
8. Uridine and deoxyuridine stocks should be prepared and stored in separated tubes.
9. Toxins as 6-OHDA or MPP^+ are generally used to mimic PD in vitro. The dopamine transporter specifically takes up both toxins. Once inside the cells, these toxins deplete catecholamines and inhibit mitochondrial oxidative metabolism. Therefore, their toxicity range will depend on the density of the dopamine transporters on the cells and also on the cellular density itself. For example, identical concentrations of 6-OHDA or MPP^+ will be more toxic in low-density cultures than in more dense ones. The concentrations listed here are intended to cause 40–60% cell death. It is important to work within this range for testing neuroprotective compounds. Also, excess concentrations of the toxins will drive nonapoptotic as well as apoptotic cell death.
10. These toxins inhibit mitochondrial complexes and deplete cells of ATP. In doing so, they also generate damaging reactive oxygen species.
11. Both 6-OHDA and MPP^+ should be freshly prepared just before use. Both are light sensitive, and therefore, their solution tubes should be wrapped in aluminum foil.
12. Always wear gloves when weighing out and working with these toxins.
13. For viability assays that involve counting nuclei, resuspend the contents of the well with a pipette before withdrawing 10 μL for placement in the Neubauer chamber. Nuclei tend to sink to the bottom of the well. Cultures to which the nuclear counting solution has been added can be stored for several days. In this case, seal the plate with Parafilm to prevent evaporation and store it at 4°C for 5 days at maximum.
14. As noted above, at least 100 nuclei should be counted per culture for viability assays. If the numbers of nuclei are excessive, more lysis buffer can be added to dilute the nuclear suspension. Keep track of the volume in final calculations of cell numbers.

Acknowledgments

This work was supported by grants from the NIH-NINDS, American Parkinson's Disease Association, and Parkinson's Disease Foundation.

References

1. Greene LA, and Tischler AS (1976) Establishment of a noradrenergic clonal line of rat adrenal pheochromocytoma cells which respond to nerve growth factor. *Proc Natl Acad Sci USA* 73, 2424–2428
2. Greene LA (1978) Nerve growth factor prevents the death and stimulates the neuronal differentiation of clonal PC12 pheochromocytoma cells in serum-free medium. *J Cell Biol* 78, 747–755
3. Greene LA, and Rein G (1977) Release of (3H)norepinephrine from a clonal line of pheochromocytoma cells (PC12) by nicotinic cholinergic stimulation. *Brain Res* 138, 521–528
4. Walkinshaw G, and Waters CM (1994) Neurotoxin-induced cell death in neuronal PC12 cells is mediated by induction of apoptosis. *Neuroscience* 63, 975–987
5. Nakamura K, Bindokas VP, Marks JD, Wright DA, Frim DM, Miller RJ, et al (2000) The selective toxicity of 1-methyl-4-phenylpyridinium to dopaminergic neurons: the role of mitochondrial complex I and reactive oxygen species revisited. *Mol Pharmacol* 58, 271–278
6. Hirata Y, and Nagatsu T (2005) Rotenone and CCCP inhibit tyrosine hydroxylation in rat striatal tissue slices. *Toxicology* 216, 9–14
7. Sherer TB, Betarbet R, Testa CM, Seo BB, Richardson JR, Kim JH, et al (2003) Mechanism of toxicity in rotenone models of Parkinson's disease. *J Neurosci* 23, 10756–10764
8. Yang WL, and Sun AY (1998) Paraquat-induced cell death in PC12 cells. *Neurochem Res* 23, 1387–1394
9. Zhou W, Hurlber MS, Schaack J, Prasad KN, and Freed CR (2000) Overexpression of human alpha-synuclein causes dopamine neuron death in rat primary culture and immortalized mesencephalon-derived cells. *Brain Res* 866, 33–43
10. Malagelada C, and Greene LA. PC12 cells as a model for Parkinson's disease research., in Parkinson's Disease: molecular and therapeutic insights from experimental models. , N. Przedborski, Editor. 2008, Elsevier. p. 375–389
11. Fahn S (1998) Medical treatment of Parkinson's disease. *J Neurol* 245 Suppl 3, P15–24
12. Marras C, and Lang A (2008) Invited article: changing concepts in Parkinson disease: moving beyond the decade of the brain. *Neurology* 70, 1996–2003
13. Zareen N, and Greene LA (2009) Protocol for culturing sympathetic neurons from rat superior cervical ganglia (SCG). J Vis Exp, 2009(23)
14. Malagelada C, Ryu EJ, Biswas SC, Jackson-Lewis V, and Greene LA (2006) RTP801 is elevated in Parkinson brain substantia nigral neurons and mediates death in cellular models of Parkinson's disease by a mechanism involving mammalian target of rapamycin inactivation. *J Neurosci* 26, 9996–10005
15. Schober A (2004) Classic toxin-induced animal models of Parkinson's disease: 6-OHDA and MPTP. *Cell Tissue Res* 318, 215–224
16. Ryu EJ, Harding HP, Angelastro JM, Vitolo OV, Ron D, and Greene LA (2002) Endoplasmic reticulum stress and the unfolded protein response in cellular models of Parkinson's disease. *J Neurosci* 22, 10690–10698
17. Rukenstein A, Rydel RE, and Greene LA (1991) Multiple agents rescue PC12 cells from serum-free cell death by translation- and transcription-independent mechanisms. *J Neurosci* 11, 2552–2563.

Chapter 19

Compartmented Chambers for Studying Neurotrophic Factor Action

Stephen D. Skaper

Abstract

Neurotrophic factors released by target tissues maintain the survival and differentiation of innervating neurons. The manner by which these target-derived neurotrophic proteins communicate with innervating neurons has been actively pursued for over three decades. The present chapter describes a technique for preparing and maintaining compartmented chambers for culturing neurons derived from either superior cervical ganglia (sympathetic neurons) or dorsal root ganglia (sensory neurons). This system recapitulates the selective stimulation of neuron terminals that occurs in vivo following release of target-derived neurotrophins.

Key words: Campenot cultures, Superior cervical ganglia neurons, Dorsal root ganglia neurons, Cell culture, Neurotrophic factors, Retrograde transport

1. Introduction

Many neurons, including sympathetic and sensory neurons, depend for their survival on retrograde signals to their cell bodies generated by nerve growth factor (NGF) or other neurotrophins at their axon terminals (1). Apoptosis resulting from the loss of retrograde NGF signaling contributes to the elimination of excess and misconnected neurons during development and to the death of neurons during the course of neurodegenerative diseases (2). Possible mechanisms of retrograde signaling include (1) retrograde transport of signaling endosomes, carrying NGF bound to activated TrkA, (2) retrograde transport of signaling molecules downstream of TrkA, and (3) retrograde propagation of a phosphorylation signal without transport of signaling molecules. Two or more retrograde signaling mechanisms may also exist to regulate neuronal cell survival, including recent evidence that withdrawal of NGF from distal axons produces

Stephen D. Skaper (ed.), *Neurotrophic Factors: Methods and Protocols*, Methods in Molecular Biology, vol. 846,
DOI 10.1007/978-1-61779-536-7_19,

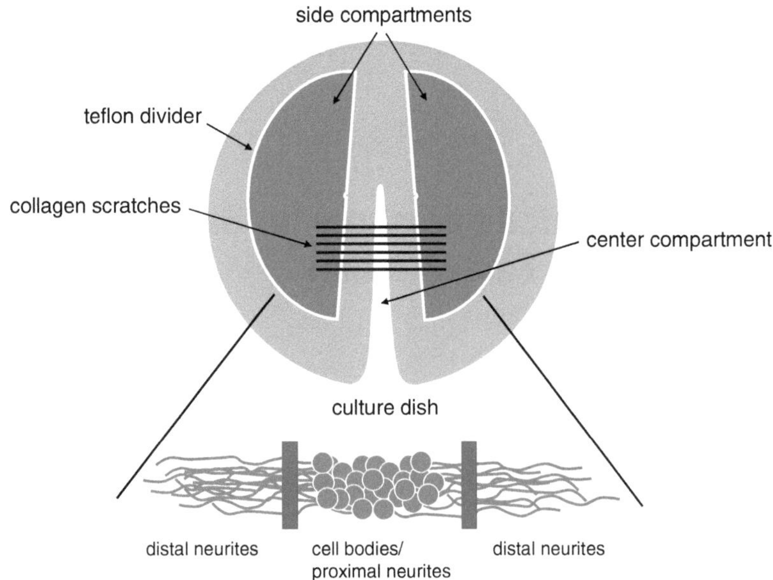

Fig. 1. Schematic diagram of Campenot chamber culture setup for isolation of distal neuronal axons. A 35-mm petri dish is divided into three chambers by a Teflon divider (7.5 mm high) that is sealed to the floor of the dish (a collagen-coated coverslip) with silicone grease. The floor of the narrow (1 × 5 mm) central compartment in which neurons are plated is transected by a series of parallel scratches which guide the growing neurites into the left and right side compartments. The lower portion of the diagram shows an illustrative enlargement of the neurons on a single track. Compartmented cultures isolate cell bodies and proximal axons and distal axons of neurons. See text for further details.

a retrograde apoptotic signal, which is transported to the cell bodies, where it initiates the apoptotic program, leading to the death of the neuron (3).

How target-derived neurotrophins communicate to the cell body of innervating neurons has been actively researched for more than three decades. In order to study retrograde transport, a culture system was originally devised by Robert Campenot, in which cell bodies are isolated from their axons (Fig. 1) (4). The technique of preparing these compartmented chambers for culturing sensory and sympathetic neurons recapitulates the selective stimulation of neuron terminals that occurs in vivo following release of target-derived neurotrophins. Since there is fluidic isolation between the compartments, treatments can be applied separately to cell bodies/proximal axons or distal axons, and cell bodies/proximal axons and distal axons can be separately harvested and analyzed. Distal axons can be axotomized, and the neurons can be studied while their axons regenerate. Compartmented cultures of rodent superior cervical ganglion (SCG) sympathetic neurons have been used in investigations of neurotrophin retrograde transport (5) and signaling (6–9), axonal transport, and axonal protein and lipid biosynthesis (10–14). Compartmented cultures of sensory neurons from rodent dorsal root ganglia (DRG) have been used in investigations of neurotrophin retrograde signaling (15, 16) and anterograde transport and axonal secretion of brain-derived neurotrophic factor (17), among other applications. Compartmented chamber cultures can support other cell types including retinal ganglion neurons (18), cortical neurons (19), and spinal cord neurons (20). Retrograde signaling events that require long-range microtubule dependent retrograde transport have important implications for the treatment of neurodegenerative disorders (21, 22). This chapter describes a procedure for preparing Campenot chambers and their use, modified from their original descriptions (4, 10).

2. Materials

2.1. Equipment and Labware

1. Stereo dissecting microscope (backlighting of stage is preferred) with fiber optic light source
2. Laminar flow biological cabinet (CL2)
3. Humidified, water-jacketed culture incubator at 37°C and 5% CO_2/95% air
4. Water bath set at 37°C
5. 0.5-mL microfuge tubes
6. 15- and 50-mL conical centrifuge tubes
7. 35-mm ∅ sterile tissue culture dishes
8. 100-mm ∅ glass petri dishes
9. 50-mL syringes

2.2. Compartmented Chamber Components

1. Teflon dividers (Camp 10, Tyler Research, Edmonton, Alberta, Canada) (see Note 1)
2. Pin rake (Camp-PR, Tyler Research)
3. Grease loader (Camp-GLSS, Tyler Research)
4. Corning high-vacuum grease (Fisher Scientific)
5. 23-gauge Luer Lock connector (Fisher Scientific)
6. Right-angled hemostatic forceps (Fine Science Tools)

2.3. Reagents

1. Tissue culture grade water, sterile (Invitrogen)
2. Dulbecco's modified Eagle's medium (DMEM) (+4.5 g/L glucose, L-glutamine, pyruvate) (Invitrogen)
3. RPMI-1640 medium (Invitrogen)
4. N2 supplement (100×) (Invitrogen)
5. Horse serum (Invitrogen) (see Note 2)
6. Fetal calf serum (Invitrogen) (see Note 2)

7. Collagen solution, type 1 from rat tail, for cell culture, sterile filtered (Sigma-Aldrich)
8. 7.5% bovine serum albumin (BSA) solution (Invitrogen)
9. Penicillin/streptomycin, 5,000 U/mL penicillin + 5,000 μg/mL streptomycin (100× stock), sterile, for cell culture (Invitrogen)
10. Cytosine β-D-arabinofuranoside (Ara-C) (Sigma-Aldrich)
11. N2-methylcellulose 400CPS (Xenex Laboratories, Coquitlam, BC, Canada)
12. Brain-derived neurotrophic factor, recombinant human (PeproTech)
13. NGF, recombinant human (PeproTech)
14. NOCHROMIX® (Sigma-Aldrich)

2.4. Culture Media and Other Solutions

1. Plating medium for newborn rat DRG neurons: Supplement 50 mL of DMEM to contain N2 supplements (500 μL from 100× stock), 0.05% BSA (334 μL from 7.5% BSA solution), 50 U/mL penicillin, and 50 μg/mL streptomycin (5 mL of 100× stock). NGF is added as appropriate (see Subheading 3.2).
2. Plating medium for newborn rat SCG neurons: Supplement 50 mL of RPMI-1640 to contain 50 ng/mL NGF (add 50 μL from a 50 μg/mL stock), 1% heat-inactivated horse serum, and 50 U/mL penicillin and 50 μg/mL streptomycin (5 mL of 100× stock). NGF is added as appropriate (see Subheading 3.2).
3. N2-methylcellulose: Weigh out 0.6 g of methylcellulose and place it in a 250-mL bottle, add a stir bar, and autoclave it for 20 min on dry (from this point, all work must be sterile). Next, add 200 mL of serum-free medium (RPMI-1640 or DMEM) supplemented with N2 components and stir in a cold room until it dissolves. Aliquot into 15-mL tubes and store at −20°C. To facilitate routine use, prepare also 1-mL aliquots and store at −20°C.
4. Ara-C is prepared as a 10-mM stock solution in phosphate-buffered saline, filter-sterilized, and stored as aliquots at −20°C for up to 6 months. Once thawed, the aliquot tube may be kept at 4°C for 4 weeks. Where used, Ara-C is diluted in culture medium to a final concentration of 1 μM.

2.5. Preparation of Collagen Substratum

1. Dilute the stock collagen solution to 1 mg/mL using 0.001 N HCl as diluent (alternative, tissue culture grade water can be used). Add 1.5 mL of the diluted collagen solution per 35-mm tissue culture plate and leave overnight in the CO_2 incubator at 37°C. Then, remove the dishes from the incubator, aspirate to collagen solution, and leave the dishes under the biological

safety cabinet with the tops removed to allow to dry. These may be utilized for the current experiment or stored at 4°C for 1 month.

3. Methods

3.1. Setting Up the Compartmented Chambers (To Be Done 1–2 Days Before Preparing Cells)

1. Fill a 50-mL syringe with Corning vacuum grease. Use the syringe to fill the grease loader, wrap it in foil, and autoclave for 45 min.
2. Make a scratch in the middle of a collagen-coated 35-mm ∅ dish with an outward motion. Scratching the substratum removes the collagen and exposes the bare plastic beneath the surface of the dish floor, which is a poor substratum for growth cone attachment. This results in axon growth being confined to the collagen-coated tracks formed between the scratches. Axons are then directed to grow to the left and right where they encounter the silicone grease barriers, grow under the silicone grease, and emerge into the distal compartments (see Note 3).
3. Place a drop of N2-methylcellulose solution on the middle of the scratch. This prevents silicone grease from adhering directly to the substratum in this zone when the Teflon divider is seated (see Note 4). Set dish(es) aside until the divider is greased.
4. Attach a 23-gauge Luer Lock connector to the grease loader. Grip the Teflon divider with a pair of 90° angle hemostatic forceps and hold it horizontally with the divider facing up under a binocular microscope. Outline the divider with grease. Each time the adapter is placed at a new starting point, insert the adapter into the grease from the previous step to assure a continuous track of grease.
5. Once grease is applied to the entire divider, remove the lid from one of the prepared 35-mm dishes, pick up the dish, and quickly invert along the axis parallel with the scratches so that the N2-methylcellulose droplet does not run.
6. While viewing the divider under the microscope, hold the dish in position over the divider, oriented so that the scratches will cross under the barriers between the proximal and distal compartments and placed so that the dish is resting gently on the silicone grease Press down on the bottom of the dish with a pair of tweezers. Make sure to press on the inside of the divider at the four corners (see Note 5).
7. Using the pair of hemostatic forceps, turn the dish over and release the divider (placing the dish on the work surface while

the divider remains clamped in the forceps can cause the position of the divider to shift) (see Note 6).

8. Place the dish with the divider under the microscope and focus on the bottom of the middle compartment. With the grease loader, make a small barrier on the floor of the dish at the opening of the proximal compartment. This is to avoid that cells placed in the middle compartment leak out (see Note 7).
9. After having set up several cultures, place the appropriate medium (for SCG or DRG cultures) in each of the side compartments and place in an incubator in which the cells will be maintained. Allow the cultures to sit for several hours and then check for leakage. If medium has leaked into the middle compartment, then the culture is unusable (see Note 8).
10. Prior studies have tested the ability of these cultures to prevent diffusion between the axon and the cell body compartment (5, 23, 24). This can easily be tested by adding low concentrations of a dye such as trypan blue to one compartment only, and look for diffusion of the dye. There should be little or no diffusion visible within 24 h (see Note 9).

3.2. Culturing of SCG or DRG Neurons

The culture media formulations used are those described in the relevant chapters of this volume, detailing protocols for preparing rodent SCG or DRG cell cultures. The original formulation developed for rat SCG neurons uses rat serum (25), which is expensive to obtain commercially and expensive and labor-intensive to prepare in the laboratory. This formulation is not used here but, as pointed out by Campenot and colleagues (10), some conditions may produce diminished axon growth and may reduce the ability of axons to cross into distal compartments. As recommended (10), methylcellulose culture media are supplemented with methylcellulose, which thickens the medium and reduces the shearing forces caused by fluid movements that can perturb attachment of neurons to the substratum. It facilitates wetting of the collagen before seating the Teflon divider, and it prevents neurons from settling in the syringe during plating.

1. Day 1: Replace culture medium in the side (distal) compartments with medium containing 100 ng/mL NGF plus + Ara-C. Add 100,000 cells to center (proximal) compartment, using medium containing 10 ng/mL NGF plus Ara-C (see Note 10).
2. Day 2: Add medium containing 10 ng/mL NGF plus Ara-C to the outside of the Teflon divider until the medium flows over the grease barrier and exchanges fluid with the center compartment (see Note 11).
3. Day 3: Replace medium in the side compartments with medium containing 100 ng/mL NGF but omitting Ara-C; replace medium in the space outside of the Teflon divider with medium containing 10 ng/mL NGF but omitting the Ara-C.

4. Day 6: Replace medium in the side compartments with medium containing 1 ng/mL NGF plus Ara-C and the surrounding space (outside Teflon divider) with NGF-free medium + Ara-C (see Note 12).
5. Day 7 onward: Use for experimentation.

3.3. Analysis of Neurons

With experience, construction of the culture dishes should require about 3 h for 48 cultures, while preparing the neurons requires 3–4 h. Compartmented cultures from both SCG and DRG provide enough cellular material for biochemical analyses such as immunoblotting, and cell bodies/proximal axons can be harvested separately from distal axons. Morphological analyses have their limitations, in that the presence of the Teflon divider produces menisci in the surface of the culture medium that interfere with the light path in inverted, phase-contrast microscopy. This can reduce the contrast, making visualization of the neurons difficult, especially in the center compartment and close to the barriers in the distal compartments. The menisci distort the ring of light produced by the phase condenser so that it is not circular and therefore cannot be aligned on the target annulus in the phase telescope. Campenot and colleagues have solved this problem by constructing a rotating shield that blocks ~80% of the ring of light, allowing only a segment of the ring of light to pass through (10). Also, they have modified the mechanical support of the light source/condenser to increase the range of movement for adjustment in the XY plane. These modifications make it possible to rotate the shield and move the light source/condenser such that the segment of the light ring that is not blocked by the shield projects nearly completely onto the target annulus, which results in a reasonable phase-contrast image. As the distortion arising from the meniscus varies with location in the culture dish, adjustments must be made whenever the field of view is moved. Also, the adjustments produce better phase-contrast images at higher magnification, as the curvature of the segment of the meniscus that must be compensated is smaller, and therefore more uniform, when the field of view is smaller. The need for this modification varies for different microscopes. Tyler Research Corporation has provided Campenot and colleagues with a modification for their Nikon Diaphot inverted microscope. It is suggested that investigators intending to perform such analyses liaise with Tyler for such modifications.

Recent studies have used microfluidic chambers instead of these compartmented chambers. Microfluidic barriers are created, which consist of narrow channels through which axons grow, such that the extracellular space is kept to a minimum (26, 27). Movement of substances across the barrier is controlled by establishing a flow through the extracellular space in the channels driven by a pressure gradient produced by maintaining different fluid levels in the compartments on each side of the barrier. As the culture substratum is on glass, this permits the use of inverted confocal microscopy (unlike traditional compartmented chambers), and shorter axons are compatible with this approach. A disadvantage is that the amount of cellular material is too small for conventional analysis of proteins and lipids.

4. Notes

1. All dividers are machined from virgin Teflon and will last for years if properly maintained. Many other types of dividers are available or can be produced relatively quickly by the company in response to an order. You can also provide a drawing that they will cut to your specifications (biomedical@tylerresearch.com).
2. Heat inactivation of fetal calf serum (and horse serum) is recommended to destroy heat-labile complement. Thaw the bottle of serum in advance, using a 37°C water bath. Next, immerse the serum bottle in the water bath after re-equilibrating to 56°C and leave for 30 min. Swirl the bottle occasionally to ensure proper mixing. Allow the serum to cool to room temperature, aliquot into 50-mL tubes, and store at −20°C.
3. The width of the collagen tracks is critical for axons to cross into the distal compartments. The optimal track width for rat sympathetic neuron cultures is about 200 μm (10). Extremely narrow tracks are not conducive to axonal growth, and if tracks are too wide, the growth cones have enough room to make U-turns away from the barrier.
4. Do not let the medium used for wetting the collagen run outside the area of the scratches before the divider is seated, as this can interfere with sealing of the perimeter of the divider to the dish. It is suggested to leave one or two tracks at each edge dry, to reduce the risk of the droplet escaping beyond the scratches (10). As the droplet does not extend the full distance along the tracks, evaporation can leave a deposit on the substratum coinciding with the edge of the droplet across the tracks, and which may interfere with axon growth. Construct each culture dish and add medium covering the full length of the tracks as soon as possible after placing the droplet.
5. It is important to press firmly enough so that the grease makes a complete seal with the dish but if too much pressure is added, the axons will not cross into the side compartments.
6. The dividers can be re-used after each experiment but must first be cleaned. Remove the divider from the plate, wipe off all

of the remaining grease, and place in NOCHROMIX® sulfuric acid solution. NOCHROMIX® is a crystalline, inorganic oxidizer. Mix one pouch (3 oz.) with 3.7 L of concentrated sulfuric acid. *Exercise caution and wear personal protective equipment.* Soak dividers for at least 1 h (new dividers should also be prepared starting at this step). After removing from the acid, rinse five times with double-distilled water, boil for 20 min, and repeat the boil-rinse steps for a total of three cycles (NOCHROMIX-sulfuric acid can be reused until it discolors). Allow the dividers to dry under a laminar flow hood, place in glass 100-mm petri dishes, wrap in aluminum foil, and autoclave. Store at room temperature until ready to use.

7. To prevent the original N2-methylcellulose solution droplet from drying out, place two drops of distal compartment medium on the scratched region of the substratum in each distal compartment, making sure that the scratched region is completely covered.
8. When first learning this technique, it is recommended to set up extra cultures to replace any that leak (most likely to occur during the initial trials period).
9. Because cultures are ordinarily maintained with higher levels of medium in the left and right compartments than in the dish perimeter, dishes in which there is a leak between compartments display a decrease in the level of culture medium in the distal compartment and an increase in culture medium in the dish perimeter. As the dishes are usually constructed a day in advance, leaking dishes are normally identified before plating the neurons by the appearance of culture medium in the dish perimeter. One can also check for leaks by omitting phenol red from the distal compartment medium when assembling cultures and checking for the appearance of phenol red in distal compartments.
10. Ara-C is provided in proximal compartments containing the cell bodies for the initial 5–7 days after plating to inhibit growth of Schwann cells and other glia. It is not included in the medium supplied to distal compartments, as these compartments are not exposed to cell suspension at plating.
11. When changing the medium, it is important to aspirate the liquid from the top of each side compartment. Also, never change the medium from the middle compartment itself, only from the surrounding space (outside the Teflon divider) and let it flow over the grease barrier into the center.
12. After 5–7 days, the distal axons of the neurons are sufficiently established in the distal compartments so that the NGF can be withdrawn from the proximal compartments containing the cell bodies and proximal axons, and neuronal survival is supported solely by the NGF supplied to the distal axons.

References

1. Campenot RB (2009) NGF uptake and retrograde signaling mechanisms in sympathetic neurons in compartmented cultures. Results Probl Cell Differ 48, 141–158
2. Levi-Montalcini R (1987) The nerve growth factor 35 years later. Science 237, 1154–1162.
3. Oppenheim, R.W. (1991) Cell death during development of the nervous system. Annu Rev Neurosci 14, 453–501
4. Campenot RB (1977) Local control of neurite development by nerve growth factor. Proc Natl Acad Sci USA 74, 4516–4519
5. Ure DR, Campenot RB (1997) Retrograde transport and steady-state distribution of 125I-nerve growth factor in rat sympathetic neurons in compartmented cultures. J Neurosci 17, 1282–1290
6. Campenot RB (2009) NGF uptake and retrograde signaling mechanisms in sympathetic neurons in compartmented cultures. in Results and Problems in Cell Differentiation, Vol. 48 Cell Biology of the Axon (ed. Koenig, E.) pp 141–158 (Springer, Berlin/Heidelberg)
7. Zweifel LS, Kuruvilla R, Ginty DD (2005) Functions and mechanisms of retrograde neurotrophin signalling. Nat Rev Neurosci 6, 615–625
8. Ginty DD, Segal RA (2002) Retrograde neurotrophin signaling: Trk-ing along the axon. Curr Opin Neurobiol 12, 268–274
9. Campenot RB, MacInnis BL (2004) Retrograde transport of neurotrophins: fact and function. J Neurobiol 58, 217–229
10. Campenot RB, Lund K, Mok SA (2009) Production of compartmented cultures of rat sympathetic neurons. Nat Protoc 4, 1869–1887
11. Posse De Chaves EI, Vance DE, Campenot RB, Kiss RS, Vance JE (2000) Uptake of lipoproteins for axonal growth of sympathetic neurons. J Biol Chem 275, 19883–19890
12. Bertrand J, Winton MJ, Rodriguez-Hernandez N, Campenot RB, McKerracher L (2005) Application of Rho antagonist to neuronal cell bodies promotes neurite growth in compartmented cultures and regeneration of retinal ganglion cell axons in the optic nerve of adult rats. J Neurosci 25, 1113–1121
13. MacInnis BL, Campenot RB (2002) Retrograde support of neuronal survival without retrograde transport of nerve growth factor. Science 295, 1536–1539
14. Eng H, Lund K, Campenot RB (1999) Synthesis of beta-tubulin, actin, and other proteins in axons of sympathetic neurons in compartmented cultures. J Neurosci 19, 1–9
15. Heerssen HM, Segal RA (2002) Location, location, location: a spatial view of neurotrophin signal transduction. Trends Neurosci 25, 160–165
16. Watson FL, Heerssen HM, Bhattacharyya A, Klesse L, Lin MZ, Segal RA (2001) Neurotrophins use the Erk5 pathway to mediate a retrograde survival response. Nat Neurosci 10, 981–988
17. Ng BK, Chen L, Mandemakers W, Cosgaya JM, Chan JR (2007) Anterograde transport and secretion of brain-derived neurotrophic factor along sensory axons promote Schwann cell myelination. J Neurosci 27, 7597–7603
18. Underhill SM, Goldberg MP (2007) Hypoxic injury of isolated axons is independent of ionotropic glutamate receptors. Neurobiol Dis 25, 284–290
19. Hayashi H, Campenot RB, Vance DE, Vance JE (2007) Apolipoprotein E-containing lipoproteins protect neurons from apoptosis via a signaling pathway involving low-density lipoprotein receptor-related protein-1. J Neurosci 27, 1933–1941
20. Sonderegger P, Fishman MC, Bokoum M, Bauer HC, Neale EA, Nelson PG (1984) A few axonal proteins distinguish ventral spinal cord neurons from dorsal root ganglion neurons. J Cell Biol 98, 364–368
21. Eide FF, Lowenstein DH, Reichardt LF (1993) Neurotrophins and their receptors–current concepts and implications for neurologic disease. Exp Neurol 121, 200–214
22. Ginty DD, Segal RA (2002) Retrograde neurotrophin signaling: Trk-ing along the axon. Curr Opin Neurobiol 12, 268–274
23. Campenot RB (1979) Independent control of the local environment of somas and neurites. Methods Enzymol *58*, 302–307
24. Heerssen HM, Pazyra MF, Segal RA (2004) Dynein motors transport activated Trks to promote survival of target-dependent neurons. Nat Neurosci *7*, 596–604
25. Mains RE, Patterson PH (1973) Primary cultures of dissociated sympathetic neurons. I. Establishment of long-term growth in culture and studies of differentiated properties. J Cell Biol 59, 329–345
26. Taylor AM, Blurton-Jones M, Rhee SW, Cribbs DH, Cotman CW, Jeon NL (2005) A microfluidic culture platform for CNS axonal injury, regeneration and transport. Nat Methods 2, 599–605
27. Park JW, Vahidi B, Taylor AM, Rhee SW, Jeon NL (2006) Microfluidic culture platform for neuroscience research. Nat Protoc 4, 2128–2136

Chapter 20

Preparation and Culture of Adrenal Chromaffin Cells

Natalia Domínguez, Miriam Rodríguez, J. David Machado, and Ricardo Borges

Abstract

Cultured chromaffin cells have been used for almost 40 years in the study of different cell functions using biochemical, electrophysiological, pharmacological, and toxicological approaches. Chromaffin cells are essentially secretory cells that are used to model sympathetic neurons or neuroendocrine cells. In this chapter, we describe the most common methods currently used to isolate and culture chromaffin cells from the animals used most commonly: cows, rats, and mice. We also provide some advice on the use of these cells in the laboratory.

Key words: Adrenal, Chromaffin cells, Culture, Collagenase, Papain

1. Introduction

Cell culture techniques are now routinely applied to modern laboratory research, chromaffin cells being no exception. Since the first seminal description of their isolation from mammalian glands (1, 2), the procedure has continually evolved. The improvements in tissue culture procedures and advances in the materials used, notably the highly purified enzymes currently available, has standardized isolation protocols and improved their reproducibility. In this chapter, we provide the "recipes" for culturing cow, rat, and mouse chromaffin cells, describing the standard protocols that we use.

Stephen D. Skaper (ed.), *Neurotrophic Factors: Methods and Protocols*, Methods in Molecular Biology, vol. 846, DOI 10.1007/978-1-61779-536-7_20, © Springer Science+Business Media, LLC 2012

2. Materials

2.1. General Supplies, Solutions, and Culture Medium

All procedures require standard culture room facilities with a laminar flow cabin, a Bunsen burner, a water bath set at 37°C (preferably with agitation), and a 37°C incubator with a water-saturated atmosphere containing 5% CO_2. An inverted microscope with phase contrast and a low vibration clinical centrifuge are also required. To plate chromaffin cells onto glass coverslips, we strongly recommend high-quality glass Marienfeld (Lauda-Königshofen, Germany). Other required materials and basic solutions include:

1. Sterile plastic Pasteur pipettes.
2. Syringes and 0.22-μm syringe filters.
3. Adjustable volume pipettes with sterile tips.
4. 15-mL conical centrifuge tubes (or 50-mL tubes for culturing cow cells).
5. Hemocytometer (Neubauer).
6. 24-well culture plates.
7. Ca^{2+}- and Mg^{2+}-free Locke's balanced salt solution: 154 mM NaCl, 5 mM KCl, 3.6 mM $NaHCO_3$, 5 mM HEPES, and 11 mM glucose.
8. Trypan blue: 4 mg/mL in Locke's solution.
9. Culture medium: We always use standard Dulbecco's modified Eagle's medium mixed 1:1 with Ham's F12 and antibiotics to culture chromaffin cells. The medium is completed by adding 10% fetal calf serum (Lonza, Basel, Switzerland, DE14-801 F, see Note 1).
10. All the solutions used are sterilized by filtering through 0.22-μm syringe filters and are supplemented to contain penicillin G (100 IU/mL) and gentamicin sulfate (40 mg/L).

2.2. Materials for Preparation of Bovine Chromaffin Cells

1. Sterile surgical material required: forceps, scissors, two scalpel blades, and one haemostatic clamp.
2. 2× 100 mm Ø Petri dishes.
3. Glass beakers: 2×≈30 mL, 1×≈100 mL, and 1×≈250 mL (for waste).
4. Two ≈35-mm Ø glass funnels.
5. Standard cotton gauzes, ≈200-μm and ≈90-μm pore-size nylon meshes.
6. 5-mL syringes and 0.22-μm sterilizing syringe filters.
7. For gradient purification, a refrigerated centrifuge capable of reaching ≈8,000 × *g* is also required, along with two sterile capped 30-mL transparent tubes suitable for centrifugation.

8. Collagenase IA.
9. Bovine serum albumin fraction V (BSA).
10. Deoxyribonuclease I (DNAase I).
11. Renografin® (ER Squibb & Sons, New Brunswick, NJ, USA) or Urografin® (Schering España, Madrid, Spain) prepared at 15% in sterile pure water (this solution is isotonic).

2.3. Materials for Preparation of Adult Rat Chromaffin Cells

1. Rats over 4 weeks of age.
2. Stereo microscope (×20 magnification).
3. Petri dishes (35 mm Ø).
4. Collagenase IA.
5. BSA.
6. DNAase I.
7. Hyaluronidase I-S.

2.4. Materials for Preparation of Young Rat or Mouse Chromaffin Cells

1. Animals of less than 4 weeks of age.
2. Stereo microscope (×20 magnification).
3. Petri dishes (35 mm Ø).
4. Papain (Worthington Lakewood, NJ).

3. Methods

General Guidelines. Cells are very sensitive to low pressure and as such, strong pipette suction and strong centrifugation should be avoided. Similarly, do not use the brake function on the centrifuge. Standard guidelines for working in culture rooms should be followed, such as wearing gloves and keeping a well-ordered hood to maintain laminar flow conditions. Unnecessary material in the working space will increase air turbulence inside the hood. As we primarily use chromaffin cells for secretion experiments, we have tailored the procedure to produce healthy secretory cells. However, it is possible that the reader may find alternative protocols designed for other purposes, although the maintenance of a healthy secretory machinery usually ensures that the cell is healthy and suitable for any experimental purposes.

3.1. Chromaffin Cell Preparation for Single-Cell Experiments

Any of the methods described below for chromaffin cell isolation and culture are appropriate for amperometry, patch clamping, or fluorescent microscopy. The following is a rule of thumb for amperometric studies using chromaffin cells of any species:

1. The isolation procedure should aim primarily to obtain healthy cells rather than a high yield. We recommend reducing the

concentration of the enzymes used for digestion in the papers cited below by 15% and minimizing the mechanical disruption during tissue digestion. For bovine chromaffin cells, we use a modified version of the original protocol using a bolus of collagenase injected through the adrenal vein, rather than perfusion (3, 4). For rat cells, we adhere to the conditions described by Gilabert (5), and for mouse cells, we use a modified version of the procedure described by Sorensen et al. (6).

2. Cells should be plated at low density (≈20,000–50,000 per cm^2).
3. Replace culture medium every 48 h using serum-free medium.
4. Although cells can be maintained for over a week after plating, optimal secretion usually occurs within 24 h of plating. However, cells may be used within a few hours.

3.2. Coverslips and Adherent Support Sterilization

Glass coverslips can be sterilized by briefly flaming both sides using a Bunsen burner before placing them in culture plates. If it is necessary to coat them with a substrate (see Note 2), place a drop of 0.01% poly-D-lysine solution (Sigma, Catalog P-1024, prepared in water) on each coverslip. Leave it for 20 min and then wash three times with sterile pure water. Sterilize the coverslip by exposing it to UV light in the fume hood for 30 min and use the treated glass within 1 week of preparation.

3.3. Cell Viability and Counting

Although many automatic systems for cell counting now exist, a standard Neubauer hemocytometer is still a valuable and cheap tool to count cells and to get a general idea of their viability. We prepare a 1:9 dilution of cells by mixing 20 μL of cell suspension + 100 μL of Locke's solution + 80 μL of trypan blue staining solution (see above). Guidelines for the correct use of hemocytometers have been published elsewhere (7).

3.4. Procedure for Bovine Chromaffin Cells

Glands may be obtained from the local abattoir. Choose only intact glands to avoid contamination and discard those with visible signs of internal blood coagulation. Do not remove the fat at the abattoir, as it provides protection against contamination. The glands do not require any special care for their transportation to the laboratory if the time from sacrifice to culturing is less than 1 h. In the case of longer intervals, it is advisable to inject 3–4 mL of Locke's solution into the glands and transport them in a plastic bag on ice. Other protocols for bovine cell culture preparation have been published elsewhere (4, 8–14).

Once in the culture room, wearing gloves spray the glands with 60% ethanol under the laminar flow cabinet, making sure that the solution does not enter the adrenal vein, and remove the surrounding fat from the glands. The general procedure to minimize contamination is to proceed progressively from a nonsterile (transport containers, external fat) to a sterile environment. As such, it is advisable to flame the surgical instruments periodically. Once the glands are clean of surrounding fat and connective tissue, spray again with ethanol and remove all the material used and discard the tissue from the hood. Clean the surface of the hood with ethanol. The procedure described below is intended for two glands but can be easily adapted for more glands. In addition, Fig. 1 can be printed and posted in the culture room as a brief guide to the procedure:

1. Prepare the enzymatic solution. Each gland will require 20 mL of a solution containing 1.5–2 mg/mL of collagenase IA and twice this amount (3–4 mg/mL) of BSA dissolved in Locke's solution. These calculations are aimed for a commercial a collagenase with an activity of 300–400 IU/mg, and should be adjusted when the enzyme has a different activity (see Note 3). The addition of 30 μg/mL of DNAase I increases the yield of the isolation procedure as it prevents cell aggregation that results from the presence of free DNA. However, given the usually high yield of bovine cultures, this step is not critical. Sterilize the solution by filtering through a 0.22-μm syringe filter into a 50-mL conical tube and place it capped in the water bath at 37°C.
2. Place the glands in a sterile Petri dish and inject each gland twice with 3–4 mL of warm (37°C) Locke's solution. Place the glands in a 100-mL glass beaker, keeping the vein orifice facing up. Cover the beaker with a dry sterile Petri dish and incubate for 5–10 min in the water bath.
3. Inject the glands with 3–4 mL of enzyme solution and return the glands to the beaker. Try to avoid accumulation of liquid in the bottom of the beaker which could contaminate the gland (see Fig. 1a).
4. After 20 min, repeat step 3. Exercise care when performing the injections as the gland has been partially digested and may rupture with excessive pressure.
5. After 20 min, the glands should be noticeably softer, indicating the success of the digestion. If this is not the case, repeat the injection and check again after 10 min. It is important to keep the collagenase incubation time to a minimum to avoid over-digestion of the chromaffin tissue.
6. Using a scalpel and clean scissors, open the gland longitudinally (along its length). It should appear as shown in Fig. 1b, with the cream-colored section representing the medulla. Make several cuts along the medulla with small scissors, and use forceps to collect the material and transfer it to a clean Petri dish. Do not collect any purple material that contains the cortical cells. This procedure should be performed swiftly.
7. Using two scalpel blades, mince the tissue into small pieces and transfer the material to a 50-mL conical tube using a plastic

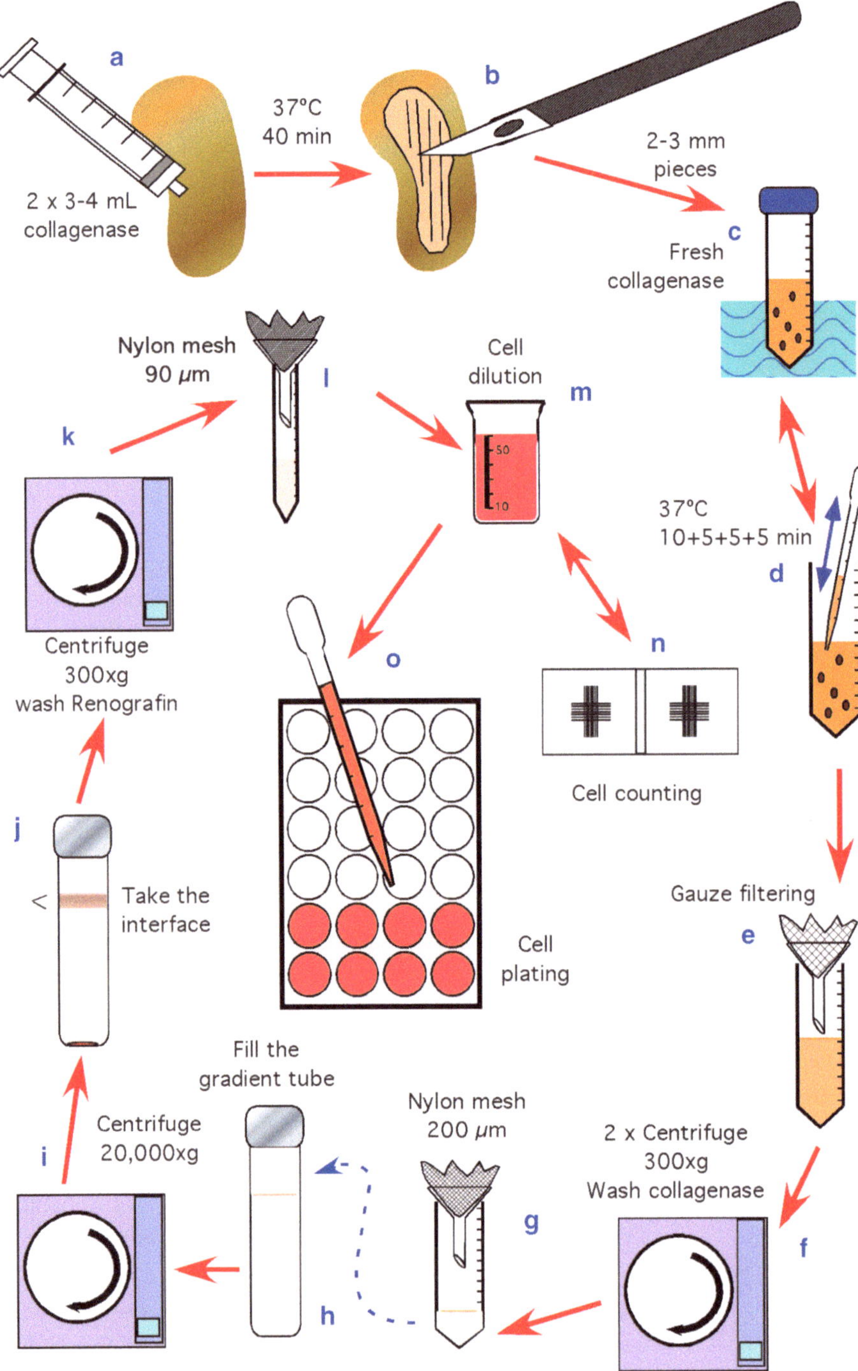

Fig. 1. Procedure to isolate and culture bovine chromaffin cells (see text for guidelines).

Pasteur pipette. Add the rest of the fresh collagenase solution and incubate this tissue at 37°C (see Fig. 1c). Every 5–10 min, carefully dissociate the material by triturating with 5–6 strokes of a Pasteur pipette, always avoiding the production of bubbles (see Fig. 1d). The total incubation time should not exceed 20–25 min.

8. Place a piece of sterile cotton gauze onto a funnel and filter the dissociated material to remove the bulk debris, collecting the cells in two 50-mL conical tubes. Distribute the filtrate between both tubes and wash the gauze with Locke's solution to maximize the yield and to dilute the collagenase (see Fig. 1e).
9. Centrifuge at ≈300 × *g* for 5 min, discard the supernatant, and resuspend the pellet by gently tapping it against the hood bench. Fill the tubes with fresh Locke's solution and repeat the centrifugation.
10. Discard the supernatant, resuspend the cells, and pass them through a 200-μm nylon mesh. Clean the mesh by rinsing with fresh Locke's solution to produce a final volume of 10 mL (see Fig. 1g).
11. The cells obtained are contaminated with debris, ruptured cells, and red and cortical cells. For many purposes, they are fit for culturing and may be diluted into culture medium at this stage. However, we prefer to purify the chromaffin cells, as outlined below (see steps 12–18).
12. Place 20 mL of 15% Renografin solution in two 30-mL centrifuge tubes of (see Note 4).
13. Mix 10 mL of cell suspension with 10 mL of Renografin to produce a final Renografin concentration of 7.5%.
14. Very slowly and carefully, add the cell suspension onto the surface of the 15% Renografin solution using a plastic Pasteur pipette. This is the most critical step! If the interface between the two Renografin concentrations is broken, the discontinuous gradient is compromised, and most of the chromaffin cells will enter the pellet (see Fig. 1h).
15. After equilibrating the tubes by weight, centrifuge at 7,500 × *g* at ≈18°C for 20 min (see Fig. 1i).
16. The chromaffin cells will appear in the interface of the gradient and can be easily recovered with a plastic Pasteur pipette (see Fig. 1j).
17. To remove the Renografin, resuspend the cells in Locke's solution in a 50-mL conical tube and centrifuge at ≈300 × *g* for 5 min (see Fig. 1k).
18. Resuspend the pellet by tapping the tube against the work surface of the hood and add 10 mL of culture medium (see Note 1). As the medium contains calcium, it is important to

avoid unnecessary manipulation of the cell suspension to prevent aggregation.

19. Pass the cell suspension through a 90-μm nylon mesh (see Fig. 1).
20. Count and plate the cells accordingly (see Fig. 1m, n). Although the described yield in the past was as large as 80–100 × 10^6 cells per gland, nowadays it is preferable to reduce this number to obtain healthier cells. Thus, 10–20 × 10^6 cells per gland is a suitable yield. As a rule of thumb, we recommend the following densities:
    - For biochemical and secretory studies in 24-well plates: 500,000 cells/well in 1–2 mL of complete medium.
    - For single-cell experiments on 12 mm Ø glass coverslips: 50,000 cells/well in 1 mL of complete medium. Bovine chromaffin cells do not usually require an adherent substrate (collagen, polylysine, etc.).
    - For collecting cells by scraping from 35 mm Ø Petri dishes: 3–4 × 10^6 cells/well in 3 mL of complete medium.
21. Change the medium using serum-free medium every 2–3 days, depending on the cell density (see Note 1).

3.5. Procedure for Adult Rat Chromaffin Cells

Animals older than ≈4 weeks produce a lower yield of cells than young rats. This is also observed with mice:

1. Prepare the dissociation solution (3–4 mL) and maintain it at room temperature: collagenase type I (250–350 IU/mL), 3 mg/mL BSA, 0.15 mg/mL DNAase I, and 0.15 mg/mL hyaluronidase I-S in Locke's buffer.
2. Sacrifice the animals according to institutionally approved ethical procedures. Place the animal on its back and spray the abdomen with 70% ethanol.
3. To access the abdominal cavity, perform an incision through the abdominal wall, cutting the skin along the linea alba and pulling it out to either side, using scissors to separate the skin from underlying tissue. The abdominal cavity should tear open, exposing the organs. Locate the two kidneys with the adrenal glands on top, which should be readily visible (see Fig. 2a).
4. Remove the glands with fine-curved forceps and place them in ice-cold Locke's buffer.
5. Under a stereo dissection microscope coupled to a cold-light source, remove the adipose tissue surrounding the gland with a stainless steel scalpel blade and decapsulate the adrenal glands. Next, remove the adrenal cortex to isolate the medullar tissue and cut this into four pieces using a scalpel. Keep the tissue wet throughout with cold Locke's buffer.

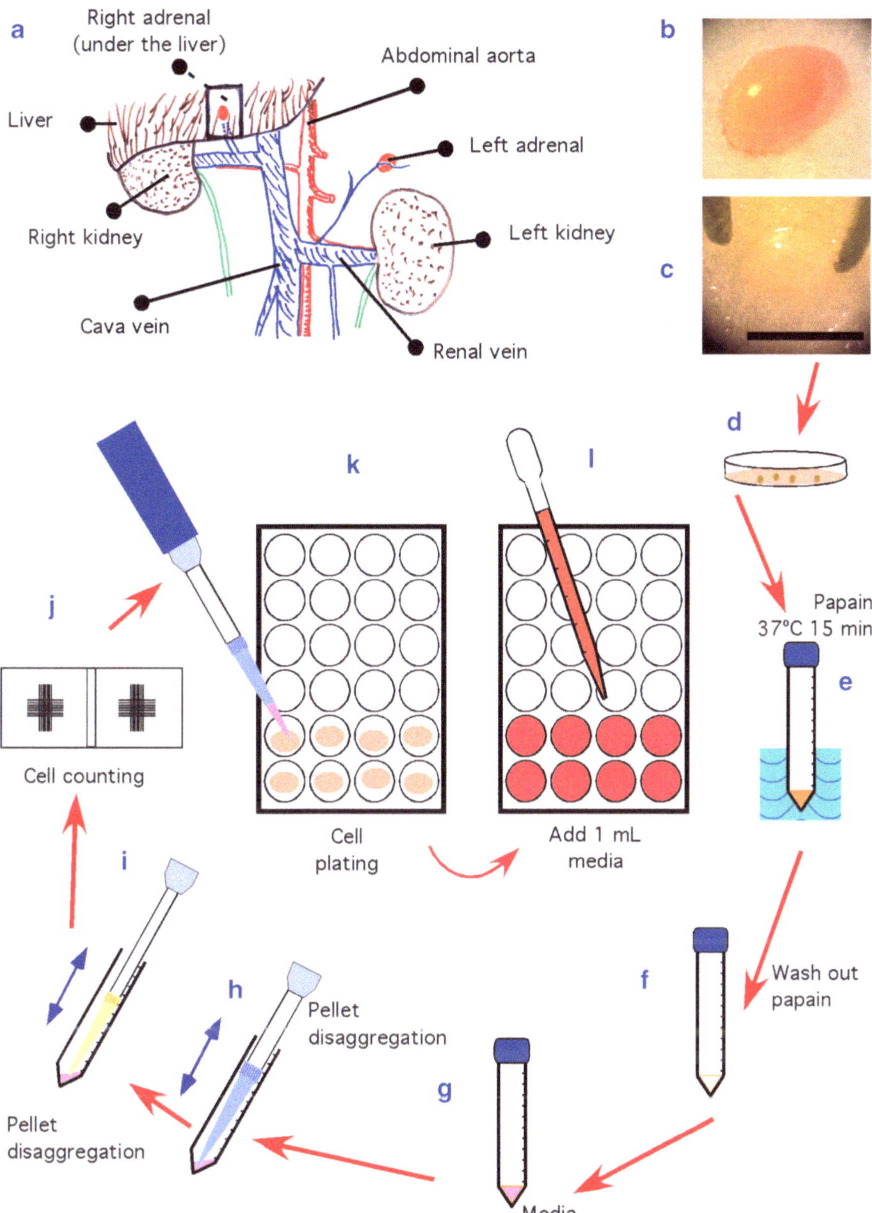

Fig. 2. Procedure to isolate and culture mouse chromaffin cells (see text for guidelines).

6. Recover the adrenal tissue using a sterile plastic pipette and transfer it to a 15-mL conical tube. Add 3 mL of the dissociation solution and incubate for 25–30 min at 37°C. The solution should be gently triturated every 10 min through a 1 mL pipette tip. During the last 5 min, the solution should be continuously triturated using a 100-μL pipette, until it becomes turbid.
7. Add cold Locke's buffer to a final volume of 10 mL to stop the enzyme reaction.

8. Place the tube containing the digested tissue in the rotor of a precooled (4°C) refrigerated centrifuge (where possible) and spin at 100 × *g* for 10 min.
9. Discard the supernatant by decanting and resuspend the pellet with 0.8 mL of prewarmed medium.
10. Add one or two drops on each poly-D-lysine-coated coverslip and place them in the incubator for 1 h to allow the cells to settle.
11. Carefully refill the culture dish with 1–1.5 mL of medium and return the plates to the incubator.
12. Replace the medium after 24 h and every 48 h thereafter (see Note 1).

3.6. Procedure for Young Mouse or Rat Chromaffin Cells

We use animals of 2–4 weeks of age. Given the small amount of collagen in these adrenal tissues, papain is used instead of collagenase, producing a considerably higher cell yield. As the quantity of starting material from the adrenal glands is very small, this procedure is intended for single-cell measurements. We usually use 12-mm Ø glass coverslips placed in 24-well plates, although other sizes can be used. Figure 1 can be printed and posted in the culture room as a brief guide to the procedure:

1. Prepare the papain solution in Locke's buffer (12–18 IU in 200 μL, sufficient for four glands), and store it at 37°C.
2. Sacrifice the animal according to institutionally approved ethical procedures, place it on its back, and spray the abdomen with 70% ethanol. Open the abdomen and remove both adrenal glands by pulling the glands away with an angled forceps (see Fig. 2a).
3. Under a stereo microscope (×20 magnification), carefully remove the adrenal capsule as much as possible of the cortical tissue (see Fig. 2b, c). Work on a sheet of clean filter paper soaked with Locke's solution. This helps to avoid drying and overheating of the tissue by the light. Transfer the medullary tissue to a sterile Petri dish (35 mm Ø) containing 1 mL of sterile ice-cold Locke's solution (see Fig. 2d).
4. Transfer the pieces to a 15-mL conical tube with 200 μL of papain solution and leave them for 15–20 min at 37°C without shaking. The incubation time should be adjusted according to the age of the animals, as younger animals will require shorter digestion.
5. Wash the tissue with 800 μL of fresh Locke's solution.
6. Remove as much of the liquid as possible, taking particular care as the tissue is almost digested, and then add 200 μL of complete medium (see Note 1).
7. Disaggregate the tissue by triturating it several times, first through a 1 mL pipette tip and then through a 100-μL pipette

tip, until the suspension becomes turbid. The degree of cell dissociation can be assessed by taking 10 μL of the suspension and observing under the inverted microscope. It is not recommended to completely dissociate the tissue, as excessive manipulation may damage the cells.

8. Take 20 μL of this cell suspension to count the number of cells in a Neubauer hemocytometer (with experience, cell counting usually becomes unnecessary).
9. Plate the cells by placing a drop of the cell suspension on 12 mm-Ø poly-D-lysine-coated coverslips and allow the cells to settle for 20 min in the incubator before adding 1 mL of medium.
10. Return the plates to the incubator.
11. The cells should be used within the first 3 days after culturing. Where necessary, change the medium (serum-free) every 48 h.

4. Notes

1. The use of a 1:1 mix of Dulbecco's modified Eagle's medium with Ham's F12 medium helps cells to maintain their round shape. This simple trick was suggested by Dr. O. Humberto Viveros and produces excellent results. We prepare 1 L of this mix from powdered medium and after adding antibiotics and filtering, we divide the contents between two bottles. Ten percent (v/v) fetal calf serum is added to one bottle (complete medium), while the second is kept serum-free (incomplete medium). The serum-free medium lasts longer than the complete, and it is used to change the medium after 48 h.
2. Other adherent substrates have been used, including collagen, poly-DL-ornithine, laminin, or fibronectin (3). We use round glass coverslips of 12 mm Ø that fit readily into 24-well culture plates. Cell adhesion improves notably when the glass is pretreated with these substrates, although we do not use any substrate for bovine cells unless they are to be used for immunocytochemistry or electron microscopy.
3. A few decades ago, the quality of commercial enzymes and sera varied considerably from one batch to another. During the 1980s, some particularly active batches of collagenase produced over 100×10^6 cells/gland, possibly due to contamination with other proteases. Although some variation occurs within different batches of these products, the controls applied by manufacturers nowadays have greatly reduced this problem.
4. Several strategies have been described for purifying chromaffin cells from other contaminants and debris. One is differential plating, which is based on the low adherence of chromaffin

cells when compared to fibroblasts or endothelial cells. Although this method gained some popularity, it is currently only performed by few, if any, researchers. Discontinuous gradients of Ficoll® or Percoll® are used by some groups for cell purification, and we have used this system in the past. However, we find that Renografin produces a purer population of chromaffin cells that are enriched in adrenaline.

Acknowledgments

We are indebted to the contributions of our predecessors working in the field of cell culture. This work was supported by the Spanish Ministry of Science and Innovation (MICINN, Grants # BFU BFU2007-64963, CSD2008-00005 and BFU2010-15822) and the Canary Islands Government (GC, Grants # C2008/01000239). N.D. is a recipient of an FPU scholarship from MICINN, M.R. received a grant from CajaCanarias, and J.D.M. is the holder of a "Ramón y Cajal" contract from the MICINN.

References

1. Kloppenborg PW, Island DP, Liddle GW, Michelakis AM, and Nicholson WE (1968) A method of preparing adrenal cell suspensions and its applicability to the in vitro study of adrenal metabolism. *Endocrinology* **82**, 1053–1058
2. Hochman J, and Perlman RL (1976) Catecholamine secretion by isolated adrenal cells. *Biochim Biophys Acta* **421**, 168–175
3. Livett BG (1984) Adrenal medullary chromaffin cells in vitro. *Physiol Rev* **64**, 1103–1161
4. Moro MA, Lopez MG, Gandia L, Michelena P, and Garcia AG (1990) Separation and culture of living adrenaline- and noradrenaline-containing cells from bovine adrenal medullae. *Anal Biochem* **185**, 243–248
5. Gilabert JA (2004) Necessary conditions to maintain rat adrenal chromaffin cells in primary culture. In *Cell Biology of the Chromaffin cell.* Borges, R & Gandia, L. Eds. La Laguna. pp 269–274. webpages.ull.es/users/isccb12/TheBook.htm
6. Sorensen JB, Nagy G, Varoqueaux F, Nehring RB, Brose N, Wilson MC, and Neher E (2003) Differential control of the releasable vesicle pools by SNAP-25 splice variants and SNAP-23. *Cell* **114**, 75–86
7. Strober W (2001) Trypan blue exclusion test of cell viability. *Curr Protoc Immunol* **Appendix 3**, Appendix 3B
8. O'Connor DT, Mahata SK, Mahata M, Jiang Q, Hook VY, and Taupenot L (2007) Primary culture of bovine chromaffin cells. *Nat Protoc* **2**, 1248–1253
9. Krause W, Michael N, Lubke C, Livett BG, and Oehme P (1996) Catecholamine release from fractionated chromaffin cells. *Eur J Pharmacol* **302**, 223–228
10. Livett BG, Boksa P, Dean DM, Mizobe F, and Lindenbaum MH (1983) Use of isolated chromaffin cells to study basic release mechanisms. *J Auton Nerv Syst* 7, 59–86
11. Fenwick EM, Fajdiga PB, Howe NB, and Livett BG (1978) Functional and morphological characterization of isolated bovine adrenal medullary cells. *J Cell Biol* **76**, 12–30
12. Almazan G, Aunis D, Garcia AG, Montiel C, Nicolas GP, and Sanchez-Garcia P (1984) Effects of collagenase on the release of [3 H]-noradrenaline from bovine cultured adrenal chromaffin cells. *Br J Pharmacol* **81**, 599–610
13. Bader MF, Ciesielski-Treska J, Thierse D, Hesketh JE, and Aunis D (1981) Immunocytochemical study of microtubules in chromaffin cells in culture and evidence that tubulin is not an integral protein of the chromaffin granule membrane. *J Neurochem* **37**, 917–933
14. Baker PF, and Knight DE (1981) Calcium control of exocytosis and endocytosis in bovine adrenal medullary cells. *Philos Trans R Soc Lond B Biol Sci* **296**, 83–103

Chapter 21

Indirect Immunofluorescence Staining of Cultured Neural Cells

Massimo Barbierato, Carla Argentini, and Stephen D. Skaper

Abstract

Immunofluorescence is a technique allowing the visualization of a specific protein or antigen in cells or tissue sections by binding a specific antibody chemically conjugated with a fluorescent dye such as fluorescein isothiocyanate. There are two major types of immunofluorescence staining methods: (1) direct immunofluorescence staining in which the primary antibody is labeled with fluorescence dye and (2) indirect immunofluorescence staining in which a secondary antibody labeled with fluorochrome is used to recognize a primary antibody. This chapter describes procedures for the application of indirect immunofluorescence staining to neural cells in culture.

Key words: Immunocytochemistry, Fluorescence, Microscopy, Immunoreactivity, Glia, Cell culture

1. Introduction

Fluorescence can be described as a luminescent light produced by certain molecules which, excited by a particular wavelength of radiation, emit a longer wavelength of light energy before returning to the equilibrium state (1). In biological terms, fluorescence is probably the most powerful contrast method available to visualize cell structures. However, there are some limits: chemical reactions reduce the lifetime of fluorophores, and thereby the visualization process and some samples can exhibit autofluorescence or photobleaching (2–5). *Immunofluorescence* is a specific example of immunocytochemistry/immunohistochemistry in which fluorochromes are linked to antibody. It is a powerful and versatile technique for the optical detection and localization of proteins, glycans, lipids, and small biological and nonbiological molecules in cells and tissues. A critical component of this technique is the fluorescence microscope,

Stephen D. Skaper (ed.), *Neurotrophic Factors: Methods and Protocols*, Methods in Molecular Biology, vol. 846, DOI 10.1007/978-1-61779-536-7_21, © Springer Science+Business Media, LLC 2012

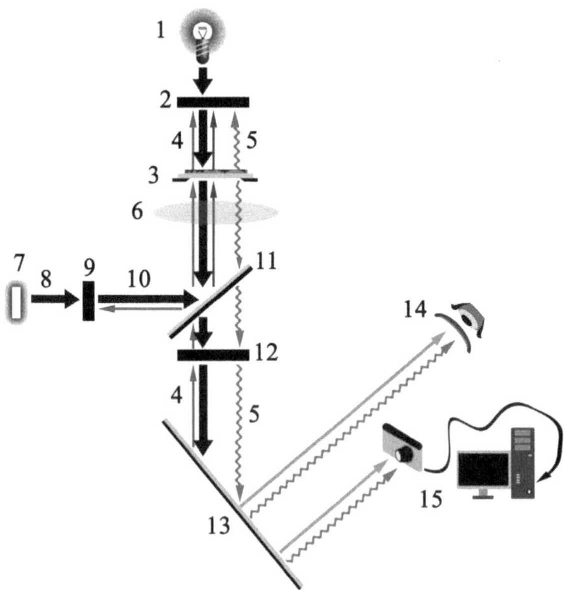

Fig. 1. Schematic diagram illustrating the principles of fluorescence microscopy. Basic components of an epifluorescence inverted microscope. In this type of microscope, a mercury or xenon low-pressure arc lamp generates a beam of full spectrum light. The excitation wavelength light is selected by a proper excitation filter, and once redirected to the specimen by a dichroic mirror, it passes through the objective lens. In this case, the objective needs to illuminate and image the sample. Fluorescent light passes an emission filter in order to select the proper wavelength and is then reflected by a mirror to the eyepiece and/or to a CCD camera (often a peltier-cooled CCD camera) to display on a monitor and storage images. Modern microscopes collect light by a cube, in which there are emission filters, a dichroic mirror, and a barrier filter, to prevent the exciting wavelengths from reaching the eyepiece and the CCD camera. Moreover, fluorescence microscopes often have a visible light lamp to permit immunohistochemical work or studies with living cells. *1*, visible light lamp; *2*, condenser; *3*, sample; *4*, visible light; *5*, fluorescent light; *6*, lens; *7*, UV mercury/xenon arc lamp; *8*, UV light; *9*, excitation filter; *10*, excitation light; *11*, dichroic mirror; *12*, emission filter; *13*, mirror; *14*, eyepiece; *15*, digital camera (or film camera) and storage or display computer system.

in which light is emitted from a UV lamp (usually a low-pressure mercury lamp) with peaks at around 253.7 and 185 nm. The UV light is filtered in order to excite different fluorochromes, while other filters are used to detect the resulting fluorescence (see Fig. 1). Like classical optical microscopy, conventional fluorescence methods have a limit of resolution due to diffraction and lens aberrations (Abbe law) (6, 7). Sharp points well below this limit appear blurry. In microscopy, diffraction is more important than lens aberration and results from the numerical aperture (NA), the type of glass of lens, and the light wavelength. In this case, the Rayleigh criterion states that two point sources are just resolved when the principal diffraction maximum of the Airy disk of one point falls in the first minimum of the other. Diffraction through a condenser-objective system of a microscope is described by

$$R = 0.61\lambda / ((\text{NAcondenser} + \text{NAobjective}) / 2),$$

where R is the angular distance, λ is the light wavelength, and NA is the numerical aperture. The factor 0.61 is derived from calculation of the position of the first dark ring surrounding the central Airy disk of the diffraction pattern, in agreement with the Bessel function. NA is described by

$$\text{NA} = \eta \sin \alpha,$$

where η is the refractive index of the medium between the lens and the sample and α is half of the lens angle. The lowest value of R, for $\lambda = 546$ nm (green light) and NA 1.25 (like the most common immersion lens), is approximately 220 nm at best.

Popular fluorescence microscopes are epifluorescence microscopes, suitable for fluorescent or phosphorescent organic and inorganic samples. In this type of microscope, UV light is diverted by a dicroic mirror and passes through the objective lens in the opposite direction. A filter interposed between the light source and the optic system selects the right wavelength to excite the sample. Before reaching the eyepiece, the light beam passes through a filter that shields it from reflected UV light. The sample image can often be detected by a digital camera for storage or display on a computer monitor in real time. Cameras can be cooled for long exposures. In the confocal microscope, fluorescently labeled samples are scanned point by point by a beam. There are essentially three types of confocal microscopes: confocal laser scanning, spinning-disk (Nipkow disk) confocal, and programmable array. All confocal microscopes scan a sample on various levels and then assemble a three-dimensional image. The resolution is higher than that obtained with an epifluorescence microscope, and all image parameters, including color balance, can be adjusted freely. However, recently, super-resolution nanoscopy techniques have broken Abbe's limit, with resolution approaching ten nanometers. These new methods include [fluorescence] phosphoactivation localization microscopy/stochastic optical reconstruction microscopy, total internal reflection fluorescence microscopy, stimulated emission depletion, structured illumination microscopy, and multimodality microscopes (8, 9).

Immunofluorescence techniques can be either direct (or primary) or indirect (or secondary antibody). In direct immunofluorescence, the antibody, which is chemically linked to a fluorophore, recognizes the target molecule and binds to it, resulting in direct detection. In contrast, indirect immunofluorescence uses two antibodies: the first (primary antibody) is specific for the target molecule, while the second (secondary) recognizes the species in which the primary antibody was raised. Because an antibody consists of a variable region (which recognizes the antigen) and an invariant region, one can generate several primary antibodies that recognize various antigens (with different variable regions), with all sharing the same invariant region. For this reason, only the secondary antibody is tagged to a fluorochrome or to biotin. In the latter case, one needs to add a fluorophore linked to the streptavidin that recognizes biotin. Direct immunofluorescence uses a single antibody for labeling and detection, has a relatively fewer steps, and reduces cross-reactivity or nonspecificity. Although the indirect fluorescence antibody (IFA) protocol is more complex and time-consuming, it offers up to 20-fold greater signal sensitivity. In addition, difficult-to-make dye-coupled antibodies can be reused in multiple experiments. Moreover, the secondary antibodies can bind to multiple sites on the primary antibody and thus produce a brighter signal since more dyes are brought to a single location. On other hand, IFA is more subject to cross-reactivity and aspecificity and requires more controls to avoid cross reactivity between primary and secondary antibodies. This chapter will detail a protocol that is useful in the application IFA to the characterization of neural cells in culture.

2. Materials

2.1. Apparatus and Labware

1. 1-L beakers
2. 1-L Cylinder
3. Micropipettes (20–1,000 μL)
4. Stirrer hotplate
5. Forceps
6. 12-mm-diameter glass microscope slides (Thermo Fischer Scientific)
7. Staining boxes or 24-well plates
8. Analytical balance
9. Microscope, Leica DMI4000 B (Leica Microsystems GmbH) (or comparable model)
10. Digital camera, Leica DFC 480 (Leica Microsystems GmbH) (or comparable model)

2.2. Reagents

1. Paraformaldehyde (PFA) (Sigma-Aldrich)
2. Poly-L-lysine hydrobromide, mol wt 70,000–150,000 (Sigma-Aldrich)
3. Bovine serum albumin (BSA) (Sigma-Aldrich)
4. Gelatin
5. Chrome alum (potassium aluminum sulfate)

6. Goat serum (Sigma-Aldrich)
7. Glial fibrillary acidic protein (GFAP) primary antibody (MAB3402, Millipore)
8. Alexa Fluor® 555 goat anti-mouse secondary antibody (Invitrogen)
9. 4′,6-Diamidino-2-phenylindole (DAPI) (Sigma-Aldrich)
10. Citifluor mounting medium (Citifluor Ltd., UK)
11. Mowiol 4-88 (Merck KGaA, Germany)
12. Fluoromount-G (Southern Biotech, USA)
13. Vectashield (Vector Laboratories)

2.3. Solutions

1. *Tris, 200 mM, pH 8.5*: Dissolve 24.22 g of Tris base in 800 mL of distilled deionized water in a beaker. Stir the mixture, and adjust pH to 8.5 by adding concentrated HCl under a chemical fume hood. Adjust the volume to 1 L by adding distilled deionized water. Sterilize by autoclaving (see Note 1).
2. *Phosphate buffer (PBS), 10×*: 1.369 M NaCl (80 g/L), 26.827 mM KCl (2 g/L), 100 mM Na_2HPO_4 (14.2 g/L, pH > 9.5), and 14.697 mM KH_2PO_4 (2 g/L). Add NaOH (1 N) to adjust pH to 7.2 ± 0.2 at 25°C. Use diluted 10 times (1×).
3. *4% PFA in PBS*: *Warning*—see Note 2. Dissolve 4 g of PFA in 50 mL of distilled deionized water in a beaker, under a chemical fume cupboard. Add 1 mL of 1 N of NaOH. On a stirring hotplate, stir and heat (~65°C) the solution until the PFA is completely dissolved, always under a chemical fume cupboard. Add 10 mL of 10× PBS and allow the solution to cool at room temperature. Check the pH (after cooling) using pH papers (never the pH meter!), and, in case, adjust to pH 7.4 (at 25°C) using 1 M HCl (~1 mL). Adjust the volume to 100 mL with distilled deionized water. Filter the solution through a 0.45-μm membrane filter (it is possible that some particles of the power of PFA do not solubilize). The solution can be stored at −20°C for several months. However, it is better to aliquot the prepared stock solution. After use, do not freeze/thaw again. Once thawed, the solution can be kept for up to 1 week at +4°C. After this time, discard the aliquot.
4. *Mowiol*: Mowiol is a mounting medium. Weigh 6 g of Mowiol 4-88 and mix in a solution of 6 mL of glycerol and 6 mL of distilled deionized water. Shake for 2 h at room temperature. Add 12 mL of 200 mM Tris pH 8.5, and heat at 50°C for about 3 h. Shake occasionally. Filter the solution through a 0.45-μm membrane filter, aliquot, and store at −20°C for 2 months (or at +4°C for 2 weeks).
5. *"Antibody" buffer*: 150 mM NaCl (2.25 g/200 mL), 50 mM Tris base (1.5 g/200 mL), 1% (w/v) BSA (2 g/200 mL),

100 mM L-lysine (3.6 g/200 mL), and 0.04% azide (add 2 mL/200 mL of a 4% stock in water). Adjust pH to 7.4 (use pH paper) adding 1 N HCl (about 6 mL). This solution can be stored at 4°C indefinitely.

2.4. Coating Glass Microscope Slides with Gelatin

1. Heat 400 mL of distilled water to 60°C on stirring hotplate.
2. Dissolve 5 g of gelatin.
3. Add 200 mg of chrome alum and cool.
4. Dip slides individually and allow to air dry by leaning them vertically against a test tube rack.
5. Store in a slide box 4°C until used.

2.5. Coating Glass Microscope Slides with Poly-L-Lysine

1. Dissolve 5 mg lyophilized powder of poly-L-lysine hydrobromide, mol wt 70,000–150,000 in 10 mL of sterilize distilled deionized water.
2. Add the poly-L-lysine solution to coverslips in a staining box or in a 24-well plate to cover the entire surface (1 mL should be sufficient).
3. Place the staining box or the plate at 37°C, in a cell culture CO_2 incubator, for at least 2 h.
4. Discard the poly-L-lysine solution and allow the surfaces to dry.
5. Store the box or the plate at +4°C for few months. The poly-L-lysine solution can be stored at +4°C for several months.

3. Methods

3.1. Coverslip Transfer to Staining Box

1. Using a coarse pair of forceps (keep cell side face up), transfer coverslips from the 24-well plate to a staining box (a homemade box with 24 flat-topped pedestals glued to its bottom to support coverslips and keep them raised above the bottom which will have some water for humidification). The box should also have a top to prevent evaporation and keep slides humidified.
2. In the following steps, all volumes are 50–100 μL unless otherwise specified. However, you can use as little as 25 μL if necessary.

3.2. Fixation and Permeabilization

1. Place coverslips in the 4% PFA solution for 10 min at room temperature, or 30 min at +4°C (except for bromodeoxyuridine staining or other fixative-sensitive antigens, in which case use 60–90 s maximum) (see Note 3). PFA cross-links proteins so that they do not solubilize when the cells are permeabilized (see below). Otherwise, fix in methanol or ethanol for 15 min

at +4°C or in an acetone solution (acetone-methanol 1:1 or acetone-acetic acid 2:1) for at list 10 min (until acetone dries) at room temperature. Alcohol and acetone solutions, in contrast to PFA, work by precipitating proteins and allow permeabilization without detergents (step 2 below).

2. After the above fixation step, pick up the coverslip with forceps and rinse four times in PBS/0.05% Triton X-100. This step can be combined with Subheading 3.3 by adding the Triton (0.1%) to the goat serum solution (see Note 4).

3.3. Block of Nonspecific Binding

1. Incubate the coverslips for 1 h at room temperature or overnight at +4°C in a buffer consisting of serum of the secondary antibody species diluted 1:10 to 1:100 in PBS or in "antibody" buffer. The serum contains high levels of antibodies and proteins which bind noncovalently to nonspecific sites and also covalently to reactive aldehyde groups created by the PFA. Without this step, the primary antibody would bind both to these reactive aldehyde groups and the nonspecific binding sites and create high levels of background staining (see Note 5). A less expensive nonspecific binding site block can be achieved by incubation in a 3% BSA/PBS solution.
2. After the blocking step, drain the coverslip (it is not necessary to rinse in PBS).

3.4. Primary Antibody Incubation

1. Incubate the coverslip in the primary antibody solution for 2 h at room temperature or overnight at 4°C (see Note 6). To prepare this solution, dilute the primary antibody with 2% serum in PBS or in "antibody" buffer. The best dilution depends on the concentration of primary antibody and the density of the antigen. Generally supernatants are used neat or diluted up to 1:10. Ascites are used at 1:100 to 1:1,000, as are polyclonal antisera. (The optimal dilution can be determined in a separate experiment.) (see Note 7).
2. Rinse the coverslip five times in PBS.

3.5. Secondary Antibody Incubation

1. Incubate the coverslip in the secondary antibody solution for 1–2 h. To prepare this solution, dilute the secondary antibody with 2% serum in PBS or "antibody" buffer. The dilution depends on the lot and concentration of the secondary antibody, the density of the antigen, and, to some extent, the concentration and the antigen affinity of the primary antibody that was used.
2. A separate experiment can be used to establish the proper antibody dilution. As a rule of thumb, the secondary antibodies will generally be used at a final concentration of 20 μg/mL. We suggest using a dilution of about 1:100 (1:500–1:1,000) for secondary antibodies. For single-label staining, use fluorescein

(e.g., FITC) or rhodamine (e.g., TRITC) dyes. It is better to use stabilized fluorochromes, but they are more expensive because they are patented, although generally they are brighter and less sensitive to pH than common dyes. For a double-label staining, use a combination of antibody labeled with different fluorochromes. In these cases, check that the emission curves do not overlap or that the overlap is as limited as possible. Remember that hereafter all steps need to be carried out under low light conditions to avoid fluorophore bleaching.

3. Rinse three times in PBS.
4. Incubate the coverslips with a counterstain solution to stain nuclei (or other cell compartments). It is important to ascertain the total number of cells on the coverslip, in particular if all cells do not express the antigen. For nuclei, use a DNA stain like DAPI. DAPI binds strongly to DNA of live and fixed cells (it easily crosses intact cell membranes). When bound to double-stranded DNA or RNA, DAPI is excited by light at 358 nm and emits at 461 or 500 nm, respectively (blue). However, its emission spectrum overlaps with green fluorescent dyes (like fluorescein) or red fluorescent dyes (like Texas Red), but this can be easily averted using an appropriate filter set. Other fluorescent dyes, like Hoechst dyes, can be used in place of DAPI. Intercalating DNA compounds like DAPI or Hoechst can exert mutagenic properties, and appropriate care should be exercised in their handling and disposal.
5. See Note 8.
6. Mount coverslips in glycerol mounting medium. However, as glycerol does not harden, you must handle slides carefully in order not to move them. If a mounting medium which hardens is desired, use Mowiol or a mounting medium such as Fluoromount-G or Citifluor mounting medium (or Vectashield) by inverting the coverslip onto a drop of the mounting medium on a 25 × 75-mm glass microscope.
7. Allow to air dry for 10 min, and clean around the edges before examining under the microscope.
8. Slides can be stored for several months in the dark at 4°C. If using glycerol or Mowiol as mounting medium, it is better to seal by nail polish in order to prevent drying during storage.
9. See Note 9.

3.6. Immunostaining Controls

1. Never perform immunostaining without appropriate controls. Negative controls must be set up for every labeling experiment; often other controls are also required.
2. *Negative controls*: When the primary antibody is omitted or replaced with an irrelevant primary antibody of the same type and concentration, be certain that there is no staining.

3. *Positive controls*: If no specific staining is observed, it is essential to demonstrate that the labeling protocol used would have worked where than antigen is known to be expressed.
4. *Preabsorption controls*: Particularly, when staining with antibodies to a peptide or other antigen available in purified form, it is important to show that preabsorption of the primary antibody with a 100 (or greater) molar excess of the purified antigen eliminates staining.
5. *Double-label controls*: When carrying out double labeling, show that the secondary antibodies each recognize specifically only the appropriate primary antibody. You should show also that the filter set used is able to discriminate several fluorochromes with minimal overlap of the emission spectra.

3.7. Troubleshooting

1. *No staining*:
   - Fixative sensitivity (wrong fixative used)
   - Antibodies not working (e.g., contaminated by microorganisms or freeze–thaw leading to misfolded antibodies)
   - Wrong secondary antibody used (e.g., anti-mouse antibody to detect a rabbit primary)
2. *High background staining*:
   - Failure to block nonspecific binding (either too little goat serum or too short an incubation). Change the type of serum or use a 3% BSA solution.
   - Secondary antibody concentration too high.
   - Primary antibody concentration too high.
   - Secondary antibody may specifically detect antigens on the tissue stained.
   - Primary or secondary antibody may be binding to Fc receptors expressed by some types, particularly macrophages or microglia.
3. *Specific staining is weak*:
   - Primary antibody concentration is too low.
   - Secondary antibody concentration is too low.
   - Fluorophore has been exposed to light during the staining.
   - Density of antigen is low (see Note 10).

3.8. Staining of Cortical Astrocyte Cultures

The procedure below is intended as a "generic" illustration and should be adapted to the cells and antigen of interest.

1. Transfer coverslips to the staining box with a coarse pair of forceps.

2. Fix with 4% PFA (about 100 μL per coverslip) for 10 min at room temperature. Wash three times with PBS.
3. Permeabilize the coverslips with 50% goat serum/0.1% Triton for 30 min.
4. Incubate with primary antibody for 1 h at room temperature. Prepare 1 mL of the GFAP antibody solution in an Eppendorf tube. GFAP is a rabbit polyclonal antibody which should be diluted in the BSA/Tris buffer to 1:1,000. After 1 h, rinse the coverslips three times with PBS.
5. Secondary antibody incubation for 1 h at room temperature. Prepare 1 mL of each antibody solution in an Eppendorf tube. Use FITC-anti-rabbit at 1:500 (dilute into BSA/Tris). After the 1 h staining, rinse each coverslip three times with PBS.
6. Mount in Citifluor mounting medium on a microscope slide. Gently blot excess mounting medium. You can use nail polish to seal. Allow slips to air dry for 30 min before examining on the microscope.
7. See Fig. 2 for a GFAP stained culture.

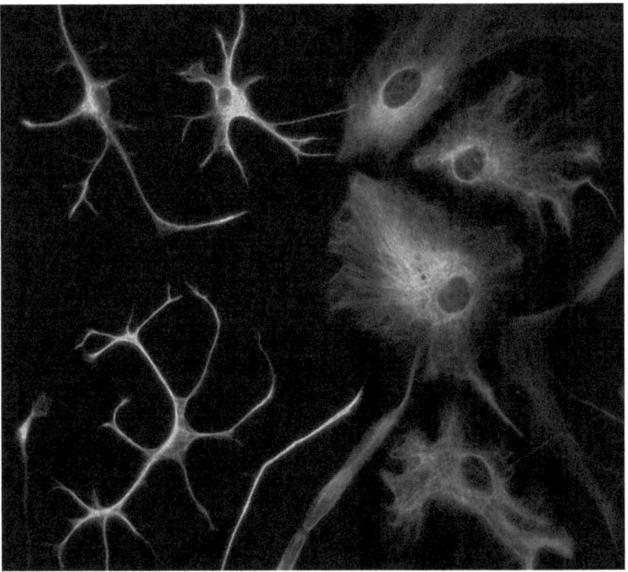

Fig. 2. Immunostaining of cortical astrocytes for GFAP. Astrocytes cultured from 2-day-old neonatal rat cortex on poly-L-lysine-coated coverslips were fixed with paraformaldehyde. Staining was performed using a mouse anti-rat GFAP monoclonal primary antibody, followed by an Alexa Fluor® 555 goat anti-mouse secondary antibody. Images were captured using a Leica DMI 4000B fluorescence microscope equipped with a Leica DFC480 camera, analyzed, and stored using Leica Application Suite software (version 2.8.1). Magnification: ×400.

4. Notes

1. Caution: The pH of the Tris solution is temperature-dependent and decreases about 0.03 pH unit for each increase of 1°C.
2. PFA is a potential carcinogen. It is a skin and eye irritant and is poisonous if ingested. Handle only under a fume cupboard and wear gloves. PFA waste cannot be disposed of down the drain and must be collected in a special waste container labeled clearly and only used to collect this waste.
3. For intracellular antigens such as GFAP, you can use instead a 10 min fixation in –20°C acid-alcohol (5% glacial acetic acid plus 95% ethanol) or in 100% methanol; whenever alcohol fixation is used, the Triton permeabilization step can be skipped.
4. Goat serum is used because the fluorochrome-conjugated secondary antibodies are raised in goats; this minimizes any unwanted binding of the secondary antibody to the goat antibodies that bind to the nonspecific binding sites.
5. Do not permeabilize cells when staining for surface-expressed antigens, particularly for galactocerebroside (a glycolipid) or A2B5 (a ganglioside)—these lipid antigens are solubilized by Triton.
6. Incubation with primary antibody can be extended from 60 min to overnight at +4°C if necessary.
7. Typical supernatant antibody concentration is 1–10 μg/mL, and ascites concentrations are 1–10 mg/mL, as are polyclonal antisera. Thus, typical final primary concentrations are about 1–10 μg/mL. (When the best concentration is not known and a guess must be made, use the supernatant undiluted, the ascites at 1:100, and polyclonals at 1:500. However, beware of the "prozone" effect—especially with IgM monoclonal antibodies.)
8. This is optional: Postfix in acid-alcohol for 10 min—particularly helpful for high-density cultures to stabilize the antibody binding and to decrease the likelihood that the cells will detach during coverslipping. This would not hurt (unless using phycoerythrin-coupled secondaries, which it destroys) and generally improves the appearance. Rinse three times in PBS.
9. Citifluor media, like AF1 and AF2, are mixtures of PBS and glycerol and contain a proprietary substance that helps retard fluorescein bleaching but is not so helpful for rhodamine or Texas Red bleaching. Vectashield works for both. *N*-paraphenylenediamine (0.1%) is useful for FITC bleaching too (but is carcinogenic). *N*-propylgallate (0.4% or higher for confocal studies) retards rhodamine and Texas-Red bleaching but does not help for FITC. Note that FITC has a steep pH

optimum, being maximally bright between pH 8 and 9. Outside this range, its intensity is many times lower.

10. If using a monoclonal antibody, consider using a polyclonal if possible or move to an immunostaining technique with a higher amplification: biotin-conjugated secondary antibodies followed by streptavidin-FITC or an immunoenzymatic technique such as peroxidase methods (secondary antibody conjugated to horseradish peroxidase or use the avidin-biotin-peroxidase complex or peroxidase anti-peroxidase methods).

References

1. Lakowicz JR (2006) Principles of fluorescence spectroscopy (3rd edition). Springer, New York, USA
2. Monici M (2005) Cell and tissue autofluorescence research and diagnostic applications. Biotechnol Annu Rev 11, 227–256
3. Heikal AA (2010) Intracellular coenzymes as natural biomarkers for metabolic activities and mitochondrial anomalies. Biomark Med 4, 241–263
4. Benson DM, Bryan J, Plant AL, Gotto AM Jr, and Smith LC (1985) Digital imaging fluorescence microscopy: spatial heterogeneity of photobleaching rate constant in individual cells. J Cell Biol 100, 1309–1323
5. Eggeling C, Volkmer A, and Seidel CA (2005) Molecular photobleaching kinetics of Rhodamine 6G by one- and two-photon induced confocal fluorescence microscopy. Chemphyschem. 6, 791–804
6. Abbe E (1873) Archiv fuer Mikroskopische Anatomie und Entwicklungs mechanik. 9, 413
7. Liang R (2010) Optical design for biomedical imaging. Spie Press, Washington USA
8. Toomre D, and Bewersdorf J (2010) A new wave of cellular imaging. Annu Rev Cell Dev Biol 26, 285–314
9. Petibois C (2010) Imaging method for elemental, chemical, molecular, and morphological analyses of single cells. Anal Bioanal Chem 397, 2051–2065

Chapter 22

Neurite Outgrowth Assessment Using High Content Analysis Methodology

Nicholas M. Radio

Abstract

High content analysis of neurite outgrowth enables the rapid and comprehensive phenotypic assessment of individual neurons in a multiwell format amenable to high throughput assays. The resulting data are considered "high content" because multiple measurements of neuronal outgrowth and morphometric data are calculated from hundreds of individual cells within each image. This approach has been widely adopted by the pharmaceutical industry to accelerate neurological drug discovery and in vitro safety assessment. High content technology utilizes automated fluorescent and/or brightfield microscopy for image acquisition. The acquired images are then quantified using mathematical algorithms to measure pertinent neurobiological morphometric information, including neurite length, count, and extent of branching for each cell within the images. Furthermore, evaluation of the individual cell-level measurements enables the detection of subpopulations of cellular responders not apparent when examining well-level averages. Using this technology, neurite outgrowth can be quantified in each well, derived from hundreds of cell measurements in a 96-well microplate in approximately 30 min.

Key words: Neurite outgrowth, Immunofluorescence, High content analysis, High content screening, Neuronal differentiation, In vitro neurotoxicity, PC12 cells

1. Introduction

Neurite outgrowth is a critical cellular event underlying the development and functionality of the nervous system and has been widely studied to evaluate neuronal health and assessment of neuronal differentiation (1). Neurite outgrowth is a process that results from the differentiation of precursor cells to a neuronal phenotype and the initiation of sheetlike lamellipodia that ultimately condense into short processes (2). As the cells mature, the processes will increase in length and complexity (see Fig. 1). In some neuronal cell models, including primary cells, neurons will become polarized by developing a single long axon and several secondary processes

Stephen D. Skaper (ed.), *Neurotrophic Factors: Methods and Protocols*, Methods in Molecular Biology, vol. 846,
DOI 10.1007/978-1-61779-536-7_22, © Springer Science+Business Media, LLC 2012

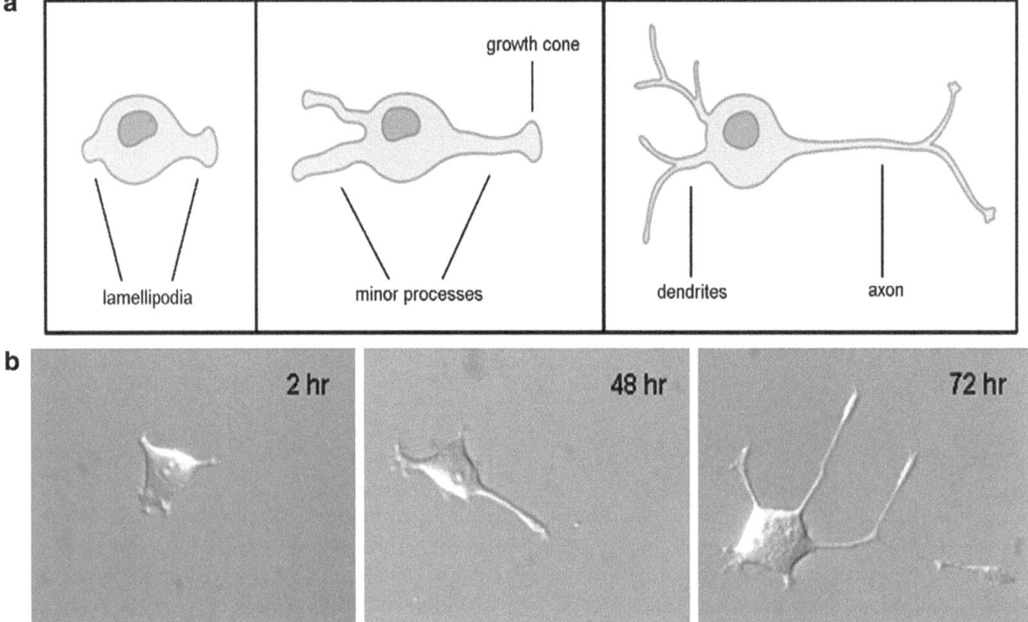

Fig. 1. Early events in neurite outgrowth. (**a**) Diagram of early events in neurite outgrowth illustrating a cell body with lamellipodia, the development of minor processes (tipped with a growth cone), and transformation of the processes into an axon and dendrites. (**b**) Hoffman modulation contrast photomicrographs of neurite outgrowth in PC12 cells 2 h, 48 h, and 72 h after treatment with 100 ng/mL NGF to induce differentiation. Reproduced with permission from Elsevier (5).

shorter in length. The growth of axonal and dendritic processes (collectively called neurites) is a critical determinant of neuronal connectivity, and disruption of this process can induce human cognitive deficits (3, 4). For example, deficits in the regulation of the dendritic cytoskeleton affect the functionality of dendrites and synapses believed to underlie some cases of mental retardation (4). Additionally, novel compounds that facilitate neurite outgrowth formation can potentially accelerate treatment of a wide variety of neurological diseases and traumas that result in nerve injury, including stroke, Parkinson's, and Alzheimer's diseases.

There are a number of well-characterized neuronal cell models that have been used to characterize chemical effects on neurite outgrowth (5). Of these models, pheochromocytoma (PC12) cells are a rodent noradrenergic cell line that has been widely utilized in neurobiological studies to examine chemical perturbations of neurite outgrowth (6). Following exposure to 10–100 nanograms per milliliter (ng/mL) nerve growth factor (NGF), PC12 cells differentiate into a sympathetic-like neuron and develop extensive neuritic processes. Unlike primary neuronal cell cultures, neurites produced from PC12 cells do not give rise to definitive axons or dendrites (7). Many studies have utilized PC12 cells to evaluate the effect of environmental compounds on neurite outgrowth (8, 9). More recently, a PC12 cell clone, Neuroscreen-1™ (NS-1) cells, has been used as an in vitro model for neurite outgrowth evaluation (1).

In comparison with the parent PC12 cell line, NS-1 cells are less prone to cellular aggregation, allowing for the evaluation of neurite outgrowth in individual cells, while retaining many of the properties of PC12 cells (1). Previous studies using more traditional, manually derived methodologies demonstrate the ability of a cell model to detect changes in neurite outgrowth; they were mostly restricted to a single chemical analysis due to throughput limitations (8, 9). For large volume chemical screening, fully automated techniques amendable to high-throughput analysis are necessary to facilitate the rapid assessment of compound libraries.

Advances in automated microscopy combined with algorithm-based quantification metrics have evolved into a cell-based measurement approach called high content analysis. High content analysis integrates quantitative microscopy with automated technology controlling image acquisition, quantified cell-level data analysis, and data collection (10, 11). High content platforms are designed to track phenotypic changes at the individual cell level in multiwell plates using fluorescent labels and/or bright field imaging. As a result of individual cell assessment, high content approaches enable the detection of subtle heterogeneous effects not apparent in average population responses (12). Furthermore, the increased throughput abilities of high content combined with the microplate formats enable the use of comprehensive dose-response curves to characterize a concentration-dependent profile. For example, analysis in a 96-well plate enables a 12-point dose-response curve for eight different compounds (see Fig. 2).

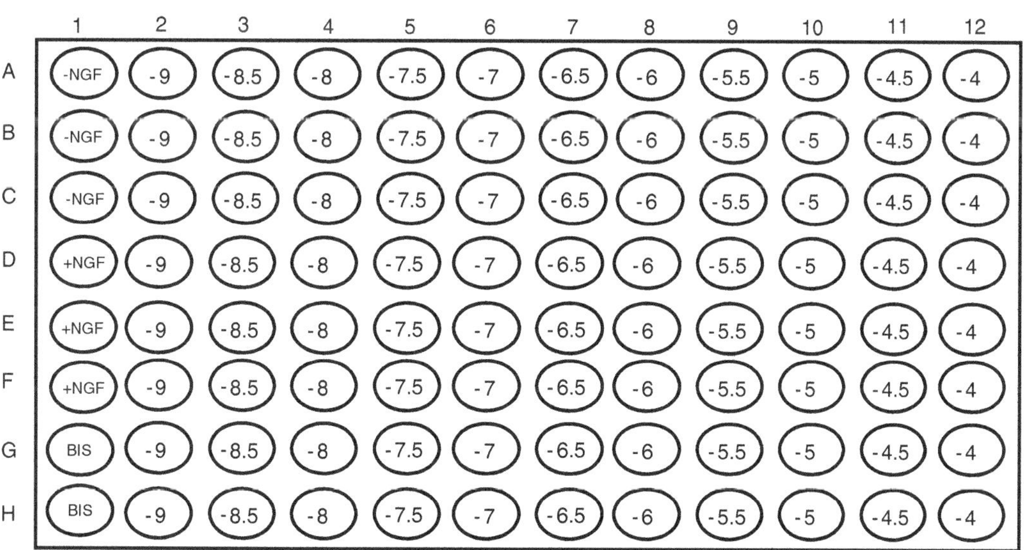

Fig. 2. Ninety-six-well plate chemical exposure format for neurite outgrowth evaluation. NS-1 cells are exposed to an 11-point dose-response curve (semilogarithmic concentration range from 1 nM to 100 μM) from eight separate chemicals (*Rows A–H*). Column 1 serves as a control column, with three replicates each of undifferentiated NS-1 cells (–NGF), cells incubated with 100 ng/mL NGF (+NGF), and cells incubated with 100 ng/mL NGF coadministered with 3 μM bisindolemaleamide-I (BIS-I) to serve as an internal pharmacologic control to evaluate expected neurite outgrowth inhibition. All treatments contain 0.1% DMSO.

The complex nature of the regulation of neurite outgrowth provides a wide range of potential targets for chemical perturbation. Chemically mediated interference with gene expression, membrane receptors, ion channels, or intracellular signaling can affect neurite initiation and growth. For example, neurite outgrowth is supported by a set of extracellular cues, including trophic factors, extracellular matrix molecules, and activity-dependent depolarization (13). Because of these many potential molecular targets, the use of neurite outgrowth as a "nodal indicator" of neuronal cell health and differentiation status has recently been investigated using high content methodologies. This approach may be more suitable for hazard identification and screening for chemical-induced changes in neurite outgrowth when the site of chemical action is unknown. These studies have employed a variety of cellular models, including cell lines, primary cultures, and stem cell models. Furthermore, over the last 2 years, the algorithms have evolved to examine additional neurobiological end points captured in punctate detection, including synaptogenesis (14). Taken together, high content analysis allows the comprehensive evaluation of neurite outgrowth in an efficient manner amenable to screening assays, as will be detailed in this chapter.

2. Materials

2.1. Cell Culture Reagents

1. NS-1 cells, a PC12 subclone (Thermo Fisher Scientific), are maintained at 37°C in a 95% humidified incubator containing 5% CO_2.
2. NS-1 cells are cultured in RPMI 1640 medium (Lonza BioWhittaker, Walkersville, MD), supplemented with 10% equine serum (HyClone, Logan, UT), 5% heat-inactivated fetal bovine serum (HyClone), 1% L-glutamine (Lonza BioWhittaker), 1% penicillin/streptomycin (Lonza BioWhittaker), and 100 ng/mL NGF (Sigma-Aldrich). After adding all components of the medium, adjust the pH to 7.4 with HCl (see Note 1).
3. Human recombinant NGF (Sigma-Aldrich) is supplied as a 0.1 mg lyophilized powder. Dilute the powder with 500 μL of 0.1% bovine serum albumin in phosphate buffered saline (PBS). Aliquot the 500 μL volume into 10-μL × 50-μL vials and store at −80°C until time of use. On the day of use, prepare NGF-supplemented medium (200 ng/mL) by adding a 50 μL aliquot to 50 mL of RPMI medium.

4. Trypsin-Versene (EDTA) solution (Lonza BioWhittaker), stored at –20°C.
5. Culture cells on BD BioCoat® Collagen Type IV, 75-cm² culture flasks with vented cap (BD Biosciences).
6. Use Falcon 15- or 50-mL sterile conical tubes (BD Biosciences) for cell transfer.
7. Transparent 96-well polystyrene microplates that have been precoated with collagen IV (BD Biosciences) are used for cell plating. Store plates at 4°C (see Note 2).
8. Bisindolemaleamide-I (Bis-I) (EMD, Darmstadt, Germany), a potent protein kinase C-inhibitor, is used as a pharmacologic positive control for neurite outgrowth inhibition and is dissolved in dimethyl sulfoxide (DMSO) (Sigma-Aldrich) and stored in single use aliquots at –20°C. The final concentration to selectively inhibit neurite outgrowth without affecting cellular viability is 3.0 μM.

2.2. Immunocytochemistry

1. Paraformaldehyde (Electron Microscopy Sciences, Hatfield PA) is prepared at a 4% (v/v) final concentration in fresh PBS. Hoechst 33342 (Thermo Fisher Scientific) is included in the fixative solution at a concentration of 5.0 μg/mL to selectively label cell nuclei.
2. PBS 10× stock (Invitrogen). Prepare 1× working solution by dilution of one part 10× with nine parts water. Store 1× and 10× solutions at 4°C.
3. Immunocytochemical staining buffer (ISB): prepare 10× stock containing 9 mM $CaCl_2$, 26.8 mM KCl, 14.7 mM KH_2PO_4, 4.9 mM $MgCl_2$, 1.4 mM NaCl, 80.6 mM Na_2HPO_4, 0.1% saponin, 5% BSA, and 0.5% NaN_3. Prepare a 1× working solution by dilution of one part 10× with nine parts water. Store 1× and 10× solutions at 4°C.
4. A mouse anti-βIII-tubulin primary antibody (1:800) (Thermo Fisher Scientific) is used to label cell bodies and neurites. Dissolve the antibody in ISB and combine with a DyLight Fluor 488-conjugated goat anti-mouse secondary antibody (1:500) (Thermo Fisher Scientific) dissolved in ISB.
5. Optical adhesive film seal (Thermo Fisher Scientific) is used for sealing plates.

2.3. Image Acquisition and Analysis

1. Automated image acquisition is performed using the ArrayScan VTI high content imaging platform (Thermo Fisher Scientific).
2. Neurite outgrowth analysis is performed using the Neuronal Profiling bioapplication (Thermo Fisher Scientific).

3. Methods

3.1. Cell Culture

1. A single passage number of NS-1 cells (i.e., ten after receipt from supplier) should be used in every experiment to avoid potential interpassage variability (15).
2. Grow NS-1 cells in T-75 flasks and trypsinize for use when they reach 70–80% confluency (see Note 3).
3. To dislodge cells, aspirate the growth medium from the T-75 culture flask.
4. Add 3.0 mL of prewarmed (37°C) Trypsin-Versene solution to the T-75 culture flask.
5. Gently tilt the flask so that the Trypsin-Versene solution covers the cell surface. The first addition of Trypsin-Versene serves as a wash step to remove the existing medium/serum and allows for a smaller volume of Trypsin-Versene to be added.
6. After covering the cell surface of the flask with the 3.0 mL of Trypsin-Versene, immediately aspirate the Trypsin-Versene solution.
7. Add 1.0 mL of prewarmed (37°C) Trypsin-Versene solution to the T-75 culture flask. Gently tilt the flask so that the Trypsin-Versene solution covers the cell surface.
8. Incubate the cells in the Trypsin-Versene solution for 5 min at 37°C.
9. Add 9.0 mL of prewarmed complete medium directly to the growth surface of the flask to dislodge the cells. Add the medium to the growth surface of the flask several times to ensure the cells are dislodged.
10. Transfer the cells into a 15-mL conical tube suitable for centrifugation.
11. Centrifuge the cells at 300 × *g* for 5 min.
12. After centrifugation, remove the supernatant containing the Trypsin-Versene solution and resuspend the pellet of cells in 3 mL of complete medium (see Note 4).
13. Perform a cell count using a hemocytometer to calculate the cell density of the harvested cells.
14. Subdivide the harvested cells into two groups: –NGF-treated cells (3 wells total) and +NGF-treated cells (93 wells total). For each group, dilute the cells so the final density is 2,000 NS-1 cells per 90 μL. For the –NGF-treated cells, dilute the cells in growth medium. For +NGF-treated cells, dilute the cells in medium containing 100 ng/mL NGF.
15. Add 90 μL per well of the harvested cell suspension (containing 2,000 NS-1 cells) into a transparent, 96-well collagen IV–coated

microplate. Ensure the cells are homogeneously mixed by triturating five times before plating the cells.

16. Leave the cells undisturbed on a solid bench top for 30 min following cell plating before moving the cells to the incubator. This will facilitate an even cell seeding for optimal neurite identification.
17. Wait a total of 2 h to allow the cells sufficient time to adhere to the bottom of the wells.
18. Add to the cells 10 μL of the 10× intermediate concentration of the test compound(s) of interest. The positive control to inhibit neurite outgrowth, Bis-I should be prepared at a 10× intermediate concentration of 30 μM. Negative control wells should only be treated with DMSO vehicle at a 10× concentration of 1.0%. After compound addition to the cells, the final volume is now 100 μL, with the positive control group exposed to 3 μM Bis-I and the negative control group exposed to 0.1% DMSO (see Note 5).
19. Expose cells to the compounds of interest and return to the incubator (37°C, 5% CO_2) for 96 h; no medium exchange is necessary.

3.2. Immunocytochemistry

1. Prepare a 2× fixative solution containing 8% paraformaldehyde and 10 μg/mL Hoechst 33342 in 1× PBS. Prewarm the fixative to 37°C before adding to the cells. Prewarming the fixative is critical to maintain the cell and neurite integrity.
2. Add 100 μL of the 2× fixative solution to the existing 100 μL of cell medium for a final concentration of 4% paraformaldehyde and 5 μg/mL Hoechst 33342. Add the fixative solution using a low-velocity fluid dispension to the side of the wells. Incubate for 20 min.
3. During the fixation, prepare 1:800 solution of the mouse anti-βIII-tubulin primary antibody in 1× ISB.
4. Following the 20-min fixative incubation, carefully aspirate the paraformaldehyde and Hoechst 33342–containing solution.
5. Fill the wells with 100 μL of ISB, and then drain each well with gentle aspiration. Wash two additional times with 100 μL of 1× ISB (see Note 6).
6. Aspirate the ISB buffer and label the cell bodies and processes with 50 μL of mouse anti-βIII-tubulin primary antibody (1:800 dilution) for 1 h.
7. Aspirate the primary antibody and wash three times with 100 μL of 1× PBS.
8. Aspirate the 1× PBS buffer and label with 50 μL of DyLight™ 488-conjugated goat anti-mouse secondary antibody (1:500 dilution) for 1 h. Protect from light.

9. Aspirate the secondary antibody and wash three times with 100 μL of 1× PBS, leaving the buffer from the final wash in the plate.
10. Seal the plate with a plate seal. Store the plate at 4°C in the dark until use. Plates can be safely stored in this manner for at least 6 months.

3.3. Image Acquisition

1. The following instructions are based on use of the ArrayScan VTI high content platform in conjunction with the iDEV™ Assay Development Workflow software in conjunction with the Neuronal Profiling bioapplication. The method describes automated measurement of neurite outgrowth in differentiated cells in 96-well plates stained using the immunocytochemical procedure described above. The assay parameters cited in this method are associated with an optimized image analysis protocol for measuring neurite outgrowth in differentiated NS-1 cells. The notes associated with this method provide guidance for adjustment of default settings and assay parameters for adapting the protocol to measure neurite outgrowth in other neuronal cell types (i.e., other cell lines, primary neuronal cultures, etc.). **Bold** text refers to a menu or heading in the Neuronal Profiling bioapplication. *Italicized* text refers to an option or value input by the end-user (see Note 7).
2. Turn on the ArrayScan VTI HCS Reader, open the vHCS:Scan Software, select the *iDEV™ Assay Development Workflow* mode, and log in.
3. From the **Select Protocol** menu, click on the *Change* button to select the *NeuronalProfiling.V4* bioapplication.
4. Proceed to the **Configure Acquisition** parameters: Select the ×*10* imaging **Objective** from the associated drop-down menu. Set the **Camera Configuration** to an option that is compatible with the ArrayScan system in use. Set **Acquisition Camera Mode** to *Standard*. Set the number of channels to *2*. The field dimensions using these settings are 660.48 by 660.48 μm.
5. Specify the correct fluorophores to be utilized in the **Image Formation** subbox: Use the drop-down menu to select *XF100 – Hoechst* as the Channel 1 dye and *XF100 – FITC* as the Channel 2 dye.
6. In the *XF100 – Hoechst* Channel 1, select *Fixed* from the **Exposure Type** drop-down menu. Set the **Initial Exp Time (s)** to *0.01*. In the *XF100 – FITC* Channel 2, select *Fixed* from the **Exposure Type** drop-down menu. Set the **Initial Exp Time (s)** to *0.03*. These exposure values serve as initial exposure settings that will used during initial image acquisition. Acquired images deemed representative to the investigator should be used to finalize exposure times, so nuclear images

use 25% of the camera's dynamic range and the cell body/neurite uses 55–65% (see Note 8).

7. Select **Autofocus** from the **Autofocus Camera Mode** drop-down and specify **Autofocus interval** *3* **fields** (see Note 9).
8. Click on a representative undifferentiated (–NGF) control well from the plate layout map. In cellular nuclei Channel 1, select the **Acquire Image** button and then select **Autofocus**. Set the final exposure value interactively by clicking the **Autoexpose** button. Repeat this process for Channel 2 for cell body and neurite image acquisition. Once suitable images are acquired for both channels of the negative control well, select **Save Field** so the image set is available for subsequent analysis.
9. Click a representative differentiated (+NGF) control well from the plate layout map. Acquire images for both the nuclear (Channel 1) and cell body/neurite (Channel 2) fluorophores. Select **Save Field** so the image set is available for analysis. Select **Next** to proceed to **Configure Assay Parameters**.
10. Within the **Protocol Optimization Task List**, proceed to the **Image Preprocessing** step. Select the **Object Type** as *Bright* for both Channels 1 and 2. Specify and enable the *LowPassFilter* as the **Background Removal Method** for both Channels 1 and 2. The **Value** for both channels should be determined by measuring the diameter of typically large valid objects (nucleus in Channel 1 and cell body in Channel 2). The diameter of these objects can be measured by using the **status tray** at the bottom of the window. Due to a greater chance of cell body size heterogeneity, double the diameter of the measured cell body size.
11. Proceed to the **Nucleus Identification Ch1** step. Enable **Smoothing**, select the **Uniform Method**, and specify a **Value** of *1*.
12. For the **Field Nucleus Identification Ch1**, enable **Thresholding**, select the **Fixed Method**, and determine a **Value** by determining a difference ("delta") between the brightest background pixels and the dimmest foreground fluorescence. Use half of this delta for a starting Value, and adjust as needed, to produce accurate borders on the nucleus. **Enable Segmentation**, selecting the **Shape Method**, and enter a **Value** corresponding to half the diameter of a typical nucleus. **Object Cleanup** should also be enabled to clean up the object mask and remove small objects from identification.
13. Proceed to the **Nucleus Validation Ch1** step. To be included in analysis, any identified nucleus can be validated based on size, shape, and fluorescence intensity measurements. Enable the **Nucleus.BorderObject.Ch1**, **Nucleus.Area.Ch1**, and **Nucleus.ShapeLWR.Ch1** selection features. To exclude

aggregated cells and noncellular particles from analysis, specify the following ranges for each of these features: **Nucleus.Area. Ch1**: minimum *20*, maximum *400* pixels; **Nucleus.ShapeLWR. Ch1**: minimum *1.0*, maximum *3.0*.

14. Proceed to the **Cell Body Identification Ch2** step. Do not enable **Smoothing**.
15. For the **Field Cell Body Identification Ch2**, enable **Thresholding**, select the **Fixed Method**, and repeat the instructions listed in **step 20** for determining an appropriate thresholding value to delineate neuron detection relative to image background. Enter a value of *10* for the **Minimum Cell Body Nucleus Overlap** to specify the minimum percentage overlap between the cell body and the nucleus required to associate the cell body with the nucleus. Enable **Segmentation** and select the **Use Seeds Method**. Next, specify a **Cell Body Demarcation** of *2* to specify the degree in pixels in which the cell body excludes neurites. The main purpose of this parameter is to eliminate neurites identified as part of Cell Bodies. Increasing values should retract the Cell Body overlays extending down neurites. Enable the **Mask Modification** with *3* so neurite origins are clear.
16. Proceed to the **Cell Body Validation Ch2** step. Enable the **Cellbody.BorderObject.Ch2**, **Cellbody.Area.Ch2**, and **CellbodyLWR.Ch2** selection features. To exclude aggregated cells and noncellular particles from analysis, specify the following ranges for each of these features: **Cellbody.Area.Ch2**: minimum *10*, maximum *1,000* pixels; **CellbodyLWR.Ch2**: minimum *1.0*, maximum *3.0*.
17. Proceed to the **Neurite Identification Ch2**, enable **Smoothing**, select the **Uniform Method**, and specify a **Value** of *1*.
18. For the **Field.Neurite.Detection Ch2 Assay Parameters**, enable **Detection**, select the **Binomial Method**, enter a **Value** of *2*, and enter a **Neurite Identification Modifier** value of *−0.97*.
19. Under the **Field.Neurite.Identification Ch2**, enter a **Direction Length** value of *2*, a **Point Resolution** value of *1*, and do not enable the **Rejection of Multiply-traced Neurite** (see Note 10). If increased sensitivity to neurite detection is desired, enable the **Aggressive Tracing** feature. Enter a value of *0* for **Trace Within Cell Body ZOI**.
20. Proceed to the **Neurite Validation Ch2** step, enable the **Neurite.Length.Ch2,** and enter a minimum and maximum value range of *0* and *1,000*, respectively.
21. Modification of the Neuronal Profiling bioapplication is now complete. Using these settings, a representative composite

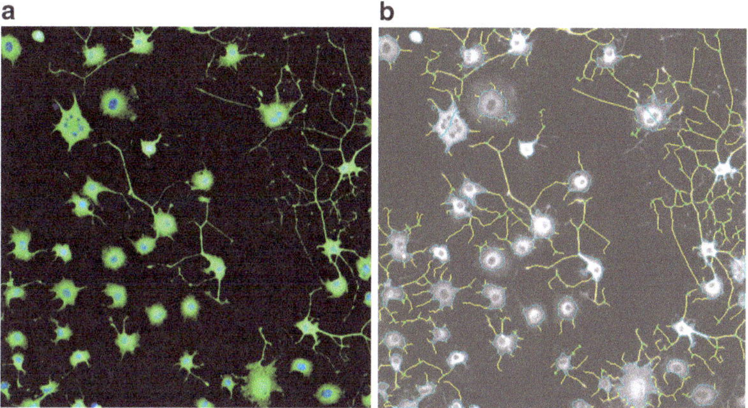

Fig. 3. Differentiated NS-1 cells (100 ng/mL NGF, 96 h) image acquired with a ×10 objective lens and a ×0.63 coupler using the ArrayScan VTI. (**a**) Nuclei are stained blue with Hoechst 33342 (5 ng/mL), and cell bodies and neurites are stained green using an anti-βIII-tubulin primary antibody (1:800) and DyLight™ 488-congjugated antibody (1:500). The resulting image was analyzed using the Neuronal Profiling bioapplication (**b**) for morphometric data including cell body area, number of neurites, neurite length, and branch point for each cell measured.

image of the NS-1 cell and applied algorithm overlay can be generated as shown in Fig. 3.

22. Click on **Next** to proceed to the **Subpopulation Characterization** task. Within the **Event Subpopulations** field, define the **Type 1 Event** so that **NeuriteTotalLengthCh2** or **NeuriteTotalCountCh2** are designated. Two additional subpopulation events can be specified if desired by the user. Within the **Scan Limits field**, enable the **Min Objects for Well** to *200* objects per well. Alternatively, this value can be adjusted according to the statistical requirements of the user.
23. Click on **Next** to proceed to the **Select Features to Store** task. Specify all available features to be stored for the **Cell Features**, **Field Features**, and **Well Features**. Select the **MEAN_NeuriteTotalLengthCh2** as the **Default Well Feature** and set the **Lower Extent** as *1* and the **Upper Extent** as *20*μm.
24. Click on **Go to Scan**, enter a **Plate ID** (if not using a barcode), such as the scientist's initials, experiment date, and plate number (i.e., *NMR12.11.2010Plate1*). Enter a **Plate Name**, such as a description of the assay contained on the plate (i.e., *Bis-I dose-response curve in NS-1 cells*). Additional notes can be electronically recorded if desired within the **Scan Comments** field.
25. Press the **Play** icon to start the high content scan of the assay plate.

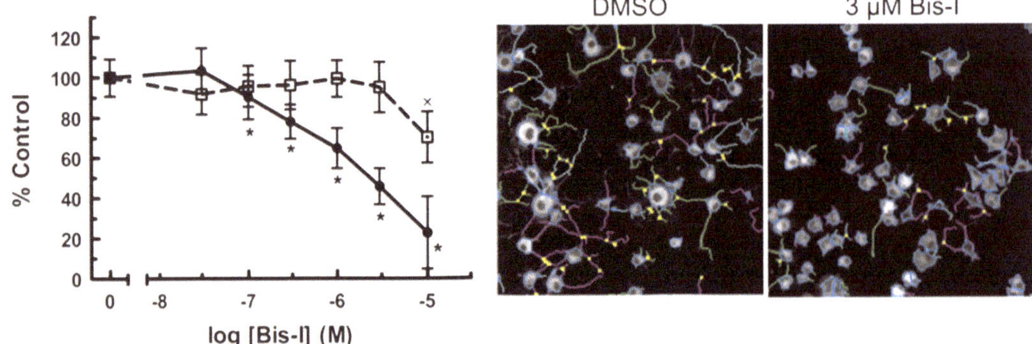

Fig. 4. Effects of neurite outgrowth inhibitors on NS-1 cells. NS-1 cells were plated at 2,000 cells per well in 100 ng/mL NGF and evaluated for either total neurite length (*closed circle*) or viability (*open square*) following 96-h exposure to Bis-I. *Inset* shows representative images of either control (0.1% DMSO) or 3 μM Bis-I-treated cells following the 96-h exposure. Data are expressed as percent of the 0.1% DMSO control and are presented as means ± standard deviation from six total wells analyzed across two independent experiments. Treatments that are Significantly different from control for total neurite length (*asterisk*) or viability (*multiplication sign*) (one-way ANOVA followed by Dunnett's test, $p < 0.05$). Reproduced with permission from Oxford University Press (1).

26. The time requirements for the immunofluorescent labeling of the cells are approximately 4 h. Time requirement for the image acquisition (three fields per well) and analysis per 96-well plate is approximately 30 min.
27. Typical results of neurite outgrowth in NS-1 cells exposed to a dose response of Bis-I are represented in Fig. 4.

4. Notes

1. Store the L-glutamine and penicillin/streptomycin mixture as 10 mL aliquots at −20°C. On the day of use, thaw and add to 1 L of RPMI medium. Store the supplemented medium at 4°C until use.
2. Alternatively, plates can be manually coated with collagen IV substrate. While this method is less expensive, quality control issues pertaining to batch-to-batch plate preparation is generally avoided when using precoated plates.
3. Approximately 1.5 million NS-1 cells are harvested per 75% confluent T-75 flask. The viability (as assessed by trypan blue exclusion is at least 95%).
4. For parent PC12 cell models, it is critical that the cells be trypsinized and triturated three times through a 20-G needle prior to plating to avoid excessive cell clumping.

5. Statistically significant decreases in total neurite length will be evident when final DMSO solvent concentrations ≥1.0% (v/v) are used.
6. Regardless of manual or automatic-mediated additions, post-fixative wash steps are potentially significant causes of cell loss and neurite integrity disruption. To minimize these effects, ensure that all wash additions are employed at low velocity and added indirectly to the sides of the well.
7. For detailed descriptions of terminology and assay parameters associated with this method, the reader is referred to the Thermo Scientific ArrayScan VTI HCS Reader User Guide and the Neuronal Profiling V3.5 Bioapplication Guide.
8. Exposure times in each fluorescent channel are determined by presampling of untreated control wells. Exposure times are calculated in a manner that utilizes a predetermined percentage of the dynamic range (0–4096) of the 12-bit CCD-camera (**Target** parameter in **AutoExposure** options). Pixel saturation should be avoided in order to avoid imposing an artificial ceiling on the fluorescent intensity values recorded by the instrument. For neurite outgrowth, it is recommended that the dynamic range of the camera (**Target** parameter) be set to between 55% and 65%.
9. The Autofocus interval determines the points within a well where a focus adjustment is performed. Conservative suggested autofocus intervals based upon the magnification of image acquisition are 4 fields using ×5, 3 fields using ×10, 2 fields using ×20, and 1 field using ×40 magnification.
10. The Reject Multiply Traced Neurites parameter concerns what happens in the event of neurites contacting several cell bodies. If Rejection of Multiply-traced neurites is selected, then multiply traced neurites will be rejected from analysis. If this feature is not selected, then one of the multiple neurite copies will be kept and the rest will be rejected from analysis. The preserved neurite is therefore assigned to a single cell body.

Acknowledgments

The author wishes to thank Theresa Freudenrich, Joshua Harrill, and William Mundy for their advice and technical assistance, as well as Neil Durso and Theresa Freudenrich for their comments and suggestions on an earlier version of this book chapter.

References

1. Radio N, Breier J, Shafer T, Mundy W (2008) Assessment of chemical effects on neurite outgrowth in PC12 cells using high content screening. Tox Sci 105, 106–118
2. Craig A, Banker G (1994) Neuronal polarity. Annu Rev Neurosci 17, 267–310
3. Webb S, Monk C, Nelson C (2001) Mechanisms of postnatal neurobiological development: Implication for human development. Dev Neuropsychol 19, 147–171
4. Ramakers G (2002) Rho proteins, mental retardation and the cellular basis of cognition. Trends Neurosci 25, 191–199
5. Radio N, Mundy W (2008). Developmental neurotoxicity testing in vitro: models for assessing neurite outgrowth. Neurotoxicology 29, 361–376
6. Vaundry D, Stork P, Lazarovici P, Eiden L (2002) Signaling pathways for PC12 cell differentiation: Making the right connections. Science 296, 1648–1649
7. Greene L (1977) A quantitative bioassay for nerve growth factor (NGF) activity employing a clonal pheochromocytoma cell line. Brain Res 133, 350–353
8. Das K, Barone S (1999) Neuronal differentiation in PC12 cells is inhibited by chlorpyrifos and its metabolites: Is acetylcholinesterase inhibition the site of action? Toxicol Appl Pharmacol 160, 217–230
9. Parran D, Mundy W, Barone S Jr (2001) Effects of methylmercury and mercuric chloride on differentiation and cell viability in PC12 cells. Toxicol Sci 59, 278–290
10. Smith C, Eisenstein M (2005) Automated imaging: Data as far as the eye can see. Nat Methods 2, 547–555
11. Dragunow M (2008) High-content analysis in neuroscience. Nat Rev Neurosci 9, 779–788
12. Giuliano K, Gough A, Taylor D, Vernetti L, Johnston P (2010) Early safety assessment using cellular systems biology yields insights into mechanisms of action. J Biomol Screen 15, 783–797
13. Neely M, Nicholls J (1995) Electrical activity, growth cone motility and the cytoskeleton. J Exp Biol 198, 1433–1446
14. Harrill J, Robinette B, Mundy W (2011) Use of high content image analysis to detect chemical-induced changes in synaptogenesis in vitro. Toxicol In Vitro 25, 368–387
15. Heumann R, Valet G, Maison D, Kemper J, Reiser G, Hamprecht B (1977) Influence of the time in culture on cellular and neuronal properties of neuroblastoma x glioma hydrid cells. J Cell Sci 27, 141–155

Chapter 23

Dissociated Cell Culture for Testing Effects of Carbon Nanotubes on Neuronal Growth

William Lee and Vladimir Parpura

Abstract

Cell culture has emerged as an important research method for studying the effects of carbon nanotubes (CNTs) on primary neurons. We describe the procedure for preparation of dissociated mixed cell culture from postnatal rat hippocampi. Based on morphological criteria and specific neuronal cell markers, neurons can be selected within this mixed cell culture and studied. We present the procedure for the assessment of neuronal cell morphology based on intracellular fluorescence of the vital dye calcein that accumulates in live neurons. This procedure encompasses fluorescence imaging and measurement of the following parameters: neurite number, total neurite length, mean neurite length, number of growth cones, number of branches, and number of branches per neurite. These combined cell culture and fluorescence microscopy approaches can be successfully used for assessment of the effects that CNTs, as water-soluble agents, have on neuronal cell growth and neurite outgrowth.

Key words: Carbon nanotubes, Cell culture, Imaging, Morphology, Neurons, Growth, Neurite outgrowth

1. Introduction

One of the challenges in the field of biomedical engineering and regenerative medicine is to employ materials that could help restore neural function at the site of injury, which may stem from traumatic insults or neurodegenerative diseases in the central nervous system (CNS) (1). Many experimental approaches aim to explore the effects of biocompatible materials in the form of injectable compounds and/or scaffolds to spur neuronal cell growth and neurite outgrowth (2). Carbon nanotubes (CNTs), as water-soluble agents or as cell-permissive scaffolds/substrates, are emerging as a material that holds great promise in such applications [for a review see ref. (3)].

Stephen D. Skaper (ed.), *Neurotrophic Factors: Methods and Protocols*, Methods in Molecular Biology, vol. 846, DOI 10.1007/978-1-61779-536-7_23,

A prerequisite for any medical use of CNTs in vivo is to understand their effects and interactions with the cell type(s) of interest. Here, we describe the procedure in which we use a dissociated cell culture system to characterize the effect of CNTs, as water-soluble agents, on modulation of neuronal cell growth and neurite outgrowth (4–7). The advantage of using a dissociated cell culture model over slice culture or in vivo models is the ability to control for experimental conditions in a well-defined manner.

We describe the procedure for preparing rat hippocampal mixed dissociated cell cultures, which is followed by the description of a procedure using the fluorescent vital dye, calcein, to reveal neuronal cell morphology. Lastly, we describe the procedure for imaging and analysis of neuronal cell morphology to assess the effects that CNTs exert on these neural cells.

2. Materials

2.1. Coverslip Preparation

1. Polyethyleneimine (PEI) (500 mg/mL; Sigma-Aldrich): Dilute the commercially available 50% w/v PEI solution 1:10 in sterile water to yield a 50 mg/mL solution. Sterile filter the solution through a 0.2-μm nylon filter (see Notes 1 and 2).
2. Round glass coverslips (see Note 3), plastic Petri dishes (35 × 10 mm and 100 × 15 mm), circular sterile filter paper (90 mm in diameter, qualitative; Whatman), and UV sterilization lamp (see Note 4).

2.2. Carbon Nanotubes

1. Single-walled carbon nanotubes (SWNTs), water-soluble. We have been using custom-made (provided by Robert C. Haddon, University of California, Riverside, CA) chemically functionalized SWNTs with polyethylene glycol (PEG) or poly-*m*-aminobenzene sulfonic acid (PABS) (see Fig. 1) as previously described (7, 8). These water-soluble materials are commercially available (Carbon Solutions, Inc., Riverside, CA; cat. no. P7-SWNT and P8-SWNT for PEG and PABS functionalized SWNTs, respectively).

2.3. Cell Culture

1. 0- to 2-day-old Sprague-Dawley rat pups (see Note 5).
2. Dissection tools (from Fine Science Tools): 10-mm angled spring scissors, 2.5-mm angled spring scissors, Graefe curved forceps (serrated), fine Dumont forceps (#55), and a blunt curved glass probe (see Note 6). Dissection tools encased within a Petri glass dish (Pyrex brand, 150 × 15 mm) (see Note 7) need to be sterilized prior to use (autoclave; 121°C; 20 psi; 45 min).

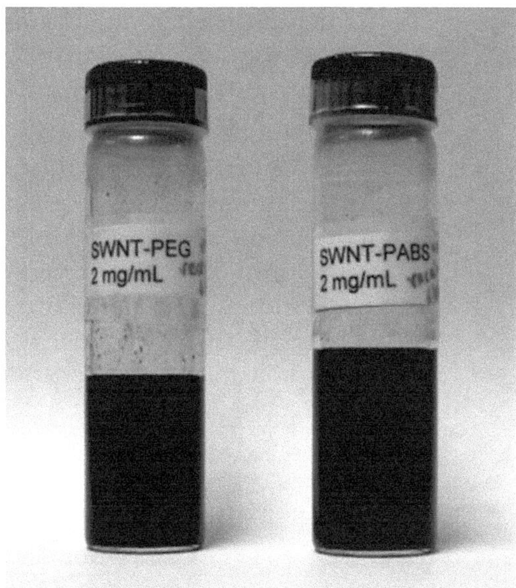

Fig. 1. Water-soluble CNTs used to investigate their effects on neuronal growth. Image of two vials containing functionalized water-soluble SWNTs, SWNT-PEG (*left*) and SWNT-PABS (*right*) at a stock concentration of 2 mg/mL, which appears as *black liquid.*

3. Hanks' balanced salt solution (HBSS) (500 mL; Invitrogen). Store at room temperature (20–24°C).
4. Minimal essential medium (MEM) (without phenol red; cat. no. 51200-038, Invitrogen). Store in the refrigerator (4°C).
5. D-glucose stock solution (2 M): Dissolve 18.02 g of D-glucose in MEM and adjust to a final volume of 50 mL. Make 1 mL aliquots and store at –20°C until use.
6. 10 mM HEPES/HBSS: Dissolve 1.1915 g of HEPES in 500 mL of HBSS and adjust the pH to 7.35 with 5 M NaOH. Filter-sterilize through a Corning 0.22-μm cellulose acetate (CA) membrane filter (Mfg. no. 4306) and store at room temperature.
7. Papain solution: 20 IU/mL papain, 0.2 mg/mL L-cysteine, 100 IU/mL penicillin, 100 μg/mL streptomycin, 20 mM D-glucose, and 10 mM HEPES in HBSS. Calculate the volume needed to dissolve 100 mg of papain (Sigma-Aldrich) to yield 20 IU/mL (see Note 8). Knowing the final volume needed, calculate the appropriate amount of D-glucose (1:100 dilution from a 2 M D-glucose stock solution), penicillin/streptomycin (1:100 dilution penicillin-streptomycin stock solution, 10,000 IU/mL to 10,000 μg/mL, respectively; Invitrogen), and L-cysteine. Add these ingredients into an appropriate amount of 10 mM HEPES/HBSS to make a calculated volume of solution. Adjust the pH to 7.35 with NaOH and sterile filter

through a 0.22-μm CA membrane filter. Add 100 mg of papain to the solution. Put the mixture on a stirring plate until the papain dissolves. Make 1 mL aliquots and store at −20°C until use.

8. Trypsin inhibitor (100 mg/mL) stock solution: Make this solution in a laminar flow hood with aseptic techniques by adding 10 mL of 10 mM HEPES/HBSS to a vial containing 1 g of Type II-O trypsin inhibitor (Sigma-Aldrich). Let the bottle sit for approximately 2 min to fully dissolve all of the contents of the bottle. Then, make 500 μL aliquots in 1.5-mL sterile microcentrifuge tubes and store at −20°C until use.
9. Mito-MEM containing MEM supplemented with 2 mM L-glutamine, 20 mM D-glucose, 1 mM sodium pyruvate, 100 IU/mL penicillin, 100 μg/mL streptomycin, 14 mM sodium bicarbonate, 0.1% (v/v) Mito + serum extender, and 5% (v/v) fetal bovine serum. Make this medium solution in a laminar flow hood. To make 200 mL of this cell culturing medium, mix 181.8 mL of MEM with 2 mL each of the following stock solutions: penicillin/streptomycin, D-glucose stock, sodium pyruvate (100 mM; Invitrogen), and L-glutamine (200 mM; Invitrogen). Add 200 μL of Mito + serum extender (Collaborative Biomedical Products, cat no. 50006) and 235.2 mg of sodium bicarbonate. After all the above supplements have been mixed and thoroughly dissolved (see Note 9), sterile filter the solution through a 0.22-μm pore CA membrane filter and add 10 mL of sterile fetal bovine serum (see Note 10) to make a total volume of 200 mL. Store at 4°C.

2.4. Cell Imaging

1. An imaging chamber with a circular recess at its bottom to accommodate for mounting of a coverslip (see Note 11). Seal the coverslip to the bottom of the chamber using Dow Corning® High Vacuum Grease (see Note 12).
2. External solution: Prepare a saline solution containing 140 mM NaCl, 5 mM KCl, 2 mM $MgCl_2$, 2 mM $CaCl_2$, and 10 mM HEPES in water (pH = 7.4). Filter through a 0.2-μm filter and store at 4°C. Immediately prior to use, warm it to room temperature and add 5 mM D-Glucose (22.5 mg of glucose per 25 mL of solution) to complete the external solution.
3. Calcein-AM stock solution: To make stock solution of calcein acetoxymethyl (AM) ester (Invitrogen), add 50 μL of dry dimethyl sulfoxide (DMSO) (see Note 13) to an individual vial containing 50 μg of calcein-AM to yield a 1 μg/μL stock solution. Store aliquots (1–10 μL) at −20°C.
4. Pluronic acid stock solution (25% w/v): Warm dry DMSO to 37°C. Weigh out 50 mg of pluronic acid F-127 (Invitrogen) and add 200 μL of dry DMSO to it in a microcentrifuge tube

to make a ~25% w/v solution. Vortex, then incubate the mixture at 37°C until it completely dissolves, which may take 30–60 min. Store at room temperature wrapped in aluminum foil. At lower temperatures, pluronic acid may precipitate out of solution and should be heated in a water bath again to redissolve it into solution. Prepare this stock solution fresh once a month.

5. An inverted microscope equipped with differential interference contrast and epifluorescence illumination, 60× oil immersion objective, and a standard fluorescein isothiocyanate (FITC) filter set for imaging calcein.
6. A camera for image acquisition and a shutter inserted in the fluorescence light path; both devices are computer/software interfaced.
7. Image analysis software.

3. Methods

3.1. Coverslip Preparation

1. Sterile technique needs to be used for coverslip preparation, which should be done in a laminar flow hood. One hippocampus can provide dissociated cells for about 20 coverslips. Place ten coverslips onto a circular sterile filter paper (90 mm in diameter) inlayed into an inverted lid of a sterile 10-cm Petri dish with maximal spacing between them. Prepare multiple sets (four sets of ten coverslips for one rat pup dissection) pending on the number of rat pups used. Sterilize the coverslips and filter paper using UV lights (see Note 3).
2. Mix 100 μL of 50 mg/mL PEI stock solution with 4.9 mL of sterile water in a 15 mL plastic tube to obtain a 1 mg/mL PEI solution.
3. Apply 100 μL of 1 mg/mL PEI solution to each coverslip with care to avoid spilling the liquid off of the top of the coverslip and onto the filter paper. Cover inverted lids with their bottoms facing up. After the PEI solution has been on the coverslip for 3 h, aspirate the solution and wash the top of the coverslip two times with autoclaved/sterile water and allow the water to rest on top of the coverslip for 3 more hours. Change the water and incubate it on the coverslip for 3 h; repeat this step. After completion of two incubation periods, aspirate the water and air-dry them inside the laminar flow hood.
4. Using a stainless steel tool such as a 5-mm flat-head screw driver, with the tip heated to glow by a Bunsen burner, emboss crossed grooves at the bottom of 35 × 10-mm Petri dishes to generate four segments. Sterilize dishes with UV light. Using tweezers, place one coverslip with the PEI-coated surface up into each segment of the dish (see Notes 14 and 15).

3.2. Cell Culture

The described procedures below, performed on rats, must be approved by the Institutional Animal Care and Use Committee prior to performing them. Hippocampal dissociated cell cultures are prepared from 0- to 2-day-old Sprague-Dawley rats using previously described procedures; this cell culture approach yields a mixed cellular population of neurons, astrocytes, and microglia, and it has been used to study neuronal cell morphology (4–7, 9). Since excellent photographs and video clips of rodent hippocampal dissections are described elsewhere (10, 11), we only briefly and textually address the isolation of hippocampi.

3.2.1. Preparatory Tasks

1. Prior to beginning a primary cell culture, place 50 μL of mito-MEM onto each coverslip inlayed within 35-mm Petri dishes using sterile technique in a laminar flow hood. Incubate dishes with their lids on overnight inside the 95% air/5% CO_2 atmosphere incubator at 37°C. This step represents a convenient way to allow the coverslip surface to wet properly and consequently allow good cellular adhesion. If shorter lead time is desired, we also have good results with cultures when medium was applied onto PEI-coated coverslips ~4 h prior to application of dissociated cells; this is roughly ~2.5 h before the initiation of a the dissection and culturing procedure (see Subheading 3.2.2).
2. Prepare the dissection area. Wear surgical gloves which should be disinfected by spraying with rubbing alcohol (70% isopropanol) as needed. Take great care to clean the area (including the dissecting scope, light source, and table top) thoroughly with rubbing alcohol, which reduces the likelihood of contamination. Place a terry cloth towel on the surface of the dissecting area. This will protect the surface of the dissecting table from debris; it will also prevent accidental sliding of dishes and tools/beakers. Place two Kimwipe™ sheets at the bottom of each of the two 200-mL beakers. This will protect tips of dissections tools. Fill beakers with 70% isopropanol up to the 75 mL mark. In one beaker, place your gross dissecting tools (10-mm angled spring scissors, Graefe curved serrated forceps, and blunt curved glass probe), while in the other, place your fine (Dumont forceps #55 and 2.5-mm angled spring scissors) dissecting tools, with their tips (covered with pipette tips) immersed in alcohol. Remove protecting pipette tips and shake alcohol off any tool prior to its use on animal tissue.
3. Attend the media, solutions, and dishes. (a) Place a bottle of fresh mito-MEM medium in an incubator to warm; this will take ~3 h. Open the lid (unscrew ~½ of turn), but do not remove it. This will allow pH equilibration. Medium can be used for up to 2 weeks; however, the fresher the better. (b) Also, remove a tube of papain solution (previously prepared in 1 mL aliquots) from a freezer and place in a hood to thaw.

Once thawed, place it in an incubator. (c) Fill a 50-mL centrifuge tube with fresh HBSS, which will be the source of HBSS for the following steps. Some of this HBSS may end up being unused and, if so, it will be subsequently discarded. Be sure to flame both the tube and the bottle before and after removing the HBSS, a practice that needs to be used will all culturing media/solutions. Place 5 mL of HBSS (from a 50-mL centrifuge tube) into a 15-mL centrifuge tube, which will be used to collect hippocampi to be cultured. Also, add 1 mL of HBSS into 35-mm culture dishes (2 per brain). Place the 15-mL centrifuge tube and the dishes on ice for later use. In addition, fill a 10-mL syringe (no needle) with HBSS.

4. Take a small transfer box with the lid (approximate box dimensions are 15×9×4 cm; with the bottom of the box covered with 3–4 layers of Kimwipe sheets) to the animal room and retrieve 1–2 pups to be dissected (see Note 16).

3.2.2. Dissection and Culturing

1. Begin at a dissecting bench. Use sterile technique. Remove the pup from the transfer box, and spray its skin with alcohol. Holding the pup's front legs with your index finger and thumb, position the pup with its back turned toward you. Using 10-mm curved dissecting scissors, sever the spine/spinal cord at the base of the skull to euthanize the pup. Then, make an incision just under the skin and almost completely around the circumference of the head. Peel the skin back to expose the skull. Inserting the scissors through the hole created during the euthanasia, cut the skull around the entire circumference of the head. Be careful to keep the tips of the scissors pointed out so as not to damage the brain beneath.
2. Lift the skull "cap" from the head using the Graefe forceps. This should expose the brain. There may be a substantial amount of blood bathing the brain at this point. Wash this blood away using a few milliliters of HBSS from the 10-mL syringe. While continuing to hold the pup, use the blunt probe to dislodge the olfactory bulbs. This can be accomplished by gently scooping (using the head of the hockey stick like glass probe) the anterior end of the brain. Again using the blunt probe, gently scoop the posterior end of the brain. This will separate the brain from the spinal cord and allow it to be removed to a culture dish containing 1 mL of cold HBSS.
3. Place the entire brain into a culture dish containing HBSS. Separate the brain into two hemispheres by using the heel of a fine dissecting scissors to "glide" from the midpoint of the cerebral cortex anteriorly through the front of the cortex between the olfactory bulbs. With fine forceps, gently pry the hemispheres apart to reveal the thalamus below. Again using the forceps, pinch the remaining tissue between each hemisphere and the thalamus to free them.

4. Remove each hemisphere to another culture dish, medial side up, containing fresh HBSS. Use forceps to grip the meninges and peel them away from the cortex. It is not necessary to remove all the meninges, just those over the tissue of interest. In addition, the choroid plexus may cover a portion of the hippocampus and may need to be removed.
5. The hippocampus will appear in the shape of the letter "C" at the posterior end of each hemisphere. The "C" begins near the dorsal surface rostrally and curves ventrally and caudally. The outer edge of the hippocampus is continuous with the rest of the cortex. The inner edge remains free with the lateral ventricle lying beneath it. Make an incision along the outer edge of the hippocampus. Incisions must also be made at the anterior and posterior ends of the hippocampus. This should completely free the tissue from the cortex. Using forceps, transfer the hippocampi to a 15-mL tube containing cold HBSS. Repeat steps 1–5 for the second pup (if dissecting 2 pups at the sitting) (see Note 16).
6. Wash the tissue by decanting the HBSS from the centrifuge tube and adding 5 mL of fresh HBSS.
7. Replace the HBSS with papain solution and leave for 1 h at 37°C in a 95% air/5% CO_2 incubator to enzymatically treat the tissue.
8. Decant the papain solution from the tube and again wash the tissue with HBSS.
9. Acutely prepare trypsin inhibitor solution (10 mg/mL) by adding 0.5 mL of 100 mg/mL trypsin inhibitor stock to 4.5 mL of 10 mM HEPES/HBSS. To stop the enzymatic action of papain, apply a trypsin inhibitor solution to the tissue for 5 min at room temperature.
10. Remove the trypsin inhibitor and wash the tissue with 5 mL of HBSS.
11. Decant HBSS and add 1.0 mL of warm culture mito-MEM medium per hippocampus.
12. Gently triturate (~40–60 strokes) the tissue through a 5-mL serological glass pipette until no visible clumps remain. Add warm culture medium so that the final dilution is 1.0 mL per hippocampus.
13. Plate the dissociated cells onto glass coverslips inlayed in 35-mm dishes at 50 μL per each coverslip, precoated with PEI and primed with 50 μL of medium.
14. Place the dishes with their lids on into an incubator for 3 h to allow cells to settle and attach to the substrate.

15. Wash debris from each dish after the 3 h. Fill each dish with 1 mL of mito-MEM. Remove this mito-MEM with suction and replace it with 1.0 mL of mito-MEM. For cultures that examine the effect of water-soluble SWNTs, replace the dish with 1.0 mL of mito-MEM with appropriate concentration of SWNTs dissolved in the medium.
16. Maintain the cultures in an incubator (37°C; 95% air/5% CO_2 atmosphere) for 3 days to allow for adequate growth, at which point neuronal cell morphology can be assessed.

3.3. Cell Imaging

3.3.1. Calcein Loading

1. Inside a laminar flow biosafety hood, place a dish taken from the 37°C incubator. Transfer one coverslip from a culture dish into a 35-mm Petri dish containing loading solution, composed of 1 μg/mL of calcein-AM and 0.025% of pluronic F-127 in external solution. Incubate the coverslip for 15 min at room temperature (see Note 17). To transfer a coverslip, use a sterile fine forceps with the tip flamed using an alcohol burner. Return the Petri dish containing unused coverslips in culture media back to the 37°C incubator immediately after transfer of an experimental coverslip (see Note 18).
2. After the end of the 15-min incubation period in loading solution, transfer the coverslip into another 35-mm Petri dish containing 1 mL of external solution to allow for the dye to de-esterify for 15 min.

3.3.2. Imaging and Analysis

1. Prepare a clean imaging chamber for attachment of the coverslip. After placing the chamber up-side down, apply a streak of sealing grease at the recess of the chamber. Using sterile forceps, take the coverslip containing cells out of external solution and place it centered onto the chamber recess with the cell-side facing toward the recess and lightly press it with forceps to loosely adhere the coverslip to the grease. Flip the chamber so that the open bath faces up, while the coverslip is at the bottom. Press the chamber down against a Kimwipe™ to seal the coverslip (see Note 19). Add ~400 μL of external solution (room temperature) into the chamber. Check for leaks (see Notes 20 and 21). Aspirate the external solution and replace it with 400 μL of external solution.
2. We use an inverted microscope (Nikon TE300) equipped with differential interference contrast and epifluorescence illumination (xenon arc lamp, 100 W) to image the cells (see Note 22). To visualize the calcein-loaded cells, we use a 60× plan apochromatic oil immersion objective (numerical aperture, 1.4; Nikon) and a standard FITC filter set. We use a cooled charge-coupled device camera (CoolSNAP *HQ*; Photometrics, Tucson, AZ)

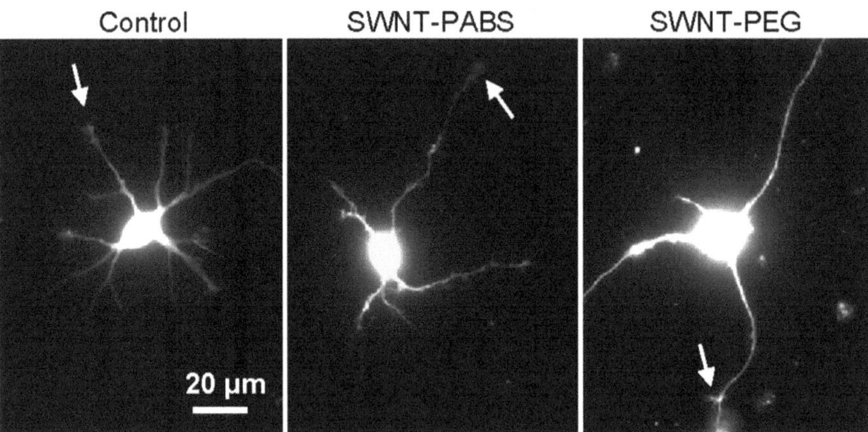

Fig. 2. Calcein-loaded neurons grown on PEI-coated glass coverslips and treated with chemically functionalized water-soluble SWNTs added to the culture medium. Fluorescence images of live neurons, accumulating the vital stain, calcein. Neurons grown on PEI-coated glass coverslips (control, sham treated) can be treated with CNTs (each at 1 μg/mL), either SWNT-PABS (*middle*) or SWNT-PEG (*right*) to affect their growth characteristics [consult the original work (7) for details]. *Arrows* indicate growth cones. Scale bar, 20 μm. Modified from ref. (7).

driven by V++ imaging software (Digital Optics Ltd., Auckland, New Zealand) or MetaMorph™ software (Molecular Devices, Chicago, IL) to acquire images and control an electronic shutter (Vincent Associates, Rochester, NY) in the excitation pathway. Neurons are identified based on their morphological features using differential interference contrast and fluorescence microscopy (see Notes 23 and 24). For neurons that extend processes beyond the field of view, acquire multiple images and merge together using Adobe Photoshop CS2 (Adobe Systems Inc., San Jose, CA) (see Note 25). The images of cultured hippocampal neurons grown on PEI substrates and exposed to water-soluble CNTs (SWNT-PEG and SWNT-PABS) are shown in Fig. 2.

3. We analyze cell morphology using the OLYMPUS MicroSuite ™ Basic software to quantify the neuronal morphological characteristics. We examine six parameters: neurite number, total neurite length, mean neurite length, number of growth cones, number of branches, and number of branches per neurite (see Fig. 3). For neurite length measurements, manually trace neurites on a computer monitor using a computer mouse. After selecting the "polygon length" mode, manually trace the neurite by clicking the mouse at various loci of the neurite to assign its segments, as illustrated in Fig. 4. Sum the lengths of all individual segments for a particular neurite to calculate the full length of that neurite (see Note 26).

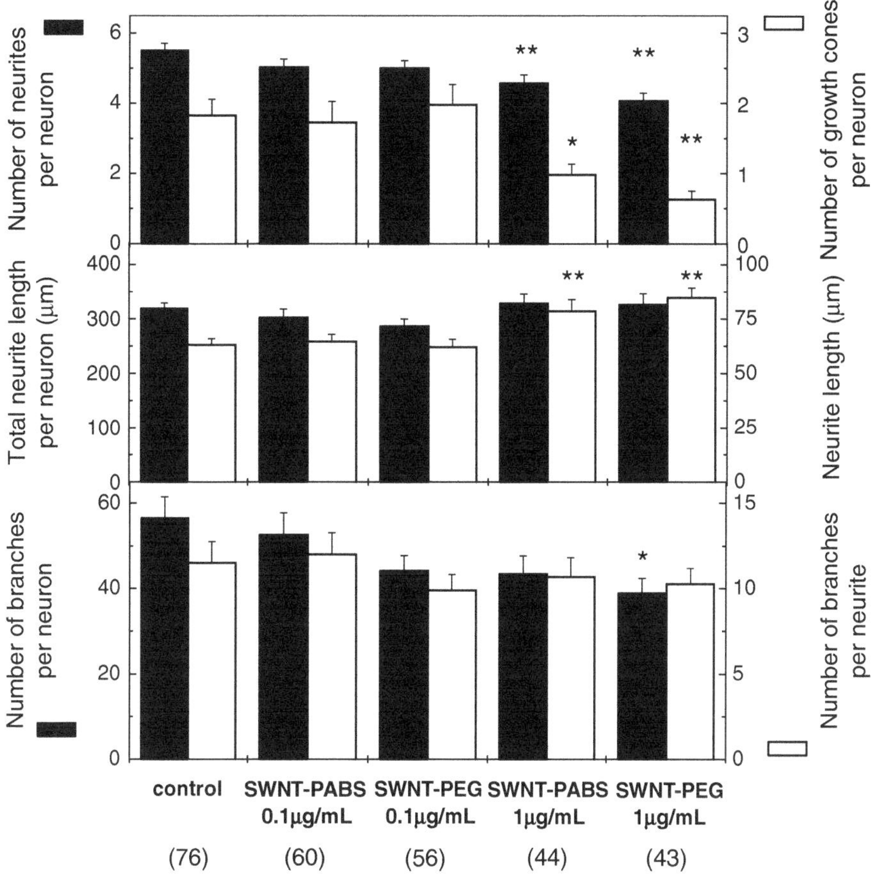

Fig. 3. Parameters of neuronal growth and morphology. Neurons grown on PEI-coated glass coverslips (control) were treated with SWNT-PABS or SWNT-PEG each at two different concentrations (0.1 and 1 μg/mL). Bars represent means ± standard errors of means. Numbers in parentheses indicate the number of neurons studied in each condition. *Asterisks* indicate a significant difference in measurements (*$p < 0.05$, **$p < 0.01$). For a description of the effects that water-soluble SWNTs exert of neuronal morphology, consult the original work (7). Reproduced from ref. (7).

4. Notes

1. Unless stated otherwise, the water used in all of our procedures is purified by the Milli-Q® Synthesis system (Millipore Corp.; http://www.millipore.com/pressroom/cp3/5khpn7). This ultra-pure water has 18.2 MΩ cm resistivity, less than 5 parts per billion of organics content, and pyrogen content less than 0.001 EU/mL. It can be sterilized by autoclaving (121°C, 20 psi, 45 min).
2. This PEI stock solution can be stored for 4–6 weeks at 4°C. Alternatively, aliquot in appropriate amounts (100 μL) and store at –20°C for up to 1 year.

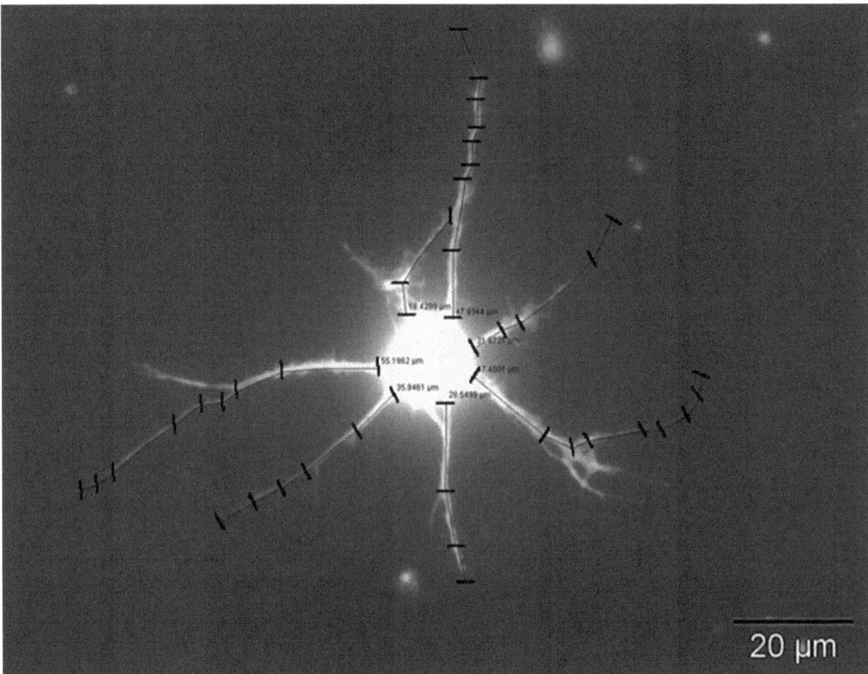

Fig. 4. Neurite tracing approach. Shown here are examples of neurite length measurements, where neurites were manually traced by an investigator marking various loci of an individual neurite (*ticks*). Lengths of individual segments (*between tick marks*) were summed to calculate the full length of a neurite (*segmental lines*).

3. The borosilicate glass coverslips used have been pretested for mammalian neural cell culture. These glass coverslips (12 mm in diameter, thickness #1, 0.13–0.16 mm; D-263 glass, Erie Scientific Company) can be purchased via Fisher Scientific (cat. no. 12-545-82-12CIR-1D). Prior to subsequent coverslip processing procedures, clean the glass coverslips (one ounce packet contains ~500 coverslips) by placing them in 2% v/v RBS 35 (Pierce) detergent dispersion in Milli-Q water and boiling for 15 min. Next, rinse the coverslips by running distilled water for 30 min and soak inside a beaker containing Milli-Q water overnight. To remove possible residual detergent film retained on coverslips, wash each individual coverslip by using a clean metal forceps to hold the coverslip and dipping it three times in Milli-Q water in a beaker. Repeat this step for two additional times in two separate beakers. Air-dry the washed coverslips in a laminar flow hood by placing them at a slanted angle onto a drying scaffold, which ensures that both sides of the coverslips are air-dried properly. The drying scaffold can be made by making accordion folds (~1 cm) of a 90-mm diameter filter paper. Store the dried washed coverslips in Petri dishes with sheets of filter paper between them to prevent scratching. Sterilize the washed coverslips under UV light prior to PEI coating. The time for washing one packet of coverslips takes ~3 h.

4. UV sterilization is done using the GS Gene Linker™ UV Chamber (Bio-Rad; Power set at Str, 2×90 s). Alternatively, one can use UV lamps in the laminar flow hood, but the duration of exposure needs to be adjusted according to the manufacturer's recommendation.
5. Rat pups can be obtained from a breeding colony or from timed-pregnant female Sprague-Dawley rats that are commercially available (e.g., Charles River Laboratories).
6. A blunt glass probe can be made using a Pasteur pipette with the tip sealed and curved (like a hockey stick) by heating it on a Bunsen burner.
7. We use appropriate plastic pipette tips to cover the tips of all tools in order to protect them from mechanical damage before placing tools into the glass dish.
8. This calculation is needed because various batches of papain may have different enzyme activity leading to variable total content of international units (IU) in 100 mg of papain. It takes about ~1 h at room temperature to dissolve papain, a process which can be assessed visually as the turbid mixture becomes a clear solution.
9. The amount of bicarbonate added to the medium is adjusted so that medium reaches the appropriate pH (7.35) when maintained in a 95% air/5% CO_2 atmosphere incubator. The pH of medium needs to be checked (measured) and the level of bicarbonate readjusted if needed. The addition of glutamine to the mito-MEM is recommended even though MEM already contains glutamine. However, glutamine has a limited storage life in the refrigerator and it breaks down more quickly at higher temperatures.
10. We use Thermo Scientific* HyClone USDA Tested Fetal Bovine Serum (cat. no. SV30014) sterile filtered by the manufacturer. We test each serum batch for various functional properties of neurons (including neurite outgrowth) and astrocytes. Based on our previously used batches, the company selects new serum with matching properties (various parameters used) and puts a desired amount of serum on reserve for 4–6 weeks until we thoroughly test it in our experimental paradigms. Presently, we use the batch FTM33793, while in the past 5 years, we used three other batches (in reverse chronological order): FRE26364, FPC20883, and FNS18021. We provide these batch numbers for guidance only.
11. We use a custom-made open diamond bath imaging chamber with a circular recess at its bottom to accommodate mounting of a coverslip (12 mm in diameter, thickness #1). Similar imaging chambers are commercially available (e.g., cat. no. RC-25 or RC-25, Warner Instruments).

12. We repackage the vacuum grease into a 3-mL syringe to which we attach a 1-cm-long 18-gauge blunt needle. This handheld syringe/needle is used to apply the vacuum grease at the circular recess to the bottom of the chamber in order to attach the coverslip. Note that the grease displays fluorescent properties and should be used parsimoniously without smudging the bottom of the coverslip with it, which would occur if too much grease is applied.
13. To prepare dry DMSO, open an ampoule containing 5 mL of sterile filtered DMSO (Sigma-Aldrich) and pour it into a 15-mL conical centrifuge tube filled to the 2 mL level with molecular sieve beads (sodium aluminosilicate molecular sieves, 8–12 mesh beads, cat. no. M-2635, Sigma-Aldrich) to absorb water. Keep the tube tightly capped and wrapped in aluminum foil, as DMSO is hygroscopic and light sensitive. Store at room temperature. The beads are reusable. However, note that beads contain indicator beads that change color from blue to pink when saturated with water. At that juncture, they need to be replaced.
14. If dishes containing PEI-coated coverslips are not used immediately, their lids can be sealed with Parafilm© and the dishes stored in a sterile container at room temperature for up to 1 week of PEI coating application. Do not expose these PEI-coated coverslips to UV light as this polymer is sensitive to UV light.
15. It is cost-effective to have all segments within a Petri dish populated with PEI-coated coverslips as this allows reduced usage of culture medium. However, the cell culture is also successful when number of coverslips is reduced from a maximal of 4 to 2–3 coverslips per dish.
16. For best results, it is advisable to dissect no more than one pup while learning the procedure. However, even for a proficient user, we do not recommend dissecting more than two pups at any one sitting. As the number of rats increases, the time needed for dissection increases. During the preparation of neuronal cell cultures, speed is of the essence. Once the rat pup dies, the brain tissue will begin to deteriorate. We have observed a decrease in cell viability which coincides with increases in the number of pups dissected.
17. To prepare calcein-AM loading solution, add 1 μL of pluronic acid stock solution to 998 μL of external solution. Vortex and then add 1 μL of calcein-AM stock solution. Vortex and transfer to a 35-mm Petri dish. This solution can be used for loading cells attached to multiple coverslips, up to several hours. We normally prepare this solution twice a day.
18. Minimize the transfer time to avoid fluctuations of the medium pH, which can affect the health of the cell culture.

19. When pressing the coverslip attached to the imaging chamber against the Kimwipe™, make sure that you place the wipe on a clean flat surface and that you do not apply too much pressure, otherwise the coverslip can crack. If you crack the coverslip, replace it after cleaning the chamber and reapplying the grease.
20. By initially placing the chamber-coverslip assembly on the Kimwipe™, you will notice some wetting of the wipe because some of the culturing medium will be retained at the bottom surface of the coverslip. Subsequently, lift up and place the whole assembly onto a different spot of the wipe. Leaks are easily recognized, as the chamber will quickly empty to wet the wipe.
21. Regardless of whether the chamber leaks or not, after pressing the assembly against the wipe, you may also notice a circular grease imprint on the wipe. If any, this should be minimal. If in excess, next time, mount the coverslip with somewhat less grease, but not an insufficient amount necessary for sealing. Do not slide the assembly across the wipe at any time as you will cause smudging of the grease onto the coverslip. This will affect the quality of images since (1) the grease does not mix with the oil used on the objective and hence it would generate the distortion of images, and (2) the grease fluoresces and thus increases the background.
22. We place our microscopes on antivibration isolation tables. We find this approach necessary in order to prevent possible movement artifacts.
23. We confirm morphological identification of neurons by labeling with the neuron-specific markers: anti-tubulin III (also referred to as Tuj1), anti-neuron-specific enolase, and the FITC-conjugated C-fragment of tetanus toxin (6, 7).
24. When acquiring images of calcein-loaded cells, one should be aware of possible photobleaching of the fluorescent dye which can be minimized by the use of neutral density filters inserted in the excitation pathway and by adjusting the camera integration time. Obtained images must have fluorescence intensity values within the dynamic range of the camera (0-4095 for a 12-bit camera) and prominently display the full extent of neurites and their branches.
25. It is sometimes necessary to take several images to capture the entire morphology of the cell when the cell's processes extend beyond the field of view. If multiple image acquisition is required, make sure that the focus of the cell remains unchanged during the entire image acquisition process as even a minute change in focus might affect the proper merging of images, especially when smaller cellular structures such as fine cell processes are in question. Merging images can be done by using the automated photomerge function in the File tab of Photoshop CS2. If Photoshop CS2 software is not available,

an alternative free image merging software that can be used is ImageJ (NIH, USA) with the stitching plug-in installed.

26. Alternatively, analysis of cell morphology can be automated by using the Neurite application module of MetaMorph™ software ver. 6.1 (Molecular Devices) to quantify the neuronal cell morphological characteristics after neurons were grown on CNT scaffolds/substrates (9). Using this approach, we examine eight parameters: neurite number, mean neurite length, maximum neurite length, neurite straightness, total neurite outgrowth, number of growth cones, number of branches, and cell body area (see Fig. 4 of ref (9)).

Acknowledgments

We would like to thank Yingchun Ni and Hui Hu for providing Figure 4, and Randy F. Stout, Jr. for comments on a previous version of the manuscript. This work was supported by the National Science Foundation (CBET 0943343).

References

1. Silver J, and Miller JH (2004) Regeneration beyond the glial scar. Nat Rev Neurosci 5, 146–156
2. Silva GA (2006) Neuroscience nanotechnology: progress, opportunities and challenges. Nat Rev Neurosci 7, 65–74
3. Lee W, and Parpura V (2009) Chapter 6 - Carbon nanotubes as substrates/scaffolds for neural cell growth. Prog Brain Res 180, 110–125
4. Hu H, Ni Y, Mandal SK, Montana V, Zhao B, Haddon RC, et al (2005) Polyethyleneimine functionalized single-walled carbon nanotubes as a substrate for neuronal growth. J Phys Chem B 109, 4285–4289
5. Hu H, Ni Y, Montana V, Haddon RC, and Parpura V (2004) Chemically Functionalized carbon nanotubes as substrates for neuronal growth. Nano Letters 4, 507–511
6. Malarkey EB, Reyes RC, Zhao B, Haddon RC, and Parpura V (2008) Water soluble single-walled carbon nanotubes inhibit stimulated endocytosis in neurons. Nano Lett 8, 3538–3542
7. Ni Y, Hu H, Malarkey EB, Zhao B, Montana V, Haddon RC, et al (2005) Chemically functionalized water soluble single-walled carbon nanotubes modulate neurite outgrowth. J Nanosci Nanotechnol 5, 1707–1712
8. Zhao B, Hu H, Yu AP, Perea D, and Haddon RC (2005) Synthesis and characterization of water soluble single-walled carbon nanotube graft copolymers. J Am Chem Soc 127, 8197–8203
9. Malarkey EB, Fisher KA, Bekyarova E, Liu W, Haddon RC, and Parpura V (2009) Conductive single-walled carbon nanotube substrates modulate neuronal growth. Nano Lett 9, 264–268
10. Gogolla N, Galimberti I, DePaola V, and Caroni P (2006) Preparation of organotypic hippocampal slice cultures for long-term live imaging. Nat Protoc 1, 1165–1171
11. Woods G, and Zito K (2008) Preparation of gene gun bullets and biolistic transfection of neurons in slice culture. J Vis Exp, pii: 675. doi: 10.3791/675

Chapter 24

High-Resolution Imaging and Evaluation of Spines in Organotypic Hippocampal Slice Cultures

Frederik Sündermann, Nataliya Golovyashkina, Christian Tackenberg, Roland Brandt, and Lidia Bakota

Abstract

Dendritic spines act as sites of excitatory neuronal input in many types of neurons. Spine shape correlates with the strength and maturity of synaptic contacts. Thus, evaluation of spine morphology is relevant for studies on neuronal development, for determination of morphological correlates of learning and memory, and for analysis of mechanisms of neurodegeneration. Here, we describe a method to determine spine morphology in an *ex vivo* model of organotypic hippocampal slice cultures prepared from transgenic or non-transgenic mice. Spines are imaged using confocal high-resolution imaging and evaluated by algorithm-based analysis. The approach permits semiautomated determination of spine density and classification of different spine types in dendritic segments from hippocampal subregions to evaluate intrahippocampal connectivity.

Key words: Dendritic spine, Hippocampus, Organotypic culture, Laser scanning microscopy, Automated morphology analysis

1. Introduction

Dendritic spines are small protrusions from dendrites and consti tute the primary site of excitatory synaptic input in many principal neurons of the brain including pyramidal cells in the neocortex and hippocampus, and Purkinje neurons in the cerebellar cortex (1). Spine volumes range between 0.01 and 0.8 μm^3 and are thought to function as microcompartments for segregating postsynaptic responses. Evidence indicates that spine alterations are related to learning, memory, and neurodegeneration and that spines show high structural dynamics (2, 3). However, it is still unclear which mechanisms are involved and what is the functional or pathological role of spine changes during development and degeneration.

Stephen D. Skaper (ed.), *Neurotrophic Factors: Methods and Protocols*, Methods in Molecular Biology, vol. 846, DOI 10.1007/978-1-61779-536-7_24, © Springer Science+Business Media, LLC 2012

Organotypic hippocampal slice cultures combine the accessibility and maintenance of in vitro culture systems with the preservation of intact hippocampal synaptic circuitry and anatomy (4). The latter is especially important for spine analysis, since spine changes can occur in a subset of neurons in a region-specific manner. Preparation of slice cultures from mice provides the further advantage that the advances in transgenic mouse technology, which resulted in the generation of a variety of different mouse models, can be used to determine the contribution of single gene products on spine morphology. An example presents the use of mouse models that are transgenic for mutated amyloid precursor protein (APP), which is known to be involved in the development of Alzheimer's disease (AD). Using this model, we have demonstrated that the density and shape of spines are altered compared to cultures from non-transgenic animals (5).

Algorithm-based analysis of spine density and morphology requires fluorescence labeling of individual neurons in a slice preparation. One way to label neurons is the use of a Sindbis virus expression system, which permits effective infection of neurons in organotypic slices with this neurotropic alphavirus (6). By using fusion constructs of green fluorescent protein (GFP) with coding sequences of interest, the effect of the synthesized protein on spines can be studied in infected neurons in an authentic environment, which also can be exploited to perform live-cell imaging of degenerating neurons (7). This approach has, for example, been taken in our lab with AD-like modified cytoskeletal proteins (5, 8). The production of Sindbis virus vectors for gene transfer has been described in many excellent papers and manuals and will not be covered in this article (see e.g., (9)). As an alternative to virus infection, algorithm-based analysis is also possible with slices that have been prepared from genetically modified mice that express fluorescent proteins in selected neurons (e.g., (10)). The following chapter describes a method to determine spine morphology in an *ex vivo* model of mouse organotypic hippocampal slice cultures using confocal high-resolution imaging and algorithm-based analysis.

2. Materials

2.1. Components for Slice Preparation, Culture, Infection, and Fixation

1. Animals: 6–8 days old mouse pups (e.g., mouse lines with C57BL/6 background; see Note 1).
2. MEM: Minimum Essential Medium Eagle, HEPES Modification without L-glutamine (Sigma-Adrich).
3. BME: Basal Medium Eagle with Earle's, without L-glutamine (Gibco BRL Life Technologies).
4. NB: Neurobasal Medium without L-glutamine (Gibco BRL Life Technologies).

5. Horse serum.
6. Glutamine, 200 mM.
7. Pen-Strep: Penicillin/Streptomycin, 100× (PAA Laboratories).
8. Fungizone Gibco Fungizone® Antimycotic, liqid (Invitrogen).
9. N1-supplement: N1 medium supplement, 100×, for neural cell cultures (Sigma-Adrich).
10. Virus: Virus constructs encoding fluorescent protein (e.g., SinRep5-eGFP) that can be optionally tagged with other protein of interest.
11. Confocal matrix : Micro-Tech-Lab (Graz, Austria).

2.2. Media, Supplements, and Other Solutions

Prepare all media under sterile conditions using sterile filter such as 50 mL Steriflip or 500 mL Steritop-GP 0.22 µm ExpressTM Membrane (Millipore), depending on the volume of medium needed.

1. Heat-inactivate horse serum at 56°C for 45 min. The aliquots can be stored at −20°C for at least 1 year. (see Note 2).
2. Dissolve glucose powder in double distilled H_2O (dd H_2O) to 20% w/v and sterile-filter using a 0.22 µm filter. Store at 4°C.
3. *Dissection medium*: 1% glutamine and 1% Pen-Strep in MEM. To 98 mL MEM, add 1 mL glutamine and 1 mL Pen-Strep. The medium can be stored at 4°C for 24 h. (see Note 3).
4. *Culture medium*: 25% horse serum, 25% BME, 3% glucose, 1% glutamine, 0.5% Pen-Strep, and 0.5% fungizone in MEM. In a 50 mL sterile conical tube add 12.5 mL horse serum, 12.5 mL BME, 1.5 mL glucose, 0.5 mL glutamine, 0.25 mL Pen-Strep, and 0.25 mL fungizone; bring volume to 50 mL with MEM and filter sterilize. Preheat only the amount of medium that is needed for a medium change on the same day (count 2–4 mL additionally for pH adjustment). Adjust the pH value to 7.2 with HCl/NaOH. If not preheated, culture medium can be stored at 4°C for at least 1 month. (see Note 4).
5. *NB-N1 medium*: 3% glucose, 1% glutamine, 1% N1-supplement, 0.5% horse serum, 0.5% Pen-Strep, and 0.5% fungizone in MEM. In 50 mL sterile conical tube, add 1.5 mL glucose, 0.5 mL glutamine, 0.5 mL N1-supplement, 0.25 mL horse serum, 0.25 mL Pen-Strep, and 0.25 mL fungizone; bring volume to 50 mL with NB and filter sterilize. Preheat only the amount of medium that is needed for a medium change on the same day (count 2–4 mL additionally for pH adjustment). Adjust the pH value to 7.2 with HCl/NaOH. If not preheated, NB-N1 medium can be stored at 4°C for at least 1 month. (see Note 4).
6. *Fixation solution*: preheat 25 mL of phosphate buffed saline (PBS) to 70°C. Add 1 g paraformaldehyde, mix and let it cool down. Add 1 g sucrose. Store at −20°C.

2.3. Culture and Dissection Dishes

1. Cell culture inserts. Millicell cell culture inserts, 30 mm, organotypic hydrophilic PTFE, 0.4 μm. #PICM0RG50, Millipore. (see Note 5).
2. Multidish 6-well plates (Nunclon™). Alternatively, single 3.5-cm culture dishes can be used.
3. Bacterial dish (10 cm) as a dissection dish.

2.4. Dissection Tools (see Note 6; Fig. 1a)

1. Large surgical scissors. Type: standard. Alloy: stainless steel. Tip shape: sharp/blunt. Tip angle: straight. Length: 13 cm. Effective cutting edge: 42 mm. (#14001-13).
2. Small scissors. Type: standard. Alloy: stainless steel. Feature: toughcut (this feature is not crucial). Tip shape: sharp/sharp. (#14058-11).
3. Graefe forceps. Length: 10 cm. Alloy: stainless steel. Tip shape: curved. Tips: serrated. Tip dimensions: 0.8 mm × 0.7 mm. (#11052-10).
4. Dumont #5 Forceps. Length: 11 cm. Alloy: titanium (or inox). Tip shape: straight. Tip dimensions: 0.05 mm × 0.02 mm. (#11252-40).
5. Hippocampus tool (spatula) (×2). Length: 16 cm. Alloy: stainless steel. Tip shape: straight. Tip diameter: 8.5 mm × 0.3 mm. (see Note 7).
6. Forceps. Length: 10.5 cm. Alloy: remanit 4301. Tip shape: angled. Tips: serrated. (#2854.1).
7. Spatula (×2). Length: 13 cm. Alloy: stainless steel. End shape: rounded. (see Note 7).

2.5. Large Equipment for Preparation and Culturing Organotypic Slices

1. Laminar flow CL2 safety cabinet.
2. Sterile dissection safety cabinet.
3. McIlwain tissue chopper. Standard razor blades can be used with the tissue chopper.
4. Dissection microscope (e.g., Leica, but any 5–10× magnifying dissection microscope is suitable).
5. Cell culture incubator at 37°C and 5% CO_2.

2.6. Equipment for Image Acquisition

Confocal Laser Scanning Microscope (e.g., Nikon Eclipse TE2000-U or Zeiss LSM 510) equipped with 10× and 20× air, 40× and 60×/63× oil immersion objectives (suitable for fluorescence imaging), lasers and filter sets corresponding to the fluorescent proteins (e.g., 488 nm argon laser for GFP), and image acquisition software (e.g., EZ-C1 software, Nikon or LSM 5 software, Carl Zeiss).

2.7. Software and Hardware Requirements

1. Autodeblur 9.3 software.
2. 3DMA Version 0204 software.

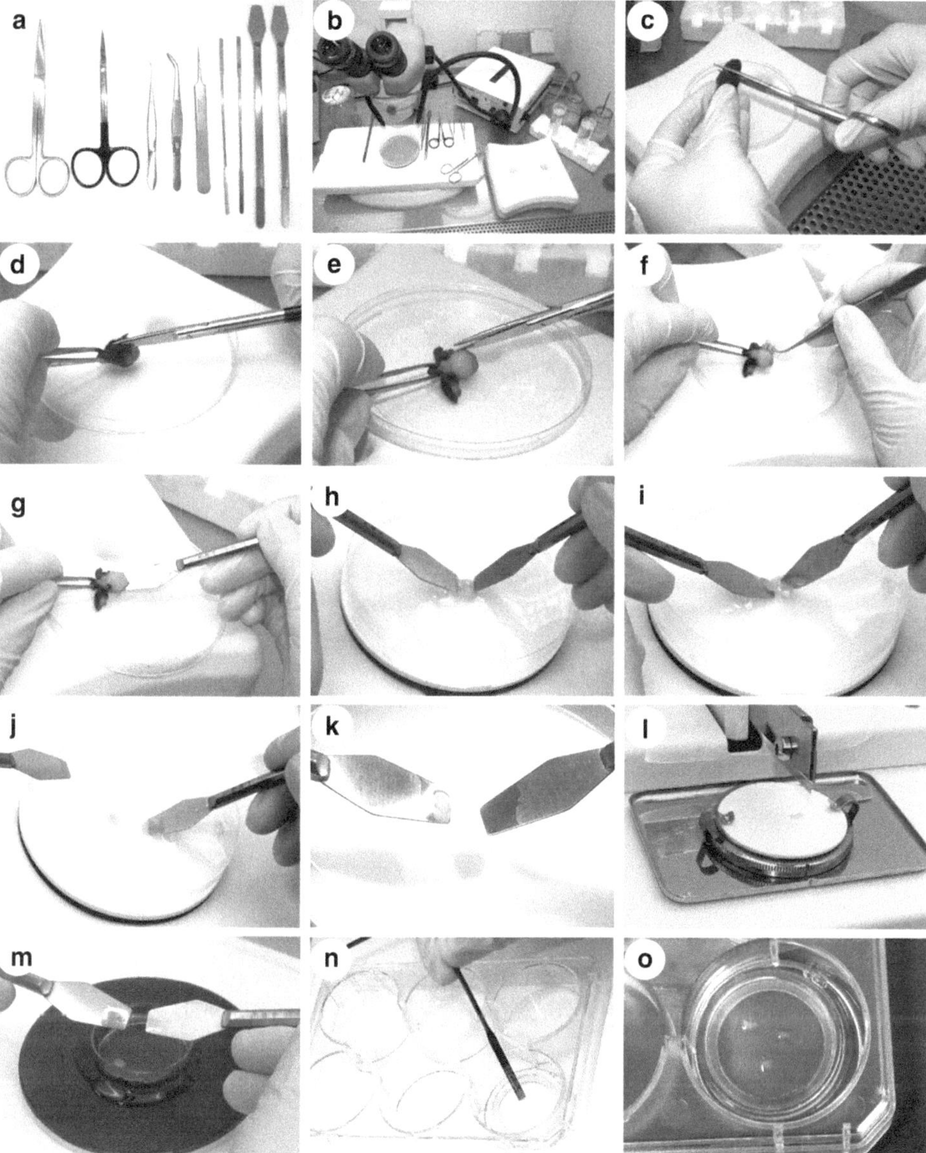

Fig. 1. Preparation of organotypic hippocampal slices from mouse brain. (**a**) Preparation tools. (**b**) Dissection setup. (**c**) Decapitation of a 1 week old mouse. (**d**) Incision of the skin along the midline of the head. (**e**) Removal of the skin to the sides with the scissors. (**f**) Removal of the skull with the curved forceps. (**g**) Removal of the brain with the spatula. (**h**) Cutting off the cerebellum. (**i**) Separation of two brain hemispheres with the spatula. (**j**) Dislodgement of the midbrain with the spatula. (**k**) Hippocampus, prepared for cutting into slices. (**l**) Cutting of the hippocampus with the tissue chopper. (**m**) Cut and separated slices. (**n**) Placing a slice onto a membrane insert within 6-well plate. (**o**) Three slices placed on the membrane insert.

3. PC with "Ubuntu Linux 6.06 LTS" as operating system and the following additional packages: Jgraph (http://www.cs.utk.edu/~plank/plank/jgraph/jgraph.html), xv (http://www.trilon.com/xv/).
4. PC with "Microsoft Windows XP" as operating system.

3. Methods

3.1. Preparation, Culturing, Infection, and Fixation of Hippocampal Slices

Organotypic slice cultures from hippocampus are cultivated according to the membrane interface technique (4) (see Note 8).

3.1.1. Dissection Preparation

The procedure takes place under the sterile culture bench.

1. Add 1 mL of culture medium to each well of a 6-well plate.
2. Place a membrane insert into each well with sterile forceps (see Note 9).
3. Place the plates and dissection medium on ice.
4. Prepare two 50-mL centrifuge tubes with sterile PBS or ddH$_2$O (see Note 10).
5. Prepare sterile eppendorf cups on ice if tissue collection for genotyping is required.
6. Cut rounds from Whatman paper, place one into each of the dissection dishes and sterilize under UV (see Note 11).

3.1.2. Dissection of the Hippocampi

The procedure takes place under the sterile preparation bench. The tools are sterilized with 70% ethanol or by autoclaving. Solid cooling pads can ensure low temperature during dissection procedure (see Fig. 1b).

1. Decapitate a 6–8 day old mouse and place the head into the lid of the sterile dissection dish (see Note 12; Fig. 1c).
2. Spray the head with 70% ethanol to avoid contamination of the samples originating from the surface of the skin.
3. Fix the head at the eyes with the angled forceps, and using small scissors, make an incision in the skin along the midline of the head (see Fig. 1d).
4. Remove the skin on both sides with the scissors (see Note 13; Fig. 1e).
5. Cut in the middle of the skull with the curved forceps pulling it neatly below the skull from rostral to caudal end (see Note 14).
6. Move away the sectioned skull with forceps (see Fig. 1f).
7. Remove the brain with spatula by gently reaching below (in between the brain and skull (see Note 15)) and lifting it up (see Fig. 1g).
8. Place the brain into the prechilled dissection medium.
9. Gently fix the brain at the position of the cerebral cortices with one of the spatulas and remove the cerebellum with the other (see Fig. 1h).

10. Further using the spatula, make a sagittal cut between the two hemispheres to separate them (see Fig. 1i).
11. Flip one of the hemispheres onto the convex side and remove the thalamus and the basal ganglia situated on top of the hippocampus under a dissection microscope (see Fig. 1j).
12. Cut out the hippocampus from the underlying cortex (see Fig. 1k).
13. With fine forceps clean away the vessels around the hippocampus (see Note 16).
14. Repeat the procedure with the other hemisphere.
15. Place the two hippocampi in a small petri dish with prechilled dissection medium on ice.

3.1.3. Preparation of Hippocampal Slices with Tissue Chopper

The procedure takes place under the sterile preparation bench.

1. Using the spatulas, place two hippocampi on a teflon stage of the tissue chopper.
2. Aspirate the excess medium with help of a sterile pipette (see Note 17).
3. Cut rapidly 400 μm thick slices with the tissue chopper (see Note 18; Fig. 1l).
4. Transfer the slices back into the small petri dish and place them on ice (see Note 19).
5. Identify intact individual slices under the dissection microscope and transfer them with the small spatulas onto the membrane inserts (see Note 20; Fig. 1n).
6. Change the culture medium below the inserts to fresh, cooled medium (see Note 21).
7. Place the 6-well plate with slices into the incubator at 37°C with 5% CO_2.

3.1.4. Maintenance and Infection of the Slices

The procedure takes place under the sterile culture bench.

1. Exchange the culture medium in the culture dishes every 2–3 days (see Note 22).
2. On day 11 post-preparation, change the culture medium to NB medium containing N1 supplement (see Note 23).
3. On day 12 post-preparation, apply the virus with the droplet method (see Note 24).

3.1.5. Fixation of Hippocampal Slices

Cultures are fixed at day 3 postinfection to ensure the highest expression of the fluorescent protein and best signal-to-noise ratio for appropriate image processing.

1. Let the slices remain attached to the culture plate membrane to preserve hippocampal structure and rinse with PBS within a 6-well plate (see Note 25).

2. Apply ~2 mL/well cold fixing solution for 2 h at 4°C.
3. Wash with PBS at least three times for 15 min each.
4. Cut out a piece of membrane with the attached slices using a scalpel and transfer the membrane with forceps onto a glass slide (see Note 26).
5. Mount the cultures with confocal matrix and cover with cover-slip (see Note 27).

3.2. Microscopy of Hippocampal Slices

1. Locate the effectively infected slices using 10× objective through eyepiece.
2. With 20× objective identify the regions of the hippocampus and the cells that will be imaged.
3. With 40× objective identify the dendritic branches of individual pyramidal neurons on the apical or basal side, respectively.
4. Image CA1 and CA3 pyramidal neurons with voxel size of 0.08×0.08×0.25 μm in the *x–y–z* directions with 60×/63× objective. Adjust the image size according to the length and shape of the imaged dendritic fragment (see Note 28).

3.3. Semiautomated Analyses of Dendritic Spines

3.3.1. Image Processing with Autodeblur

1. Open the file in the Autodeblur program.
2. Open the 3D Deconvolution menu (see Fig. 2) at Deconvolution → 3D Deconvolution.
3. Optionally you can insert a cutting step (see Note 29; see Fig. 3).
4. Choose "Adaptive PSF (Blind)" as deconvolution method.
5. Click on <change> in the "Optics Settings" to adjust the image and PSF settings. These options depend on the confocal setup (numerical aperture, magnification, refractive index) and the emission wavelength of the used fluorescent markers.
6. The "Deconvolution Settings" depend on the level of noise of the images. For confocal images an iteration number from 10 to 15 and a level of noise between low and medium is sufficient.
7. Adjust the "Output Settings" to your needs. For 3DMA analysis, a 16-bit tiff image format is recommended.
8. Click on <Start> to start the process or on <Batch> to add another file for deconvolution. The batch process can be started from "Batch processing" in the "Deconvolution" menu.

3.3.2. Analysis with 3DMA

Autodeblur is run on a PC with "Windows XP" as an operating system. The original "3DMA" software is created for operating under "Unix/Linux" systems. To perform 3DMA based analysis, a "Windows XP" PC for deconvolution and a "Unix/Linux" PC (Ubuntu 6.06 LTS) for 3DMA analysis is required. For data transfer, a "Share" folder, which can be accessed by both operating systems, is recommended (see Note 30).

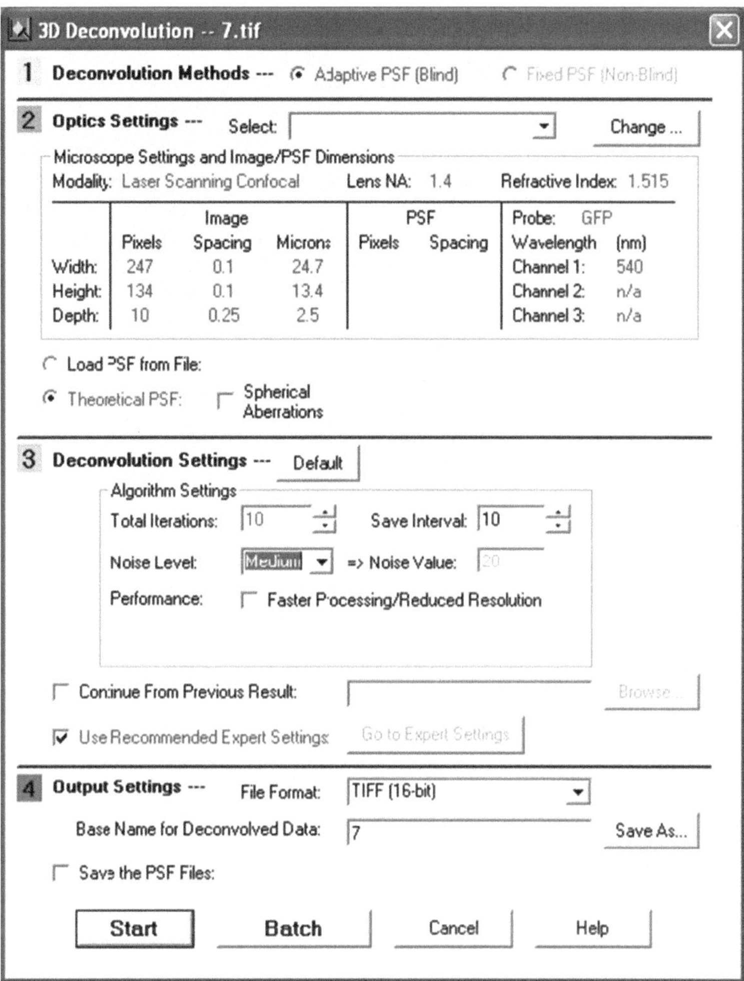

Fig. 2. Deconvolution settings dialog window. In this dialog window deconvolution settings can be adjusted as described in the protocol (refer to Subheading 3.3.1, steps 4–7). The shown preselections for spine analysis are recommended by the authors.

1. Within the “Linux” operating system, open a terminal (xterm) and switch in superuser mode (sudo su).
2. After deconvolution it is necessary to create a new folder for your analysis. Change to this folder and type “create_wdir” (see Fig. 4 line: 01) in order to automatically create the required subfolders.
3. Copy the image data (*.tif and *.aqh files) into the “raw” folder. Change to the “raw” folder (see Fig. 4 line: 02).
4. To make the files readable for “3DMA” convert them into a “rsz”-format (see Fig. 4 line: 03-05).
5. Switch to the “Recompiled Fixed Source” mode by typing “3dmaswitch” in the terminal window.
6. Convert the file by typing “aqh2mtif <filename>” (without suffix, e.g., filename.tif will be filename).

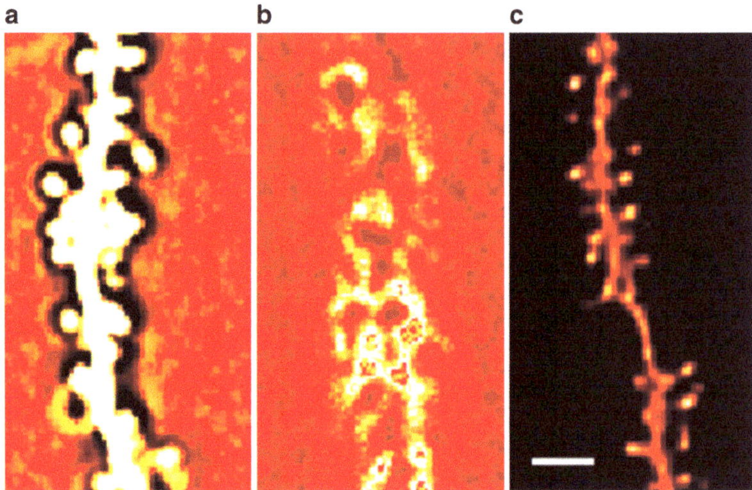

Fig. 3. Results of the deconvolution. (**a**) The deconvolution algorithms enhance edges within images and can produce artefacts like black borders around bright image compounds. (**b**) Other artefacts are shadows of brighter structures imprinted in dark planes of their neighborhood. The intensity of these imprints is slightly higher than the background. (**c**) Projection of an appropriately deconvolved image stack. Scale bar, 2 μm.

```
01:> create_wdir
02:> cd raw
03:> 3dmaswitch
04:> aqh2mtif <file_without .tif>
05:> 3dmaswitch
06:> cd ../cases
07:> 3dmaGUI_lnx
```

Fig. 4. List of command lines for 3DMA file processing.

7. When all files are converted, switch back to "Original Source" mode by typing "3dmaswitch" again.
8. Change to the "cases" folder and start 3DMA by "3dmaGUI_lnx" (see Fig. 4 line: 06-07).
9. After the welcome screen you are asked to choose an application. Choose <Dendritic Spine>.
10. In the "Application Routine Menu" (see Fig. 5a) you can choose between different routines. Start with "Spine Detection Routines."
11. Choose the options as shown in (see Fig. 5b) and click on <continue>.
12. In the "Data and Parameter Settings Section" switch to <resized> and limit the listed files to files with "rsz" extension by typing in the "Data Filename in raw directory" field "../raw/*.rsz." After clicking <Load Data> your files will appear in the file section. Choose one file for processing by clicking on it. Remove the other files from the list by pressing "space" key repeatedly.

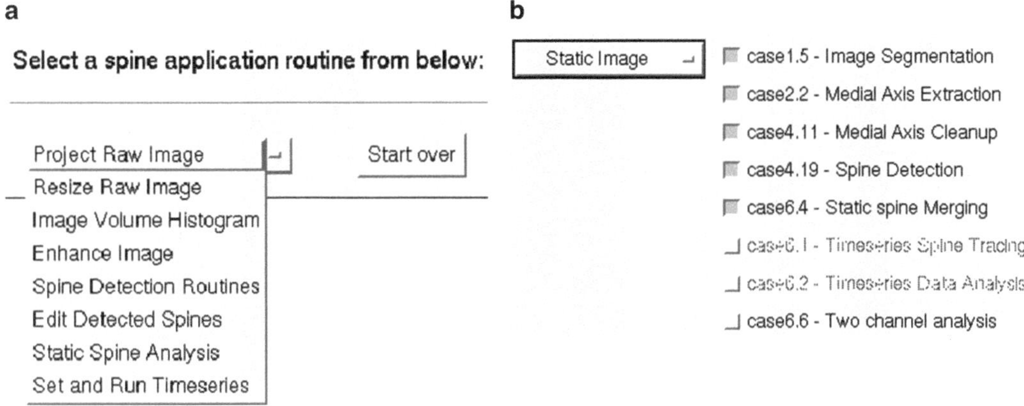

Fig. 5. Screenshot of 3DMA menus. (**a**) Applications essential for spine analysis are listed: "Spine Detection Routines," "Edit Detected Spines," and "Static Spine Analysis". (**b**) After choosing "Spine Detection Routines" default options are recommended.

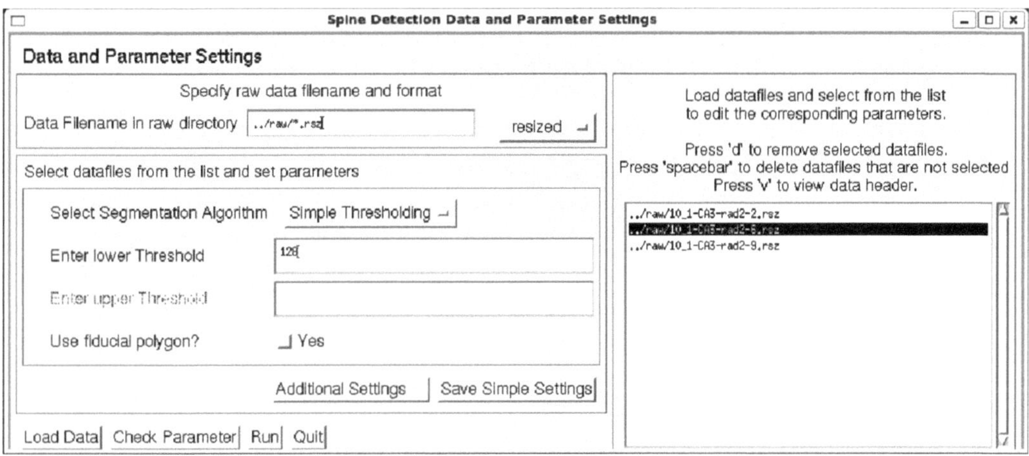

Fig. 6. Data and parameter selection for spine detection.

13. Select the options as shown in (see Fig. 6): uncheck <Use fiducial polygon>, select <Simple Thresholding>. Enter a threshold level in <Enter lower threshold>. As an alternative, choose <Indicator Kriging> and enter a lower and an upper threshold level (see Note 31; Fig. 7).
14. Click <Save Simple Settings> to save the entered settings. After confirmation start your analysis by clicking <Run>.
15. After finishing the processing of the file start a <New Run>, choose <Dendritic Spine> as application, and select the <Edit Detected Spines> routine for manual correction.
16. In the menu which appears select the file to edit by double clicking. Deselect falsely detected spines and dendrites by right mouse click (see Fig. 8). Click <Save> to finish the editing process (see Note 32).

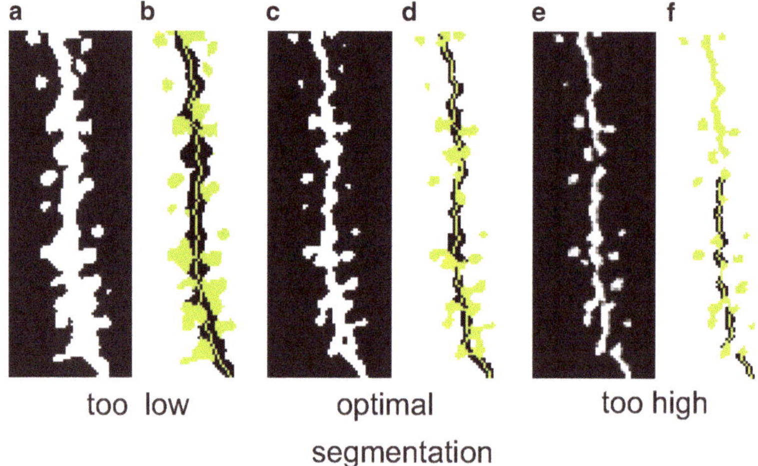

Fig. 7. Effects of thresholding. (**a**,**c**,**e**) Comparison of different threshold levels for spine detection to visualize the effects of thresholding. (**b**,**d**,**f**) Projections of thresholded images with its detected spines. (**b**) Image with threshold set too low shows a bias toward stubby spines. (**d**) Optimally thresholded image with no bias toward any spine type. (**f**) Images with threshold set too high shows segmented dendritic backbone and a bias toward thin and mushroom spine types.

Spine	Dendrite	X	Y	Z	Length	Volume	Voxels
32	1	150.809	215.085	11.553	0.306	0.255	141
3	1	127.686	111.246	9.576	0.770	0.213	118
17	1	147.568	167.885	11.649	0.567	0.267	148
16	1	130.311	140.467	10.017	0.559	0.650	360
6	1	136.308	43.744	10.795	0.685	0.070	39
7	1	126.486	122.027	10.432	0.701	0.067	37
2	1	169.219	237.656	7.531	0.595	0.058	32
1	1	147.053	132.105	8.355	0.784	0.137	76
5	1	124.569	74.569	9.877	0.837	0.117	65
11	1	164.026	211.030	9.542	1.050	0.489	271
12	1	138.077	177.606	14.891	1.008	0.513	284
13	1	161.132	203.034	12.248	0.919	0.423	234
14	1	121.598	54.201	10.265	0.806	0.368	204
15	1	143.831	153.731	9.604	0.604	0.556	308
		142.478	60.196	11.196	0.850	0.083	46
		140.550	68.550	10.150	0.517	0.108	60

Sort by Dendrite index | Save | Quit

Fig. 8. Cropped dialog window for manual editing of detected spines. On the left side in the full dialog view, an overview projection of the dendritic segment with colored spines is shown.

17. To complete your analysis and obtain the statistical data, start again a <New Run>, select the <Dendritic Spine> application, and choose the <Static Spine Analysis> routine.
18. In this menu load your data by clicking <Load Data> and set the output filename. Alternatively, change the output folder as well. The default options are "../spine/detect/analysis.txt."
19. The analysis file can be read with every text or word editor (see Note 33).

4. Notes

1. Different transgenic mouse lines can be used with C57BL/6 background, e.g., APP transgenic line, as an Alzheimer's disease model.
2. Use aliquots from the same batch of serum, since horse serum can differ from batch to batch.
3. About 25 mL are needed for the dissection dish and about 3 mL for each small dish with hippocampi. Hence, 50 mL of dissection medium are enough for ~5 pups.
4. Antibiotics can have additional effects on slices in culture. e.g., penicillin can reduce GABAergic transmission (11). If one is able to successfully culture slices without using antibiotics, they should be omitted from the medium. However, once started it is important to use antibiotics throughout the entire experimental procedure to avoid differential influence on the results.
5. Different inserts are commercially available. However, only Millipore provides PTFE membranes. The material is crucial for appropriate slice development (certain degree of flattening during the culture time) and subsequent confocal microscopy (the PTFE membrane is optically clear in contrast to membranes from other materials).
6. One can use dissection tools with the same basic features (e.g., size and material) from other providers.
7. Above-mentioned tools are purchased from Fine Science Tools GmbH, Heidelberg and Carl Roth GmbH & Co. KG, Karlsruhe.
8. Organotypic cultures are often made also according to roller-tube technique (12, 13). However, during cultivation with the roller-tube method the slices thin to a practically one-cell layer. The membrane interface culture method maintains a thickness of slices at ~100–150 μm during the cultivation period, which makes this method more suitable for the study of 3D spine analyses.

9. Take care that air bubbles do not remain under the membrane. The membranes will become completely transparent when wetted.
10. One of the tubes with sterile PBS/ddH_2O may serve to clean most of the blood and tissue from the tools before placing them into 70% ethanol, the other to rinse the tools from ethanol before continuing to use them for further dissection.
11. Preparing the whole package of dishes at once, and packing them back into the original sterile bag after sterilization together with Whatman paper ensures a ready dissection dish for many preparations. The Whatman paper provides a better contrast and a non-slippery surface during the dissection procedure.
12. Cut off an ~3 mm portion of the tail if genotyping is required. The tail should be placed immediately into a sterile eppendorf cup kept on ice. Pay attention that the dish in which the decapitation is taking place is cleaned throughout with 70% ethanol after each mouse to avoid false-positive PCR results due to cross contamination of the genetic material.
13. The upper and lateral skull should be completely clean and accessible for further preparation.
14. Care must be taken to avoid injury during this procedure. Pulling the forceps slightly upward, while sliding it might help avoid injury to the brain.
15. During this move the cranial nerves are cut so that the brain can be flipped out.
16. The remaining connective tissue might block the separation of the chopped slices.
17. Avoid injuring the slices but gently correct their position with the pipette tip to align them perpendicular to the axes of the blade.
18. Set the blade arm of the chopper in the position closest to the hippocampi to minimize the cutting time. Add a drop of cold dissection medium onto the slices immediately after chopping.
19. During this procedure gently pick up the slices with help of the big spatulas and try to separate the slices without injuring them by fine, round movements of the spatulas close to one another (see Fig. 1m). Keep the slices on ice for 30 min before continuing the procedure.
20. Up to 3 or 4 slices can be cultured on one insert (see Fig. 1o).
21. This procedure takes place under the sterile culture bench. Suck all the medium from the dish with a sterile glass pasteur pipette during each of the medium changes otherwise the new medium will increase in volume due to the residual medium resulting in leveling above the membrane and eventually to

suffocation of the cells within slices. Carefully place the fresh medium below the membrane and not on the slices by pulling it out from the pipette tip between the insert and the wall of the culture dish.

22. To avoid temperature changes within the slice during medium exchange keep the 6-well plate on a warm pad while it is out of the incubator. The culture medium during slice cultivation should always be applied pre-warmed.
23. Changing to a different medium prior to infection is required to drastically lower the serum content that might counteract the virus infection but in parallel it complements some of the serum proteins to maintain adequate culture conditions.
24. The use of Sindbis virus was chosen for this approach because it infects neurons with high efficiency and provides a fast synthesis of the fluorescence protein on levels that ensure good image quality. Preparation of Sindbis virus is carried out under safety level 2 conditions according to the Sindbis Expression System Manual (Catalog no. K750-01). A small aliquot of the virus is thawed on ice and 1 μL of the viral solution is gently applied with a pipette onto each slice without touching the surface of the slice. Silicone-coated tips have reduced surface tension that helps in dispensing the infectious particles.
25. For an extensive rinse use ~2 mL/well of PBS (one below and one above the membrane).
26. It is easier to cut the membrane if the insert is transferred onto a piece of cardboard box. Suck off the medium prior to the procedure and soak up on paper the excessive liquid.
27. Due to the thickness of the slices, place two small, round coverslips on the glass slide, flanking the membrane as elevators. Cover the sample together with the small coverslips with one bigger coverslip. Avoid air bubbles under the coverslip, as that would disturb imaging process.
28. The length of the imaged dendritic segments should be between 20 and 30 μm. Choose a dendritic segment that is not crossed by other dendrites.
29. Deconvolution is also an edge-sharpening procedure. Therefore, it can introduce some artefacts, like a darker border around brighter image components and an imprint of the brighter areas in darker planes (see Fig. 3). To reduce these artefacts cut the empty planes around the area of interest during a preprocessing step and adjust the options in the "Deconvolution Settings", like iteration number and level of noise.
30. 3DMA-Neuron as a software package for automated neuronal morphology analysis is no longer available from the homepage of the authors (W.B. Lindquist and C.M. Weaver). It has been

commercialized by Nihon Visual Science, Inc. In case of interest to purchase the software, contact Nihon Visual Science, Inc. (E-mail: info@nvs.co.jp). Other programs such as NeuronStudio, a noncommercial program created at Mt. Sinai School of Medicine by the Computational Neurobiology and Imaging Center (14), may also be applicable for spine analysis but has not been tested by us. We are also currently developing a robust and easy-to-use software tool, which we will provide free of charge upon request.

31. Choosing an appropriate threshold level is the most critical step during spine analysis. If the threshold is too low, the volume of the dendritic segment and the spines appear larger, which results in a bias toward stubby spine type. If the threshold is too high, the volume of the dendritic segment and the spines appear smaller than they are which results in a bias toward thin and mushroom spine types. A threshold chosen much too high will segment the dendritic backbone and make further analysis impossible (see Fig. 7).
32. Open the file with "rsz" extension, to look at the original image in parallel. This will help to determine falsely detected spines.
33. In the analysis file, the following parameters are listed: total number of spines, number of mushroom, stubby and thin spines, spine density, dendrite radius, and length and volume of each spine.

Acknowledgements

This work has been supported by the Deutsche Forschungsgemeinschaft (DFG BR1192/11-2) and a Lichtenberg-Fellowship within the graduate college "Membranes and cellular communication" (supported by the state of Lower Saxony; FS). We appreciate the valuable work of Neelam Shahani and Tobias Wolf during the establishment of this technique.

References

1. Harris KM, and Kater SB (1994) Dendritic spines: cellular specializations imparting both stability and flexibility to synaptic function. *Annu Rev Neurosci 17*, 341–371
2. Tackenberg C, Ghori A, and Brandt R (2009) Thin, stubby or mushroom: spine pathology in Alzheimer's disease. *Curr Alzheimer Res 6*, 261–268
3. Kasai H, Fukuda M, Watanabe S, Hayashi-Takagi A, and Noguchi J (2010) Structural dynamics of dendritic spines in memory and cognition. *Trends Neurosci 33*, 121–129
4. Stoppini L, Buchs PA, and Muller D (1991) A simple method for organotypic cultures of nervous tissue. *J Neurosci Methods 37*, 173–182
5. Tackenberg C, and Brandt R (2009) Divergent pathways mediate spine alterations and cell death induced by amyloid-beta, wild-type tau, and R406W tau. *J Neurosci 29*, 14439–14450

6. Ehrengruber MU, Lundstrom K, Schweitzer C, Heuss C, Schlesinger S, and Gahwiler BH (1999) Recombinant Semliki Forest virus and Sindbis virus efficiently infect neurons in hippocampal slice cultures. *Proc Natl Acad Sci USA 96*, 7041–7046
7. Bakota L, and Brandt R (2009) Live cell imaging in the study of neurodegeneration. *Int Rev Cell Mol Biol 276*, 49–103
8. Shahani N, Subramaniam S, Wolf T, Tackenberg C, and Brandt R (2006) Tau aggregation and progressive neuronal degeneration in the absence of changes in spine density and morphology after targeted expression of Alzheimer's disease-relevant tau constructs in organotypic hippocampal slices. *J. Neurosci 26*, 6103–6114
9. Ehrengruber MU, and Lundstrom K (2007) Alphaviruses: Semliki Forest virus and Sindbis virus vectors for gene transfer into neurons. *Curr Protoc Neurosci Chapter 4:Unit 4.22*
10. Feng G, Mellor RH, Bernstein M, Keller-Peck C, Nguyen QT, Wallace M, Nerbonne JM, Lichtman JW, and Sanes JR (2000) Imaging neuronal subsets in transgenic mice expressing multiple spectral variants of GFP. *Neuron 28*, 41–51
11. Andersen P (1983) Basic mechanisms of penicillin-induced epileptiform discharges. *Prog Clin Biol Res 124*, 3–13
12. Gähwiler BH (1988) Organotypic cultures of neural tissue. *Trends Neurosci 11*, 484–489
13. Gähwiler BH, Capogna M, Debanne D, McKinney RA, and Thompson SM (1997) Organotypic slice cultures: a technique has come of age. *Trends Neurosci 20*, 471–477
14. Rodriguez A, Ehlenberger DB, Dickstein DL, Hof PR, and Wearne SL (2008) Automated Three-Dimensional Detection and Shape Classification of Dendritic Spines from Fluorescence Microscopy Images. *PLoS ONE 3*, e1997

Chapter 25

Imaging Amyloid Precursor Protein In Vivo: An Axonal Transport Assay

Tomás L. Falzone and Gorazd B. Stokin

Abstract

Fusion of fluorescent probes to axonally transported proteins represents an established approach that enables live imaging of axonal transport. In this approach, in vivo examination of fluorescent particle dynamics provides information about the length, directionality, and the velocity by which axonally transported proteins travel along axons. Analysis of these parameters provides information about the distribution of axonal proteins and their dynamics in and between different subcellular compartments. Establishing the movement behavior of amyloid precursor protein within axons indicated that live imaging approaches offer the opportunity to significantly enhance our understanding of the biology as well as pathology of axonal transport. This chapter provides a fluorescence-based procedure for measuring axonal transport of APP in cultured newborn mouse hippocampal neurons.

Key words: Amyloid precursor protein, Axonal transport, Fluorescent probes, In vivo fluorescent imaging, Primary hippocampal cell culture

1. Introduction

Axonal transport is a complex system that guarantees the delivery of proteins and organelles to synapses, axonal compartments, and cell bodies in a timely manner. This transport system, which supplies high synaptic protein turnover with the appropriate cell-derived commands and delivers environmental signals from different presynaptic environments to cell bodies, relies on molecular motors moving cargoes along microtubule tracks at the expense of ATP hydrolysis. The force-generating molecular motors transport cargoes either toward the synapses or toward cell bodies by means of anterograde or retrograde axonal transport, respectively (1).

Amyloid precursor protein (APP) is a type II integral membrane protein (2). In neurons, APP-containing vesicles undergo

Stephen D. Skaper (ed.), *Neurotrophic Factors: Methods and Protocols*, Methods in Molecular Biology, vol. 846, DOI 10.1007/978-1-61779-536-7_25,

constitutive bidirectional axonal transport. Several observations link APP to the pathogenesis of Alzheimer's disease (AD). In brief, aberrant *App* genotypes segregate with AD phenotype (3), and aberrant accumulation of proteolytic fragments of APP in the form of amyloid peptides and plaques are considered a neuropathological hallmark of AD brains (4). Intriguingly, aberrant accumulation of APP and its proteolytic fragments has been long described within axons of AD brains (5) as well as in several other diseases ranging from traumatic brain injury to multiple sclerosis. The early pathological presence of APP in the axons points to a role of APP beyond AD, suggesting that APP may represent a bona fide surrogate marker of axonal injury (6, 7).

Early work on axonal APP focused largely on immunochemical studies. These approaches indisputably localized APP within axons and uncovered its aberrant accumulation in disease (8). Axonal transport of APP was first suggested in a sciatic nerve ligation study, which showed APP synthesized in the dorsal root ganglia accumulating proximally to the ligation site within axons (9). It was after the development of fluorescent approaches, however, when bidirectional movement of APP was first visualized, together with the assessment of its dynamic behavior within axons (10, 11). Although the function of axonal APP movement behavior and the mechanisms regulating axonal transport of APP remain largely unknown, these approaches indisputably revolutionized our understanding of APP biology and allowed the identification of APP axonal transport impairments in animal models of AD, which added a new twist to the current pathogenesis of AD (12–15). In conclusion, novel fluorescent approaches to studying axonal transport can be applied to any axonally transported transmembrane or soluble protein. The tracking and surveillance of these proteins have only started producing a wealth of new data that will significantly further our understanding of the axonal milieu (16, 17). This chapter describes a detailed protocol in which the above approach is applied to the axonal tracking of APP in cultured mouse hippocampal neurons.

2. Materials

2.1. Primary Hippocampal Neuron Cultures

1. Cold Hank's balanced salt solution (HBSS, Gibco). Store at 4°C (see Notes 1–4).
2. Hank's buffer: to 500 mL HBSS, add 0.4 g of D-glucose (Sigma-Aldrich), 0.834 g of 4-(2-hydroxyethyl)-1-piperazineethanesulfonic acid (HEPES, Sigma-Aldrich), 5 mL of 100× penicillin-streptomycin (Invitrogen) and filter-sterilize using 0.22-μm-diameter filter cartridge. Store at 4°C.

3. Disaggregation buffer: add to phosphate buffered saline (PBS, Gibco) 1.5 mM DL-cysteine HCL (Sigma-Aldrich), 0.025% bovine serum albumin (BSA, Sigma-Aldrich), and 35 mM D-glucose (Sigma-Aldrich). Store at –20°C.
4. Papain solution: add 45 U of lyophilized papain (Worthington) and 0.05% DNase I (Boehringer Mannheim) to disaggregation buffer. Once papain is dissolved, filter-sterilize by passing through a 0.22-μm-diameter filter cartridge.
5. Complete Dulbecco's Modified Eagle's Medium (DMEM, Gibco): supplement DMEM with 10% fetal bovine serum (FBS, Gibco) and 500 μM L-glutamine (Sigma-Aldrich). Store at 4°C.
6. Complete Neurobasal medium: enrich Neurobasal medium (Gibco) with 1:50 (v/v) of 50× serum-free supplement B27 (Invitrogen) and 500 μM L-glutamine (Sigma-Aldrich). Store at 4°C.
7. Boric acid buffer: to 400 mL of water, add 1.24 g of boric acid (Sigma-Aldrich) and 1.9 g of borax (Sigma-Aldrich), then filter-sterilize the solution. Store at 4°C.
8. Poly-D-lysine-coated coverslips: wash round coverslips (Fisher Scientific) three times using 100% acetone with gentle agitation, then three times with 100% ethanol, and finally three times with sterile water. When dried, coverslips are incubated with 100 μg of poly-D-lysine (MW 30,000–70,000) per mL of boric acid buffer for several hours at room temperature, then washed thoroughly with sterile water, and placed in a 24-well dish (Falcon, BD biosciences) for cell plating.
9. Stereo dissecting scope (SMZ660, Nikon).
10. Fine forceps and scissors (Roboz).

2.2. Transfection

1. Pcdna3 CMV APP YFP vector expressing a protein fusion between APP695 (APP) and yellow fluorescent protein (YFP) at the C-terminus of APP (see Note 5) (10).
2. Lipofectamine 2000 (Invitrogen). Store at 4°C.
3. 10-day-old primary hippocampal cell culture.

2.3. Axonal Transport Imaging

1. Inverted epifluorescence microscope (TE-2000U, Nikon)
2. Heat- and CO_2-controlled chamber (Harvard Apparatus, Holliston, MA)
3. Photometrics CoolSNAP HQ-cooled CCD camera (Roper Scientific, Ottobrunn, Germany)
4. Glass bottom plates (MatTek, Ashland, MA)
5. MetaMorph 6.0 (Universal Imaging Corporation, Marlow, UK)

3. Methods

3.1. Harvesting Cells from Hippocampi

1. Primary hippocampal cell cultures are harvested from postnatal day 1 mice (see Note 6). Using fine forceps under a dissecting scope, cut the skin above calvarium by a midline anteroposterior incision. Separate from the underlying bones of the calvarium both skin flaps generated by the incision. Incise the cranial bones in the top midline portion and chip away to expose the brain.
2. Detach the brain from the cranial nerves, remove from the skull, and cut in half. Use a small spoon to remove the hippocampi from the brain and meninges, and place into cold HBSS (see Note 7).
3. Under a tissue culture hood, wash the excised hippocampi with 10 mL of HBSS at 4°C and then digest in 0.22-μ-filtered papain disaggregation buffer for 20 min with moderate shaking at 37°C.
4. After digestion, wash hippocampi twice with 10 mL of complete DMEM (see Note 8), triturate 12 times with a 1-mL micropipette in 2 mL of complete DMEM, and plate on top of poly-D-lysine-coated coverslips at 60,000 cells per well in a 24-well dish (see Note 9).
5. After allowing 2 h for neurons to attach, replace the complete DMEM with 500 μL of complete Neurobasal medium. Maintain cell cultures at 37°C in 5% CO_2 incubator.

3.2. Transfection

Ten- to fourteen-day-old primary hippocampal cell cultures are transfected with pcdna3 CMV-APP-YFP expression vector using Lipofectamine 2000.

1. For each well, dilute 0.8 μg of endotoxin-free purified DNA in 50 μL of plain Neurobasal medium and gently mix. Similarly, gently mix and dilute 2 μL of Lipofectamine 2000 in 50 μL of plain Neurobasal medium and then incubate for 5 min.
2. After incubation, combine the diluted DNA with diluted Lipofectamine 2000 to a total volume of 100 μL, mix gently, and incubate for 20 min.
3. Remove 300 μL of complete Neurobasal medium from each well and add 100 μL of DNA-Lipofectamine complexes to each well containing primary hippocampal cells (see Note 10). Mix gently by rocking the plate back and forth.
4. Incubate cells at 37°C in a 5% CO_2 incubator and replace medium to complete Neurobasal medium 2–4 h later (see Note 11).

3.3. In Vivo Imaging

1. In vivo transport of APP-YFP vesicles is imaged 16 h after transfection using a heated stage at 37°C and in a 5% CO_2 chamber using an inverted epifluorescence microscope connected to a Photometrics CoolSNAP HQ-cooled CCD camera and driven by MetaMorph 6.0. (see Note 12).
2. Remove transfected cells from multiwell plates by lifting the coverslip with forceps and flipping it over a glass bottom plate containing complete Neurobasal medium equilibrated in a 5% CO_2 incubator (see Note 13).
3. Identify fluorescent (transfected) cell bodies using an oil immersion super apochromatic 100× lens. Follow axonal projections at least two fields away from the cell body view. Anterograde and retrograde axonal transport is determined depending on the orientation of cell bodies and projection tips.
4. Register axonal dynamics of fluorescent APP-YFP vesicles in 15-s stacks format (tiff) and capture at a speed of ten frames per second at ×100 magnification and 2 × 2 binning for continuous registration (see Note 14, Fig. 1). Kymographs are generated from stacks and analyzed using MetaMorph or converted to QuickTime or AVI movie format for presentations. Calibration of movie pixel magnification and number of frames per second should be established prior movie registration and used across experiments.

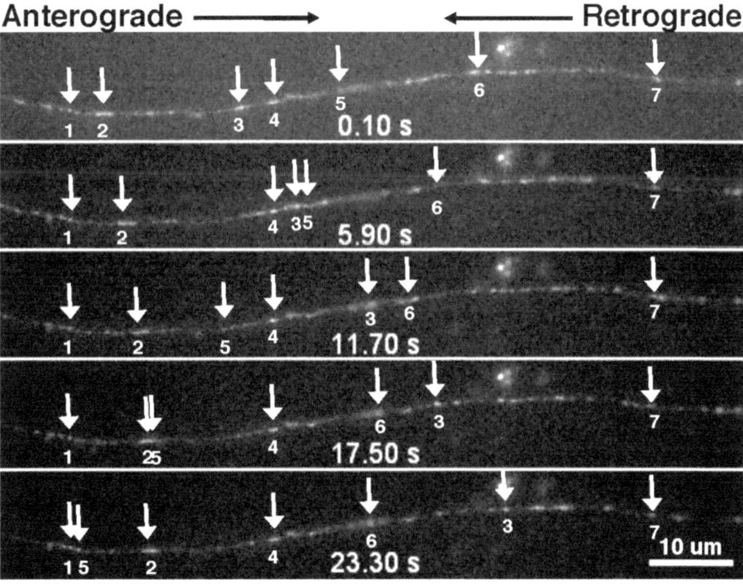

Fig. 1. Axonal transport movie registered in primary hippocampal neurons transfected with APP-YFP using an inverted epifluorescence microscope. 100-ms frame image montage of a movie registered in a transfected axon showing dynamics of APP-YFP vesicles. *White arrows* indicate stationary (1, 4, 7), anterograde (2, 3) or retrograde (5, 6) moving vesicles, respectively. Time corresponds to frames extracted from a 30-s movie. Bar equals 10 μm.

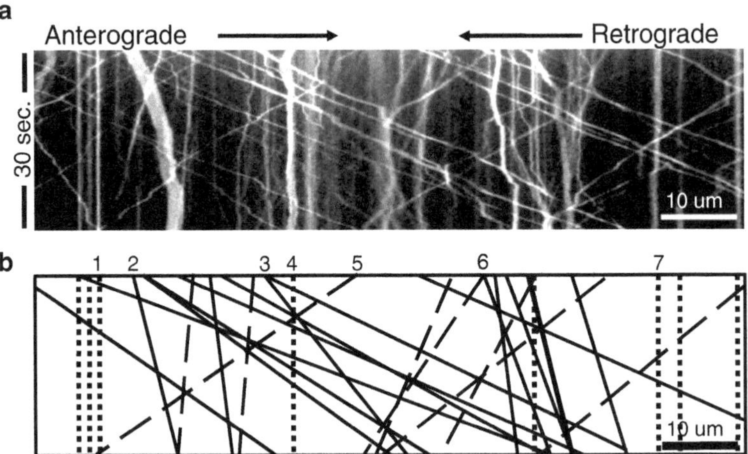

Fig. 2. Kymograph generated from movie showing APP-YFP axonal transport dynamics. (**a**) Kymograph generated from movie in Fig. 1 showing the distance displacement of anterograde, stationary, and retrograde APP-YFP vesicles along 30 s. (**b**) Schematic representation of the above kymograph in which all APP-YFP vesicles were tracked and plotted. *Straight*, *dashed*, and *dotted black lines* correspond to anterograde (2, 3), retrograde (5, 6), and stationary (1, 4, 7) particles, respectively. Bar equals 10 μm.

3.4. Kymograph Generation and Analysis

1. Open movie stack in a working station computer using MetaMorph (see Note 15). Draw a line of five-pixel width from cell body to axon tip to define the precise location of the measurements and to track the length of the axon in captured movies.
2. Using the kymograph function in MetaMorph, generate a montage image from the plotted line in which distance and time are represented by the X and Y axes, respectively (Fig. 2a).
3. Using MetaMorph, in each kymograph plot lines following all visible straight and descending fluorescent particles generated by APP vesicle dynamics along time (Fig. 2b). Transfer measurements extracted from lines such as angle, X and Y axes, and start and end points obtained from pixel identification to an Excel file and convert to microns by time for analysis (Fig. 3a). APP-YFP vesicle movement is approximated from kymograph as a percentage of clearly defined, stationary, anterogradely or retrogradely moving APP-YFP out of defined particles, which started and ended movement within the time and distance captured in the kymograph. Average speed, distance, and directionality of axonal transport are calculated in Excel files for further statistical analysis (Fig. 3b).
4. In line with scientific rigor, all kymographs are coded before the beginning of each experiment and scored blind to the identity of the sample. Only upon completion of particle plotting and of all data collection, the samples are uncoded to allow analysis.

a

	Features	Angle	Distance in pixels	Distance in um	Time (Sec)	Speed (distance/time)
	L5	110.0254	653	87.64566	29.75	2.946072605
	L11	113.0494	557	74.76054	29.75	2.512959328
	L21	113.2764	444	59.59368	23.875	2.496070366
	L14	115.5885	497	66.70734	29.75	2.242263529
	L12	121.1345	394	52.88268	29.75	1.777569076
Anterograde	L32	123.6502	332	44.56104	29.75	1.497850084
	L7	126.4692	322	43.21884	29.75	1.452734118
	L16	137.5114	218	29.25996	29.75	0.983528067
	L1	147.3438	141	18.92502	29.625	0.638819241
	L22	160.4841	84	11.27448	29.75	0.378974118
	L23	164.2774	67	8.99274	29.75	0.302276975
	L24	166.5316	57	7.65054	29.75	0.257161008
	L6	166.7595	56	7.51632	29.75	0.252649412
	L18	174.2417	24	3.22128	29.75	0.108278319
	L9	174.9364	21	2.81862	29.625	0.095143291
	L15	179.2778				
	L20	179.2778				
	L27	179.2778				
	L2	179.5165				
Stationary	L3	180				
	L4	180				
	L26	180.2407				
	L25	180.9629				
	L13	186.2606	26	3.48972	29.75	0.117301513
	L8	186.4993	27	3.62394	29.75	0.121813109
	L31	193.2269	55	7.38210	29.5	0.250240678
Retrograde	L19	202.3801	98	13.15356	29.75	0.442136471
	L30	206.7589	119	15.97218	29.5	0.541429831
	L17	216.0131	173	23.22006	29.75	0.780506218
	L29	232.2419	306	41.07132	29.75	1.380548571
	L28	232.36	271	36.37362	29.75	1.222642689

b

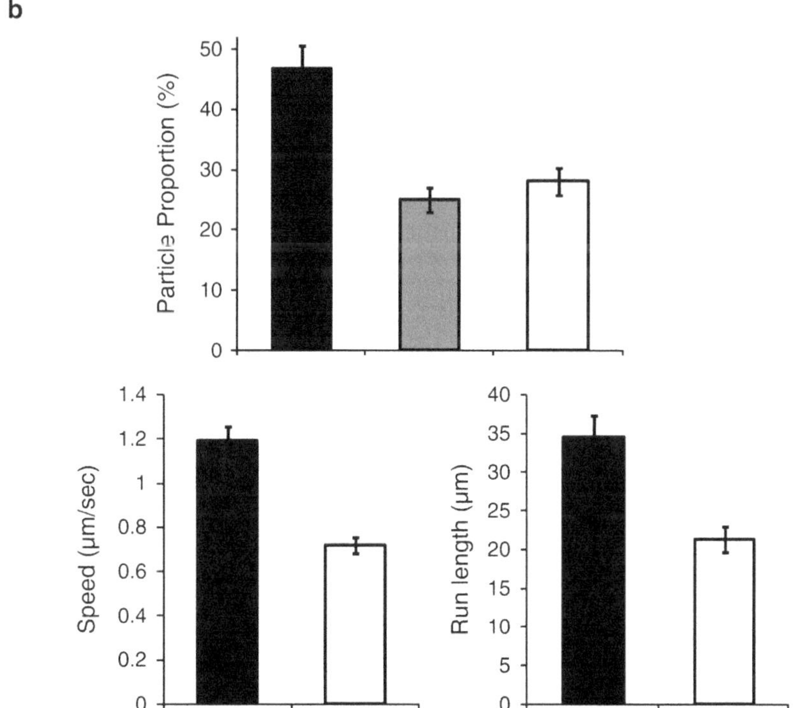

Fig. 3. Axonal transport measurements obtained from kymograph analysis. (**a**) Excel file showing calculations extracted from lines plotted on kymograph from Fig. 2b. Each line was separated by decreasing angle and corresponds to anterograde, stationary, and retrograde particles. Distance, time, and speed are calculated from the kymograph. (**b**) Results of average particle proportion (%), speed (µm/s), and run length (µm) calculated from Excel file. *Black*, *gray*, and *white* bars correspond to anterograde, stationary, and retrograde transport, respectively.

4. Notes

1. Prepare all solutions using ultrapure water by purifying deionized water to attain a sensitivity of 18 MΩ cm at 25°C.
2. Prepare, store, and manipulate all reagents at room temperature, unless indicated otherwise.
3. Follow regulations strictly for disposal of waste materials.
4. Hank's buffered salt solution is fairly complicated to make, which is why most people buy it.
5. Any APP cDNA, or other axonally transported protein cDNA, fused to a fluorescent probe can be used for the in vivo imaging of protein movement within axons. Alternatively, or simultaneously, fluorescent dyes may be used for labeling proteins or organelles. In example, mitochondria can be dynamically visualized using Mitotracker (Invitrogen).
6. Unless otherwise stated, all steps of the described method are performed at room temperature and pressure.
7. Hippocampal spoon (Fine Science Tools) can be used to facilitate detachment of hippocampi from cornu Ammonis.
8. Complete DMEM used for washing allows FBS to inactivate papain activity.
9. Pipette trituration is a key step. Do it too much and you'll kill the cells, do it too little and you'll have many clumps.
10. Little drops can be delivered at different points of the round coverslips to facilitate distribution. Make sure the medium covers the entire well surface.
11. The same medium removed before starting transfection can be used to replace Lipofectamine transfection medium. This procedure helps cells recover after transfection due to factors and proteins released by cells during their development.
12. Continued registration can be done with different camera software devices in TIFF format and then opened in a different computer station for analysis.
13. High magnifications (×100) require a short working distance from specimen to lens. The sandwich made by flipping the coverslips inside the glass bottom plate (1 mm) allows the cells to be in the lens working distance for imaging and movie registration.
14. Cooled CCD camera allows fast continuous registration at low sensitivity reaching the speed of ten frames per second at a time resolution of 100 ms per frames. 2 × 2 binning was set up to favor high-speed acquisition by reducing the recorded information.
15. Kymograph can also be generated in ImageJ software using "multiple kymograph" plug-in.

Acknowledgements

This work was supported by PICT-2008-0293 (ANPCyT)(T.L.F.), Alzheimer Association NIRG-10-172840 Grant (T.L.F.) and P3-0338 grant from the Slovenian Research Agency (G.B.S.).

References

1. Hirokawa N, and Noda Y (2008) Intracellular transport and kinesin superfamily proteins, KIFs: structure, function, and dynamics. Physiol Rev 88, 1089–1118
2. Kang J, Lemaire HG, Unterbeck A, Salbaum JM, Masters CL, Grzeschik KH, et al (1987) The precursor of Alzheimer's disease amyloid A4 protein resembles a cell-surface receptor. Nature 325, 733–736
3. Bertram L, Lill CM, and Tanzi RE (2010) The genetics of Alzheimer disease: back to the future. Neuron 68, 270–281
4. Goedert M, and Spillantini MG (2006) A century of Alzheimer's disease. Science 314, 777–781
5. Cras P, Kawai M, Lowery D, Gonzalez-DeWhitt P, Greenberg B, and Perry G (1991) Senile plaque neurites in Alzheimer disease accumulate amyloid precursor protein. Proc Natl Acad Sci USA 88, 7552–7556
6. Stokin GB, and Goldstein LS (2006) Axonal transport and Alzheimer's disease. Annu Rev Biochem 75, 607–627
7. Stokin GB, and Goldstein LS (2006) Linking molecular motors to Alzheimer's disease. J Physiol Paris 99, 193–200
8. Joachim CL, Duffy LK, Morris JH, and Selkoe DJ (1988) Protein chemical and immunocytochemical studies of meningovascular beta-amyloid protein in Alzheimer's disease and normal aging. Brain Res 474, 100–111
9. Koo EH, Sisodia SS, Archer DR, Martin LJ, Weidemann A, Beyreuther K, et al (1990) Precursor of amyloid protein in Alzheimer disease undergoes fast anterograde axonal transport. Proc Natl Acad Sci USA 87, 1561–1565
10. Kaether C, Skehel P, and Dotti CG (2000) Axonal membrane proteins are transported in distinct carriers: a two-color video microscopy study in cultured hippocampal neurons. Mol Biol Cell 11, 1213–1224
11. Goldsbury C, Thies E, Konzack S, and Mandelkow EM (2007) Quantification of amyloid precursor protein and tau for the study of axonal traffic pathways. J Neurosci 27, 3357–3363
12. Stokin GB, Lillo C, Falzone TL, Brusch RG, Rockenstein E, Mount SL, et al (2005) Axonopathy and transport deficits early in the pathogenesis of Alzheimer's disease. Science 307, 1282–1288
13. Stokin GB, Almenar-Queralt A, Gunawardena S, Rodrigues EM, Falzone T, Kim J, et al (2008) Amyloid precursor protein-induced axonopathies are independent of amyloid-beta peptides. Hum Mol Genet 17, 3474–3486
14. Falzone TL, Stokin GB, Lillo C, Rodrigues EM, Westerman EL, Williams DS, et al (2009) Axonal stress kinase activation and tau misbehavior induced by kinesin-1 transport defects. J Neurosci 29, 5758–5767
15. Araki Y, Kawano T, Taru H, Saito Y, Wada S, Miyamoto K, et al (2007) The novel cargo Alcadein induces vesicle association of kinesin 1 motor components and activates axonal transport. EMBO J 26, 1475–1486
16. Eva R, Dassie E, Caswell PT, Dick G, ffrench-Constant C, Norman JC, et al (2010) Rab11 and its effector Rab coupling protein contribute to the trafficking of beta 1 integrins during axon growth in adult dorsal root ganglion neurons and PC12 cells. J Neurosci 30, 11654–11669
17. Bilsland LG, Sahai E, Kelly G, Golding M, Greensmith L, and Schiavo G (2010) Deficits in axonal transport precede ALS symptoms in vivo. Proc Natl Acad Sci USA 107, 20523–20528

Chapter 26

The Use of Specific AAV Serotypes to Stably Transduce Primary CNS Neuron Cultures

Saafan Z. Malik, Margaret A. Maronski, Marc A. Dichter, and Deborah J. Watson

Abstract

Although primary neuronal cell cultures are a valuable source of in vitro insight for many neurobiologists, all current gene expression technologies for these cells have significant drawbacks. Some of these limitations of current gene expression protocols include toxicity, transient expression, a requirement for postnatal neurons, and/or low efficiency. To date, many types of experiments were not possible because of these limitations. Here, we outline a methodology by which primary cultured neurons can be transduced at any age, after plating, with virtually no toxicity and continued gene expression for the lifetime of the culture. This method involves the use of adeno-associated viral vectors, which have the potential to be highly useful for either upregulation or downregulation of single or multiple genes, including neurotrophins, other neuroprotective genes, and neurotoxins.

Key words: Viral vector, Gene delivery, Cultured neurons, Neurotrophins, Immunocytochemistry, Neuroprotection assays, Glutamate toxicity

1. Introduction

Reliable testing of candidate genes in primary cultured neurons is a critical technique required to identify trophic and protective factors for a variety of neurodegenerative diseases. Neuroscientists would benefit from the ability to design reliable neurotoxicity assays and test potentially neuroprotective agents on primary cultured neurons. However, common techniques such as recombinant Sindbis, SV40, or Semliki forest viral vectors as well as plasmid transfection using calcium phosphate, commercially available lipids, or electroporation suffer from many drawbacks, including toxicity, transient expression, the requirement for immature neurons, and/or low efficiency any of which could prevent an essential experiment.

Stephen D. Skaper (ed.), *Neurotrophic Factors: Methods and Protocols*, Methods in Molecular Biology, vol. 846, DOI 10.1007/978-1-61779-536-7_26, © Springer Science+Business Media, LLC 2012

We and other groups have shown that a few well-known adeno-associated viral (AAV) vectors can mediate efficient, stable, and nontoxic transduction in vivo and in vitro. (e.g. refs 1–3). Recently, we published a detailed comparative study of AAV vectors packaged with each of seven naturally occurring serotypes in addition to four engineered capsid proteins (3). One of the important findings of that study was the particular usefulness of AAV1 capsids to mediate gene expression in a very high percentage of cultured neurons. Of note, there is another potential benefit of AAV vectors: the ability to use the same vector directly in an in vivo experiment with proper planning.

Here, we detail the planning and methodology for designing and producing such a vector, testing it in vitro, using it for neuroprotection assays in vitro, and issues of note when moving into in vivo studies.

2. Materials

There is a "canonical" book in this area to which readers are referred for general methodology. (see "Rat Hippocampal Neurons in Low-Density Culture" in (4)). We have summarized the overall technique with specific modifications necessary for the assays in this chapter (see below and also (5)).

2.1. Poly-L-Lysine Coating and Preincubation of Tissue Culture Plates with or Without Glass Coverslips

1. Many cultured cells, especially neurons, will not adhere to plastic or glass without prior coating of the surface. Poly-L-lysine (Peptides International) has a positive charge which attracts the negative surface charge of the neurons to the plate or well. Poly-L-lysine is made up in borate buffer.
2. Borate buffer: add a stir bar to 500 mL of endotoxin-free water (BioWhittaker/Lonza) and set on a stir plate in the biosafety hood to stir. Add 2.38 g boric acid and 1.27 g borax to the water and continue stirring for approximately 15 min or until all the powder is dissolved. Resterilize the solution through a 0.2-μm filter and store in a labeled, sterile bottle at room temperature for up to 6 months.
3. Poly-L-lysine solution: dispense 100 mL borate buffer into a sterile 500-mL screw-cap bottle under the hood. Add one bottle (100 mg) poly-L-lysine (molecular weight 70–150 kDa, Peptides International, Louisville, KY) to the borate buffer and swirl until all the poly-L-lysine has dissolved in the buffer, about 2 min.
4. Preincubation medium: combine 500 mL of Neurobasal medium and 5 mL (one bottle) of B27 (with antioxidants) supplement.

5. Plates and coverslips are required according to the specific experimental conditions. Coverslips must be 12-mm round German glass coverslips (Fisher Scientific). Autoclave before use.

2.2. Primary Hippocampal Cell Culture

1. One liter bottle of Dulbecco's modified Eagle's medium (DMEM) containing 25 mM HEPES buffer, 4.5 g/L glucose, no glutamine (e.g., BioWhittaker/Lonza BW12-709 F).
2. Fetal calf serum (Hyclone). Store in 50 mL aliquots at −20°C.
3. Trypsin, 2.5% (Invitrogen).
4. HEPES-buffered saline solution (HBSS) and water (BioWhittaker/Lonza).
5. Trypan blue, 0.4% in saline.
6. A 37°C water bath with a shelf above the water.
7. A timer.
8. Glass Pasteur pipettes and a rubber bulb.
9. Bunsen burner.
10. 35-, 60-, and 100-mm-diameter tissue culture plates.
11. Poly-L-lysine-treated culture plates.

2.3. Construction of AAV Genomic Plasmid

1. A plasmid containing the AAV inverted terminal repeats (ITRs) and polyA sequence that has been propagated in recA- competent cells and sequenced.
2. Additional plasmids containing the expression elements of interest.
3. Standard molecular cloning reagents, including restriction enzymes and buffers, ampicillin and/or kanamycin, LB broth and plates, agarose gels, 37°C shaking incubator.
4. Competent bacterial cell lines such as SURE2 or STABL2 that are recA-. These must be kept at −80°C and thawed on ice a few minutes before transformation.
5. Primers that allow sequencing of the final plasmid, especially the ITRs.

2.4. Packaging of AAV

1. Contact information for vector preparation experts (see Subheading 3.5 for recommendations).

2.5. Concentration Response of Neurotoxic Agent

1. An AAV-green fluorescent protein (GFP) vector with a titer of at least 5× (10) GC/mL.
2. Poly-L-lysine plates or coverslips of your choice.
3. L-glutamate.
4. Phosphate-buffered saline.
5. Microcentrifuge tubes for making dilutions.

6. 10 mL sterile cell culture reservoir.
7. 10% neutral buffered formalin and appropriate waste containers.
8. VECTASHIELD with DAPI (Vector Laboratories).
9. Parafilm.
10. Microscope slides and nail polish.
11. A fluorescent microscope with a digital imaging camera, attached computer and software. For example, a Nikon 80i microscope, a CoolSnap cf2 monochrome camera, and ImagePro software.

2.6. Quantification of Neuroprotection

1. Permeabilization buffer: Tris-buffered saline (TBS) containing 0.3% Triton (TBST).
2. Blocking reagent: 5% goat or donkey serum in TBS.
3. Appropriate combinations of primary and secondary antibodies and detection reagents. For identification of neurons, rabbit polyclonal antimicrotubule-associated protein 2 (1:500), Biotin-SP-conjugated goat anti-rabbit IgG (Biotin-SP is Jackson ImmunoResearch's trade name for biotin with a 6-atom spacer positioned between biotin and the protein to which it is conjugated), and Alexa594-conjugated streptavidin (1:500 each). For astrocytes, monoclonal anti-glial fibrillary acidic protein (GFAP) antibody (Sigma, 1:500), Biotin-SP-conjugated donkey anti-mouse IgG, and Alexa594-conjugated streptavidin (Molecular Probes, 1:500 each). Other primary antibodies such as anti-NeuN or anti-GFP will also be useful.

3. Methods

3.1. Preincubation of Plates and/or Coverslips

1. Observing sterile technique, use sterile forceps to add one coverslip per well of a 24-well plate or five coverslips to a 35-mm-diameter tissue culture plate (Nunc). Alternatively, use a 24-well plate with no coverslips. The specific setup depends on the type of experiment you wish to perform.
2. Set the plates in a single layer on an incubator tray and keep inside the hood at room temperature.
3. For a 35-mm plate, add 1.5 mL of the poly-L-lysine solution and make sure the coverslips, if present, are submerged and arranged in a single layer on the bottom of the dish. Store at room temperature for 24 h.
4. The next day, aspirate the poly-L-lysine solution and add 1.5 mL of BioWhittaker water to each dish, again being sure to submerge the coverslips, and arrange in a single layer on the

bottom of the dish. It is important that the coverslips remain in the same orientation as previously (i.e., do not flip over and do not move or overlap). Store for a further 24 h inside the hood.

5. The next day, aspirate the water and add 1.0 mL of the preincubation medium for approximately 72 h. The use of German glass and this preincubation step promotes greater viability in older cultures. The reduction or lack of serum, in the culture, reduces or eliminates glia.
6. Gridded coverslips can be used if necessary (see Note 1).

3.2. Serum-Free Primary Hippocampal Neuron Cultures (see Notes 2 and 3)

1. Immediately before use, add 0.5 mL of thawed trypsin to 4.5 mL of sterile HBSS with HEPES at room temperature in a 15-mL centrifuge tube, mix and add to 35-mm dishes containing the dissected material. Place each of the 35-mm dishes into the bottom of a 100-mm dish and carry to 37°C water bath. The 35-mm dishes should have their lids on and remain sterile at all times. Place the 100-mm dishes on a shelf above the water level and cover the bath. Set timer for 15 min.
2. After 15 min, return trypsinized hippocampi to the hood. Carefully pour the mixture into sterile 60-mm plates and add ~10 mL of sterile HBSS with HEPES to dilute the trypsin. Return to the water bath. After 5 additional minutes, retrieve the 60-mm dish and remove as much HBSS with HEPES as possible without suctioning up any hippocampi. Then add another sterile 10 mL HBSS with HEPES to the dish, place it back into the 100-mm bottom dish, and place in water bath for another 5-min wash. Afterwards, dispense the hippocampi into a 15-mL centrifuge tube until all hippocampal material is in tube (~5 mL). Discard the rest of the isolation medium containing trypsin and HBSS with HEPES.
3. Constrict the opening of the sterile 9-in. Pasteur pipette with a Bunsen burner until resistance forms in the suction bulb, indicating the opening is sufficiently small. With the constricted Pasteur pipette, triturate hippocampi gently 14 times until all tissue is broken up into individual cells and cannot be seen individually any more with the solution appearing cloudy without any material floating in it. Avoid bubbles by keeping the tip of the pipette in the solution while triturating. Add 5 mL of Dulbecco's modified Eagle's medium supplemented with 10% fetal bovine serum.
4. Count the cells using 10 μL of cell suspension drawn from the center of the tube and 90 μL of trypan blue. Calculate the number of cells/mL and the percent viability.
5. For this protocol, use 100,000 cells/mL. Add additional Neurobasal medium with B27 to bring up the volume to the required concentration for plating.

6. For 24-well plates, use 0.5 mL or 150,000 cells/well. For 96-well plates, use 500 μL per well. For 35-mm plates containing coverslips, add 1.5 mL into each plate at 100,000 viable cells/mL equaling a total of 150,000 cells total per plate.
7. Store the cultures in a 37°C, 5% CO_2 humidified incubator. For reduced number of glia in cultures, use Neurobasal plus B27. Most glial cells are inhibited by lack of serum in the medium, but a small percent will survive in the Neurobasal/B27. If your experiments dictate that absolutely no glia should be present, you can add a mitotic inhibitor 24 h after plating the neurons.
8. Neuronal cell cultures are generally used from 7 days up to 5 weeks. Cultures should be fed by adding a few drops of new Neurobasal/B27 medium to each plate. There is a greater risk of contamination at the time of feeding than of the cultures metabolizing the medium in the plates, so the cells can be fed about once per week.
9. The same protocol can be used for cortical cultures.

3.3. Generation of AAV Vectors for Neuronal Cell Transduction: Design Considerations

1. AAV is a small (20 nm), single-stranded, icosahedral, nonreplicating DNA virus composed of a ~4.7-kb genome comprised of one ITR on each end of the genome and two open reading frames which generate the multiple genes necessary for the AAV life cycle and capsid production. Recombinant AAV vectors are generated by packaging an engineered AAV genome into an AAV capsid sequence that may or may not be derived from the same AAV serotype as the genome. To generate recombinant vectors, the rep/cap sequences can be delivered in trans, allowing for more flexibility in vector design. Currently, genome plasmids are generally based on the AAV2 ITRs because they can be easily cross-packaged into capsids from AAV1, 5, 6, 9, etc. Each capside sequence probably binds to a separate receptor, though not all of them are known. These vectors are called AAV2/5 or AAV2/9, etc., as specified by the user. To generate the genome of the vector, the plasmid must contain all the expression components in one continuous sequence. (The total size of the entire plasmid is not critical but must contain the antibiotic resistance gene, etc.) The usual arrangement of AAV elements is 5′ITR (usually derived from AAV2) followed by a promoter, a gene of interest, a polyA sequence, and the 3′ITR. The entire sequence between the 5′ITR and the 3′ITR should not be much larger than the size of the wild-type genome (~4.7 kb), otherwise the recombinant genome will not be packaged efficiently. Unrelated "stuffer" sequence is sometimes included if needed to bring the sequence up to 4.7 kb. Maximal packaging size is about 5.2–5.3 kb (6–8). To overcome some aspects of the size limitations, trans-splicing vectors have been developed (for more details, see (9)).

2. The length of the ITRs is fixed at ~145 bp each, so scientists have worked hard to discover, generate, or optimize short promoters that are cell-type specific, regulated in some way, or active in all cells. Short promoters would of course allow more space for longer genes, multiple genes, other regulatory elements, and/or reporter genes. Promoters that are in use currently in neuroscience research are (among many others) hCMV immediate early promoter and enhancer (620 bp), GUSB promoter (378 bp), CBA promoter (1.7 kb), and EF1α (1,194 bp). More cell-type-specific expression can be achieved by using the neuron-specific synapsin-1 promoter (470 bp) and neuron-specific enolase promoter (1.8 kb) or the glial-specific GFAP promoter (2.2 kb) (2, 10, 11). Other promoters, including various drug-regulated promoters and combination promoters, are constantly being identified, optimized, and tested. Finally, a novel avenue of approach is to use a promoter which is active only during certain physiologic states important for neuronal or glial cell biology, for example, apoptosis, excitotoxicity, or neurotrophin response. We have noted that the hCMV promoter takes a few days to generate detectable expression in cultured neurons (3). Cell-type-specific promoters may speed up the process, or faster expression of the transgene may be achieved by the use of self-complementary genomes which bypass the rate-limiting step of second-strand synthesis (12).
3. It is frequently useful to include a reporter gene if the size of the construct allows it. The internal ribosomal entry site (IRES) initiates cap-independent translation of a second cistron, with the caveat that expression of the second cistron is always weaker than the first cistron. The benefit is that, if you can detect expression of the reporter, it suggests that the cDNA in the first cistron is being expressed well. Multiple IRES elements from viruses and mammalian cell have been identified; Our recent experience suggests that the IRES from the foot-and-mouth disease virus produces reasonable cap-independent translation of GFP in 293T cells (S. Malik and D. Watson, manuscript in preparation.) The IRES-GFP component in our constructs is 1,164 bp. An alternative strategy is to include a second promoter to drive expression of the second cDNA; however, they may compete for transcription factors.
4. Finally, most AAV constructs contain the woodchuck hepatitis virus posttranscriptional regulatory element before the polyA sequence to increase transgene expression (13).
5. If you plan to extend your in vitro findings to in vivo experiments, it is essential to determine whether your serotype transduces cells in the brain region and/or in the cell type of interest. For this, a literature search or pilot experiment is necessary.

6. For this, a literature search or pilot experiment is necessary. (14, 15).
7. For a more detailed review of these issues, see (16) and references therein.

3.4. Construction of AAV Genomic Plasmid

1. To construct your genomic plasmid, begin with a plasmid containing the ITRs and as many other components as desired. Many useful plasmids are available from academic investigators and from an extensive AAV plasmid catalog available at the National Gene Vector Biorepository (https://www.ngvbcc.org/ReagentRepository.action), as well as from commercial sources. Be sure a map is supplied and a sequence if possible. It is highly advisable to sequence the AAV parts of the plasmid before continuing.
2. It is ESSENTIAL to use competent bacteria that lack recA so that the ITRs stay intact. For this, a literature search or pilot experiment is necessary. If the provider of your source plasmid has not done this, you can be sure that the ITRs are not complete and you should *not* continue to use that plasmid. Unfortunately, this is not an uncommon event. Again, this is extremely important. Appropriate bacterial strains include SURE2 and STABL2 cell lines which are recA-. Clone in your promoter and cDNA of interest, as well as other elements. The AAV genomic plasmids usually contain the polyA element already; this should be confirmed. Again, you MUST use competent cells lacking recA to propagate the intermediates. Sequence the final product, especially the ITRs.
3. Confirm that the cassette is active by transfecting it into 293T cells. Your gene of interest should be detectable by Western blot, immunostaining, or other means.

3.5. Packaging of AAV

1. Methods for the packaging and purification of recombinant AAV vectors are continually improving. For experiments using neurons in vitro, clinical grade vectors are not required; research grade vectors are sufficient. If financial resources are available, we strongly recommend working with an experienced core facility, such as the one at the University of Pennsylvania(http://www.med.upenn.edu/gtp/vectorcore/), the University of Michigan (http://www.med.umich.edu/vcore/), or many other academic vector cores that package AAV vectors for outside investigators. Alternatively, you can contact a commercial AAV production company, although it should be noted that packaging with certain serotypes is only available at some companies. We would suggest that commercial kits for packaging and purifying AAV vectors in your own laboratory should be a last resort. We have observed significant toxicity of purification by-products on neuronal cell cultures.

Also, please note that as of this writing, the National Gene Vector Laboratory no longer produces vectors.

2. Titering of AAV vectors is also critical to match titers if you are testing multiple serotypes for your application. Neuronal cell cultures do not respond well to volumes above ~50 μL in a 35-mm plate. Therefore, the more concentrated vector you can use, the better. For example, in previous studies, we have used vectors with titers in the 5 × 10 (12) or 5 × 10 (13) genome copies; more concentrated vectors will be available as purification methods become more sophisticated. Controls should include (1) cultures receiving the same dose in microliter of the vehicle only (in our case, PBS plus 5% glycerol) or (2) cultures with no treatment at all. If possible, it is also helpful to test more than one independent batch of vector to rule out any problems that might have occurred during production.
3. The issue of serotype may require some testing for your particular application. In our experience, using reporter genes, AAV1, 8, and 9 were particularly efficient and nontoxic to neuronal cell cultures, whereas AAV2 was relatively ineffective (3). AAV5 and AAV6 caused toxicity at the same high doses, which was measured on a scale of 0–3 (see below). At lower doses, these two vectors were nontoxic but less efficient. It should be noted that another group ((1) found AAV6 to be sufficient for neuronal cell transduction). AAVRh10 shows very promising expression in vivo (17, 18).

3.6. Concentration Response of Neurotoxic Agent

Before performing any neuroprotection assays, the dose-response curve must be determined. This should be performed with a GFP reporter vector of the same serotype as the experimental vectors. Vectors are stored at −80°C in small aliquots and thawed on ice before use.

3.6.1. Transduction of the Cells for Neurotoxicity Experiments

For 96-well plates:

1. Plate dissociated primary hippocampal neurons (100,000 cells/mL) in poly-L-lysine-coated 96-well plates.
2. One week after plating, transduce each well with the GFP reporter vector, at an approximate multiplicity of infection (MOI) 1.5×10^5. The volume of vector added should not exceed about 1/10 of the media volume. At 21 days in vitro (DIV), the cells should be approximately 90% GFP positive.

For 24-well plates:

1. Add a single dose of the GFP reporter vector to cultured rat hippocampal cells on DIV 7. Expression of GFP should be detectable in about 90% of the cells by 1-week posttransduction.

Check for any changes in morphology (axonal retraction or beading) indicating cell toxicity.

For coverslips in a 35-mm plate:

1. Add a single dose of 2.0–2.5 × 10^{11} genome copies of the GFP reporter vector to cultured rat hippocampal cells on DIV 7. The volume of vector added should not exceed about 50 μL in a 35-mm plate. Expression of GFP should be detectable in about 90% of the cells by 1-week posttransduction or earlier.

3.6.2. Addition of the Neurotoxic Agent

1. Two weeks after plating, add L-glutamate (0–400 μM) to the wells, which should still contain the regular medium as well. There should be a minimum of three replicates per experimental condition. As a control, transduce some cells but do not treat with glutamate. These will be processed the same as the other wells and will serve as the denominator for the calculation of percent toxicity (see below).
2. Twenty-four hours after addition of glutamate, aspirate the medium and fix all cells with 10% neutral buffered formalin for 10 min at room temperature in a fume hood. Remove the formalin and dispose of it in chemical waste containers. Gently wash the cells twice in PBS before proceeding. Add a small drop of VECTASHIELD containing DAPI to each well. At this stage, the plate can be sealed with parafilm and kept at 4°C for up to a few days before imaging. Alternatively, the coverslips should be mounted on glass slides with the same mounting medium and sealed with nail polish.

3.6.3. Quantification of Neurotoxicity by Intrinsic GFP Fluorescence

1. An overall sense of the toxicity can be gained by a quick visual inspection of the wells. We use a scale of 0–3, where 0 = healthy cultures with uniform GFP in the processes; 1 = beaded GFP expression in the processes but little or no cell loss; 2 = substantial cell loss and/or fragmented GFP fluorescence in the processes and cell body; and 3 = majority of cells dead.
2. After that, live cells are quantified. Carefully seal the plate with parafilm to prevent leakage of any fluid or mounting medium. In an initial experiment, we obtained digital images of three nonoverlapping fields at 10× with the plate inverted on the microscope stage, which was set to its lowest setting to accommodate the plate (or with slides mounted normally on the microscope).
3. Manually count cells on each image, three nonoverlapping fluorescent 10× fields per well and three wells per condition. To insure reproducibility, it helps to have multiple lab members count the cells. Calculate cell survival as percentage of the untreated wells.
4. Immunostaining is also possible (see Subheading 3.7).

3.6.4. Quantification of Neurotoxicity by an Automated Acquisition Protocol

1. Once an initial manual cell count is complete, an automated acquisition protocol can be established. Again, this should be established with a healthy culture first.
2. First, determine image acquisition settings with your software that clearly visualize all the DAPI-stained nuclei in a selected field of a healthy culture (not treated with glutamate) and save that setting. The nuclei should appear oval to round and bright enough to be seen clearly without overexposure. Use an objective of high enough power to see the fluorescence easily, but not so high that the field of interest contains less than 50 cells. We generally use a 10× objective.
3. Next, while keeping the microscope on the same field, increase the image acquisition time so that all of the GFP-positive cells in that field are clearly visualized. Save that setting as well.
4. Perform this protocol on at least five nonoverlapping fields in each channel and save the images separately, with filenames labeled by glutamate dose, fluorescent agent, and field. It is helpful to save the images by field as you acquire them so it is easy to assign fields to images as you go along.
5. Overlay the images for each field to confirm that the GFP-positive cells each contain a DAPI-positive nucleus.
6. Returning to the individual images, use the captured images to perform an automated count of the cells. It is usually easier to do this with the nuclei, which are clearly delineated. The settings for counting can be adjusted to include cells of a particular range of pixels. These settings should also be saved and used for all the wells. In our experience, it is sometimes difficult to determine whether a glutamate-treated cell should be counted or not, because of its altered morphology. Establishing imaging and counting protocols on healthy cultures and using the same parameters for glutamate-exposed cultures will help to keep all the quantification assays uniform.
7. To validate the automated imaging and counting protocols, confirm that the automated values match the manual cell counts.
8. Finally, calculate the percent of live cells for each concentration of glutamate, using the number of cells in the healthy cultures as the denominator.

3.6.5. Determine the Dose-Response Curve

1. Plot the dose of glutamate against the percentage of live cells. Choose the concentration of glutamate to use in all the following neuroprotection assays. We prefer a concentration of glutamate that represents about 30% cell survival so that we can (1) determine whether a neuroprotective response is significant; (2) determine whether there may be an unexpected result of neurotoxicity by a supposedly neuroprotective construct; and (3) have the potential to see significant intermediate degrees of

protection since it is unlikely that any given agent will protect all the cells. In other words, we try to avoid a ceiling or a floor effect, while retaining the ability to detect degrees of protection below a complete rescue.

2. If there is any indication that the GFP itself is toxic to the cultures, substitute a vector of the same serotype containing LacZ. Alternatively, if you suspect the serotype is toxic, try one or more alternative serotypes containing GFP.

3.7. Quantification of Neuroprotection

1. This should be performed with a vector of the same serotype as the reporter vector, but containing the gene of interest. This technique does not require a GFP-containing vector (although it does not preclude the use of one). Surviving cells are detected by immunostaining for the protein of interest.
2. Fix the cells as described above but do not use mounting medium.
3. A general immunostaining procedure is followed. After fixation, permeabilize with TBST and block in 5% goat or donkey serum in TBS. Expose to the primary antibody overnight at 4°C, wash and incubate with the corresponding secondary antibody, wash and incubate with the fluorescent streptavidin (1 h at room temperature each), and then wash again and invert onto glass slides with VECTASHIELD mounting medium containing DAPI to identify nuclei. If two antibodies are to be used at the same time, note that one should be directly conjugated and the other should be biotinylated. Similarly, the species of the two antibodies should be different so that the secondary antibodies do not bind to both primary antibodies. As to the fluorescent tags, they should not be in the green range (488 nm) if a GFP-expressing vector is being used and not in the blue range (350 nm) to avoid conflict with the blue signal from the DAPI. In general, we use a red fluorescent tag for the neurons and, if necessary, a fluor in the 633-nm range for the second antibody. After the last wash, apply VECTASHIELD mounting medium with DAPI, about one drop per well. This mounting medium does not set well, so it is essential to seal the plate well or to invert the coverslips (if present) on a glass slide, dry, and seal with nail polish.
4. For this protocol, neurons are identified by immunostaining with rabbit polyclonal anti-microtubule-associated protein 2 (1:500) and detected with Biotin-SP-conjugated goat anti-rabbit IgG and Alexa594-conjugated streptavidin (1:500 each). Astrocytes are identified on separate coverslips with monoclonal anti-GFAP antibody (1:500), Biotin-SP-conjugated donkey anti-mouse IgG, and Alexa594-conjugated streptavidin (1:500 each). Confirm the specificity of staining by omitting the primary antibody in each case. Immunofluorescence is colocalized

with GFP intrinsic fluorescence or immunofluorescence for confirmation.

5. After immunostaining, keep the plates and/or coverslips in the dark. To image, invert the plate on the lowered microscope stage as described earlier or mount the glass slide onto the stage. Keep the room dark to the extent possible.
6. The imaging protocol should be optimized as described above, except that the fluorescent signal from the antibody should be optimized on healthy cultures before moving on to the treated cultures.
7. Confirm the dose response of glutamate by counting the DAPI cells alone and plotting that number against the concentration of glutamate. To calculate the percent transduction of live cells, express the number of GFP-positive cells as (GFP+/DAPI+).
8. Plot the concentration of glutamate vs. the percentage of surviving cells, using the untreated cultures (untreated with vector and untreated with glutamate) as controls. A second control should be included, which is the vector alone but no glutamate. This will confirm that the gene of interest has no independent effect on neuronal cell survival. We prefer a bar graph with the healthy cultures at 100% as the first bar.

4. Notes

1. Gridded plastic coverslips may be used as an alternative to the glass coverslips for some neurotoxicity protocols. Sterilize the coverslips in ethanol in a covered container overnight at room temperature. Clean coverslips by adding and aspirating ethanol six times to multiwell plates that fit the coverslips and check the orientation under the microscope. The letters should be right side up. If not, flip the coverslip with sterile forceps but avoid scratching them. Allow the ethanol to evaporate 30 min then add 1 mL of poly-L-lysine solution to each well, making sure each coverslip is submerged, and incubate overnight at room temperature. Aspirate poly-L-lysine solution from wells, add 1 mL (BioWhittaker/Lonza) tissue culture water to each well, making sure coverslips are submerged, and store at room temperature overnight. Aspirate water and allow it to evaporate 24 h at room temperature. Recheck coverslip orientation and store in a drawer until ready for preincubation.
2. We find it convenient to preincubate the plates from Friday to Monday.
3. It is important to note that we do not include the protocol for Cesarean section of the rat at stage E18 or E19 or the procedure

for sterile isolation of embryonic hippocampi in this chapter. This is a complicated procedure that should be learned from an experienced professional, should be approved by Animal Care and Use Committee and your institution, and should conform to federal guidelines (19). Our cultures are generally derived from E19 Sprague Dawley rats.

Acknowledgments

The work in our laboratory is supported, in part, by grants from the National Institutes of Health RO1 NS040978 (DJW), RO1 NS041811 (MAD), a pilot grant from the University of Pennsylvania Institute for Medicine and Engineering (DJW), and by funds from the University of Pennsylvania Department of Neurosurgery. As always, we thank Dr. Sean Grady for mentorship and support.

References

1. Howard DB, Powers K, Wang Y and Harvey BK (2008) Tropism and toxicity of adeno-associated viral vector serotypes 1, 2, 5, 6, 7, 8, and 9 in rat neurons and glia in vitro. *Virology* 372:24–34
2. Kugler S, Lingor P, Scholl U, Zolotukhin S and Bahr M (2003) Differential transgene expression in brain cells in vivo and in vitro from AAV-2 vectors with small transcriptional control units. *Virology* 311:89–95
3. Royo NC, Vandenberghe LH, Ma JY, Hauspurg A, Yu L, Maronski M, Johnston J, Dichter MA, Wilson JM and Watson DJ (2008) Specific AAV serotypes stably transduce primary hippocampal and cortical cultures with high efficiency and low toxicity. *Brain Res* 1190:15–22
4. Goslin K, Asmussen H and Banker G 1998 Rat hippocampal neurons in low-density culture. In: *Culturing nerve cells*. Banker G, Goslin K (eds). MIT Press: Cambridge, MA
5. Cummings DD, Wilcox KS and Dichter MA (1996) Calcium-dependent paired-pulse facilitation of miniature EPSC frequency accompanies depression of EPSCs at hippocampal synapses in culture. *J Neurosci* 16:5312–5323
6. Dong JY, Fan PD and Frizzell RA (1996) Quantitative analysis of the packaging capacity of recombinant adeno-associated virus. *Hum Gene Ther* 7:2101–2112
7. Grieger JC and Samulski RJ (2005) Packaging capacity of adeno-associated virus serotypes: Impact of larger genomes on infectivity and postentry steps. *J Virol* 79:9933–9944
8. Wu Z, Yang H and Colosi P (2010) Effect of genome size on AAV vector packaging. *Mol Ther* 18:80–86
9. Yan ZY, Zhang YL, Duan DS and Engelhardt JF (2000) Trans-splicing vectors expand the utility of adeno-associated virus for gene therapy. *Proc Natl Acad Sci USA* 97:6716–6721
10. Shevtsova Z, Malik JM, Michel U, Bahr M and Kugler S (2005) Promoters and serotypes: targeting of adeno-associated virus vectors for gene transfer in the rat central nervous system in vitro and in vivo. *Exp Physiol* 90:53–59
11. Xu R, Janson CG, Mastakov M, Lawlor P, Young D, Mouravlev A, Fitzsimons H, Choi KL, Ma H, Dragunow M, Leone P, Chen Q, Dicker B and During MJ (2001) Quantitative comparison of expression with adeno-associated virus (AAV-2) brain-specific gene cassettes. *Gene Ther* 8:1323–1332
12. McCarty DM, Monahan PE and Samulski RJ (2001) Self-complementary recombinant adeno-associated virus (scAAV) vectors promote efficient transduction independently of DNA synthesis. *Gene Ther* 8:1248–1254
13. Brun S, Faucon-Biguet N and Mallet J (2003) Optimization of transgene expression at the posttranscriptional level in neural cells: implications for gene therapy. *Mol Ther* 7:782–789
14. Michel U, Malik I, Ebert S, Bahr M and Kugler S (2005) Long-term in vivo and in vitro AAV-2-mediated RNA interference in rat retinal

ganglion cells and cultured primary neurons. *Biochem Biophys Res Commun* 326:307–312
15. Xia H, Mao Q, Eliason SL, Harper SQ, Martins IH, Orr HT, Paulson HL, Yang L, Kotin RM and Davidson BL (2004) RNAi suppresses polyglutamine-induced neurodegeneration in a model of spinocerebellar ataxia. *Nat Med* 10:816–820
16. Boulaire J, Balani P and Wang S (2009) Transcriptional targeting to brain cells: Engineering cell type-specific promoter containing cassettes for enhanced transgene expression. *Adv Drug Deliv Rev* 61:589–602
17. Cearley C and Wolfe JH (2006) Transduction characteristics of adeno-associated virus vectors expressing cap serotypes 7, 8, 9 and Rh10 in the mouse brain. *Mol Ther* 13:528–537
18. Klein RL, Dayton RD, Tatom JB, Henderson KM and Henning PP.(2008) AAV8, 9, Rh10, Rh43 vector gene transfer in the rat brain: effects of serotype, promoter and purification method. *Mol Ther* 16:89–96
19. National Research Council (1996) Guide for the Care and Use and Use of Laboratory Animals. National Academy Press: Washington, DC

Chapter 27

Preparation and Characterization of Biocompatible Chitosan Nanoparticles for Targeted Brain Delivery of Peptides

Secil Caban, Yılmaz Capan, Patrick Couvreur, and Turgay Dalkara

Abstract

Here, we describe a nanocarrier system that can transfer chitosan nanoparticles loaded with either small peptides such as the caspase inhibitor Z-DEVD-FMK or a large peptide like basic fibroblast growth factor across the blood–brain barrier. The nanoparticles are selectively directed to the brain and are not measurably taken up by liver and spleen. Intravital fluorescent microscopy provides an opportunity to study the penetration kinetics of nanoparticles loaded with fluorescent agents such as Nile red, and has demonstrated that this nanomedicine formulation is rapidly transported across the blood–brain barrier.

Key words: Nanoparticles, Brain drug delivery, Blood–brain barrier, Neurotrophic factors, Caspase inhibitors, Neuroprotection

1. Introduction

The blood–brain barrier (BBB) is a diffusion barrier essential for normal function of the central nervous system (1). Only lipophilic or nonionic water soluble molecules with low molecular weight can cross the BBB. Essential compounds such as amino acids, hexoses, and peptides need specific carriers or transporters to permeate the brain. Accordingly, the majority of available pharmaceutical agents cannot efficiently reach the brain parenchyma, which is one of the bottlenecks in the treatment neurological and psychiatric disorders (2). One of the strategies developed to overcome this limitation is to conjugate the drug molecule to an antibody directed against one of the carrier proteins expressed on the luminal surface of brain capillary endothelial cells (e.g. transferrin receptor) (3–5). Although this approach has been shown to be functional for the brain delivery of several BBB-impermeable molecules, its efficiency is generally low because the ratio of drug to carrier protein is not above 1:1 (6).

Stephen D. Skaper (ed.), *Neurotrophic Factors: Methods and Protocols*, Methods in Molecular Biology, vol. 846, DOI 10.1007/978-1-61779-536-7_27,

Instead of creating chimeric drug–antibody complexes, conjugating the antibody to the nanoparticles loaded with substantial amounts of the drug molecule has recently emerged as an alternative method to improve the efficiency of receptor-mediated brain drug delivery (7, 8). In rodents, the transferrin receptor appears to be a promising target for a number of reasons. First, it is highly expressed on the luminal surface of brain capillary endothelia (5); second, less than a quarter of the receptors are saturated by the circulating transferrin (9), leaving plenty of unoccupied receptors for the antibodies to bind; and finally, it is rapidly shuttled between the luminal and abluminal membranes (turnover half time is 2–3 min) to provide high throughput transferrin delivery to the brain (10). Furthermore, using specific antibodies that target the transferrin receptor subtype-1 located mainly on the brain endothelium, nanoparticles can selectively be targeted to the brain without affinity to the transferrin receptors in other tissues predominantly expressing the subtype-2 (10–12).

Nanoparticles are generally made up of biocompatible polymers (13). They are coated with hydrophilic polymers such as poly(ethylene glycol) (PEG) to reduce rapid clearance from the circulation by the reticuloendothelial system (14) (see Fig. 1). Chitosan is a polymer obtained by deacetylation of chitin, a naturally occurring polysaccharide in shells of crustaceans. Chitosan is highly biocompatible; it does not lead to allergic reactions or rejection, and is degraded to nontoxic amino-sugars in tissues (15). Another advantage of chitosan is that it can be prepared under exceptionally mild conditions owing to its hydrophilic nature, making it particularly attractive for loading delicate compounds such as peptides and functional macromolecules (16–18).

Experimental studies suggest that neurotrophic factors provide protection against acute as well as chronic neurodegeneration (19). Similarly, the small peptide inhibitors of proteases that play important roles in cell death (such as caspases) are also neuroprotective (20, 21). One important advantage of both classes of agents

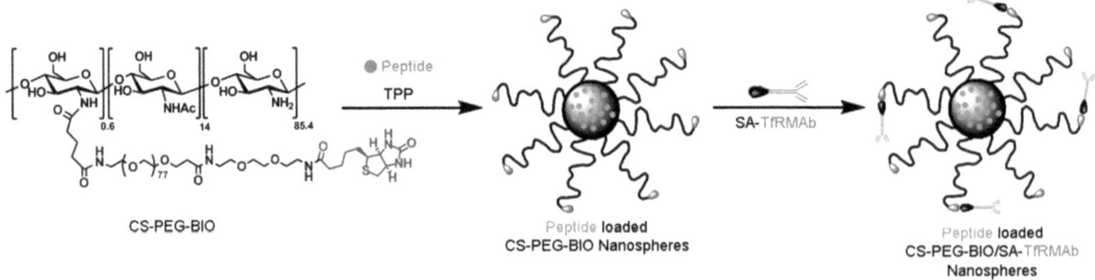

Fig. 1. Schematic outline of nanoparticle preparation. Peptides (Z-DEVD-FMK or bFGF) or Nile red are loaded into CS–PEG–BIO polymers and then nanosperes are formed by ionic gelation with tripolyphosphate solution (TPP). The nanoparticles are conjugated with anti-mouse transferrin monoclonal antibodies (TFRMAb) by means of biotin–streptavidin bonds (reprinted and modified from ref. (7) with permission).

is that they have a relatively long therapeutic time window in acute disorders such as cerebral ischemia (21–23), increasing their chance to be successfully translated to stroke treatment in the clinic. Unfortunately, however, both groups of these peptidergic agents cannot cross the BBB.

We have recently demonstrated that PEGylated chitosan nanoparticles conjugated via avidin–biotin linkage with the anti-transferrin antibody could efficiently deliver these promising agents to the central nervous system, and provide neuroprotection when given systemically (7, 24). This nanomedicine rapidly penetrates the brain, possibly by combining the affinity of the antibody for the transferrin receptor to trigger receptor-mediated transcytosis across the BBB with the ability of cationic chitosan to interact with the negative charges on the surface of the brain capillary endothelium (25). The combined systemic administration of the nanoparticles loaded with basic fibroblast growth factor (bFGF) and a caspase inhibitor (Z-DEVD-FMK) substantially reduced infarct volume as well as neurological deficit after experimental focal ischemia in the mouse (26), similar to the results obtained with their intracerebroventricular administration (27). Thus, chitosan nanoparticles open new and exciting opportunities for brain delivery of biologically active peptides including neurotrophic factors and hence hold promise for the treatment of central nervous system disorders.

2. Materials

Prepare all solutions using ultrapure water prepared by purifying deionized water to attain a sensitivity of 18 MΩ cm at 25°C. Animal housing, care, and application of experimental procedures were all done in accordance with institutional guidelines. All animal experiments described in this and the previous section were approved by Hacettepe University Ethics Committee. Modified chitosan polymers were provided by Eduardo Fernandez-Megia, Ramon Novoa-Carballal, and Ricardo Riguera from Universidad de Santiago de Compostela, Spain.

2.1. Preparation of Chitosan and Surface-Modified Chitosan Nanoparticles

1. Dissolve chitosan (CS) in water to obtain a concentration of 1.75 mg/mL.
2. Dissolve chitosan–polyethylene glycol (CS–PEG) in water to obtain a concentration of 1 mg/mL.
3. Dissolve chitosan–polyethylene glycol–biotin (CS–PEG–BIO) in water to obtain a concentration of 1 mg/mL.
4. Dissolve sodium tripolyphosphate (TPP) in water at two different concentrations: 0.4 mg/mL and 0.84 mg/mL (see Notes 1 and 4).

5. *N*-benzyloxycarbonyl-Asp(OMe)-Glu(OMe)-Val-Asp(OMe)-fluoromethyl ketone (Z-DEVD-FMK) (BACHEM, USA): Centrifuge a vial containing 1 mg of the peptide and then dissolve the contents in 75 μL dimethyl sulfoxide. Divide the solution into aliquots. Next, dilute each aliquot with water and further divide into smaller working aliquots. Store at –20°C until use (see Note 5).
6. Centrifuge a vial containing 50 μg of bFGF (Invitrogen, USA) and then dissolve the contents in 500 μL water to obtain a solution of 100 μg/mL. Divide this solution into smaller working aliquots and store at –20°C (see Note 5).
7. Nile red: prepare a 1 mg/mL solution in dimethyl sulfoxide.

2.2. Conjugation of Streptavidin and Monoclonal Antibody to CS–PEG–BIO Nanoparticles

1. Dissolve 380 mg of tetrasodium ethylenediaminetetraacetic acid in 10 mL water to obtain a 0.1 M solution.
2. Dissolve 7 mg of streptavidin (Sigma) in 250 μL water.
3. Dissolve 40 mg of Traut's reagent (2-immunothiolane, Thermo Fisher Scientific Inc, Pierce, USA) in 10 mL water to yield a 4 mg/mL solution.
4. Prepare a solution of 34 mg NaOH, 310 mg boric acid, and 373 mg potassium chloride dissolved in 100 mL water.
5. Functional grade anti-mouse monoclonal antibody to the transferrin receptor (TfRMab) (functional grade purified anti-mouse CD71, eBioscience, USA, clone R17217): prepare a 1 mg/mL solution.
6. Phosphate buffered saline (PBS, pH 7.4) is used as the release medium in release studies.

3. Methods

3.1. Preparation of Nanoparticles

1. Dissolve CS or the surface-modified CS (CS–PEG or CS–PEG–BIO) polymers in water by magnetic stirring at 700 rpm. CS is dissolved in half an hour whereas dissolving CS–PEG and CS–PEG–BIO requires 45–50 min (see Fig. 2) (28).
2. Dissolve TPP in water by magnetic stirring at 700 rpm.
3. Add Z-DEVD-FMK or bFGF or Nile red solution into the polymer solution after the polymer is dissolved. The final concentration of polymer must be kept the same with the beginning concentration mentioned in Materials (see Note 2).
4. Add TPP solution drop-wise onto the polymer solution while stirring. Volumetric ratio of the polymer to TPP, which is a cross-linker used to form nanoparticles, depends on nanoparticle type. For CS nanoparticles the ratio is 1/1 mL, whereas it

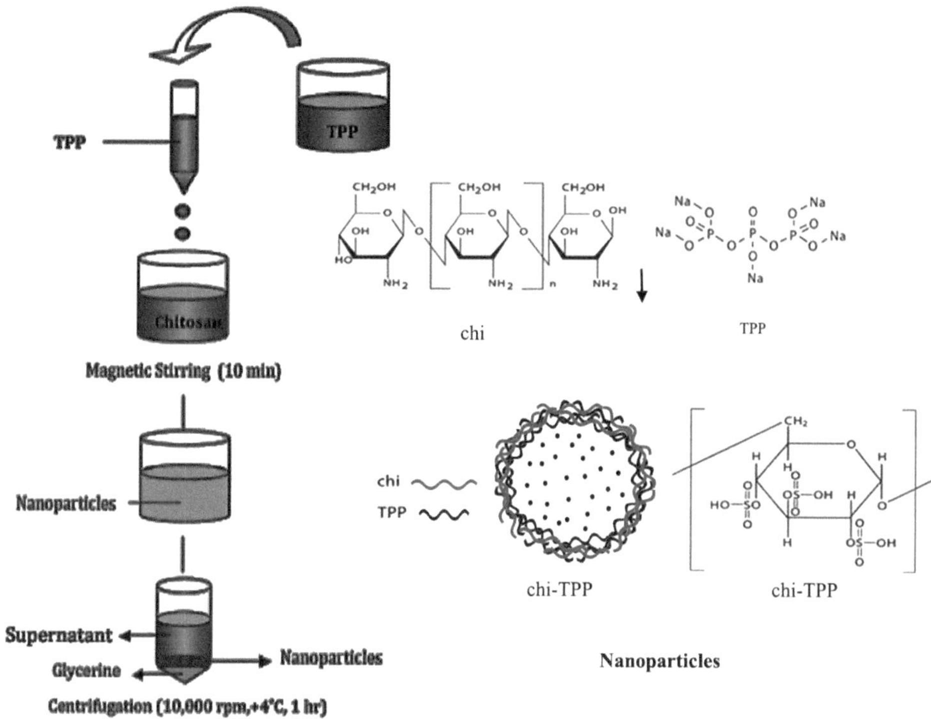

Fig. 2. Chemical structures of chitosan (Chi) and sodium tripolyphosphate, and schematic illustration of chitosan nanoparticle preparation procedure with ionotropic gelation method. (TPP: tripolyphosphate solution). (Modified from ref. (28) with permission).

is 5/2 mL for CS–PEG and CS–PEG–BIO nanoparticles (see Note 3).

5. Continue stirring the nanoparticle suspension for 15 min.
6. Centrifuge the nanoparticle suspension at 10,000 rpm (9,277 g) at +4°C for 1 h.
7. Discard the supernatant and resuspend in water for in vivo administration or resuspend in PBS for release studies (see Tables 1 and 2, Fig. 3).

3.2. Conjugation of TfRMAb to Nanoparticles

1. Add 1,900 μL of borate buffer onto 100 μL of Traut's reagent and stir for 5 min on a magnetic stirrer.
2. Take equal volumes of the above and of the streptavidin solution and mix together on a magnetic stirrer for 90 min to obtain thiolated streptavidin.
3. In parallel, stir 100 μL of TfRMAb solution (1 mg/mL) with 5 μL of m-maleimidobenzoyl-*N*-hydroxysuccinimide solution (5 mg/mL in dimethylformamide) for 30 min at room temperature. This will transform some of the amino groups into maleimide groups.

Table 1
Examples of nanoparticle size and zeta potential

Polymer type	Particle size (nm)	Zeta potential (mV)
CS (blank)	301	22
CS (Z-DEVD-FMK loaded)	650	20
CS (bFGF loaded)	424	23

Table 2
Examples of drug association efficiency and loading capacity of nanoparticles

Nanoparticles	% AE[a]	% LC[b]
CS (Z-DEVD-FMK loaded)	23	$0{,}019 \times 10^{-3}$
CS (bFGF loaded)	21	$2{,}739 \times 10^{-3}$

[a]% Association Efficiency (% AE) = 100 × (total peptide amount − free peptide amount)/total peptide amount
[b]% Loading Capacity (% LC) = 100 × (total peptide amount − free peptide amount)/nanoparticle weight

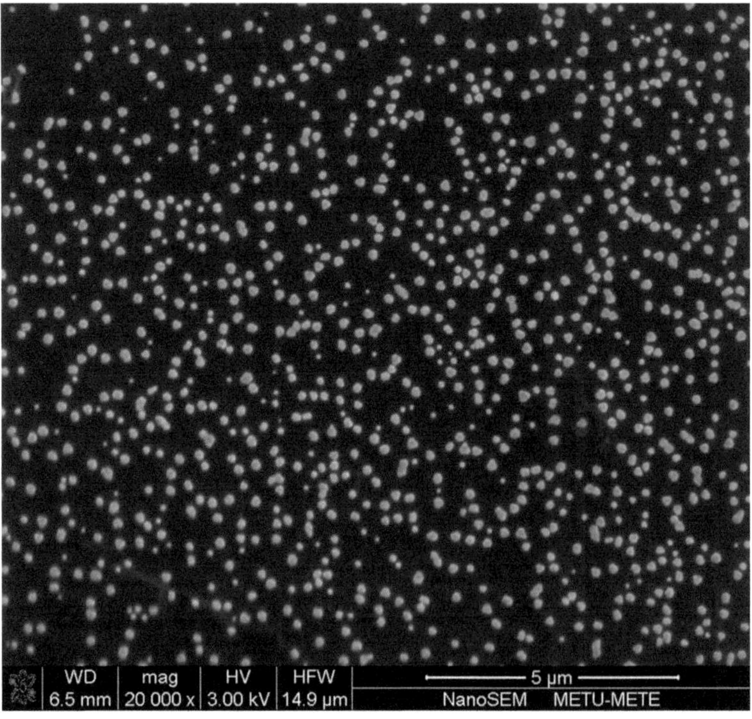

Fig. 3. Scanning Electron Microscopy (SEM) images of loaded and PEGylated chitosan nanoparticles.

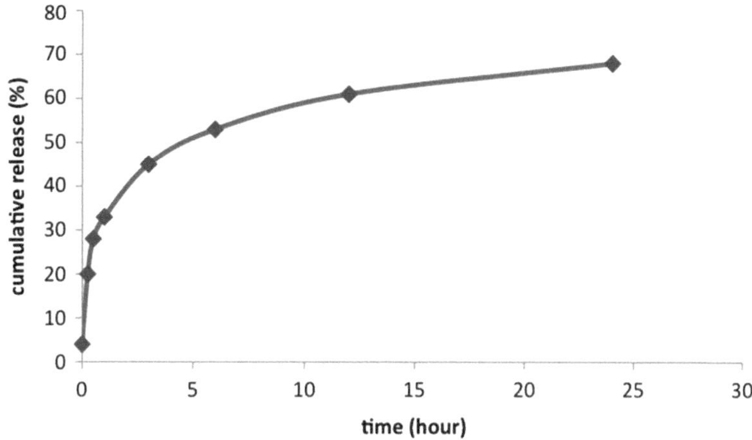

Fig. 4. A typical in vitro release profile of a peptide from chitosan nanoparticles.

4. After preparing CS–PEG–BIO nanoparticles, add 12.5 μL of streptavidin solution and 12.5 μL of antibody solution to the nanoparticle suspension and incubate for 30 min at room temperature.

3.3. Release Studies from Nanoparticles

1. Add 1 mL of PBS in each release compartment after discarding supernatant.
2. Place each eppendorf tube in a water bath at 37°C on a horizontal shaker.
3. Centrifuge nanoparticles at 10,000 rpm (9,277 g) for 10 min and then filter at predetermined time intervals (see Note 6).
4. Determine the concentration at each time point and calculate the cumulative values (see Fig. 4).

3.4. In Vivo Monitoring of Nanoparticle Penetration into the Brain

1. Anesthetize mice with isoflurane during surgery and with urethane (750 mg/kg, intraperitoneal, followed by 500 mg/kg 30 min later) during the experiment.
2. Monitor body temperature with a rectal probe and maintain at 37.0 ± 0.2°C using a homeothermic blanket control unit (Harvard Apparatus). Monitor systolic blood pressure noninvasively by using a cuff and tail probe (NIBP controller; ADInstruments). Monitor pulse rate and oxygen saturation by an oxymeter using amini Y clip on the left lower extremity (V3304 Tabletop Pulse Oxymeter; Surgivet) (see Note 7).
3. Open a cranial window of 5 × 5 mm over the parietotemporal cortex and leave the dura intact to maintain physiological conditions. Seal the window with dental acryl and then fill it with artificial cerebrospinal fluid (124 mM NaCl, 5 mM KCl, 1.25 mM NaH_2PO4, 1.3 mM $MgSO_4$, 2.4 mM $CaCl_2$, 25 mM $NaHCO_3$, and 10 mM glucose; pH 7.4) at 37°C. (see Notes 8 and 9).

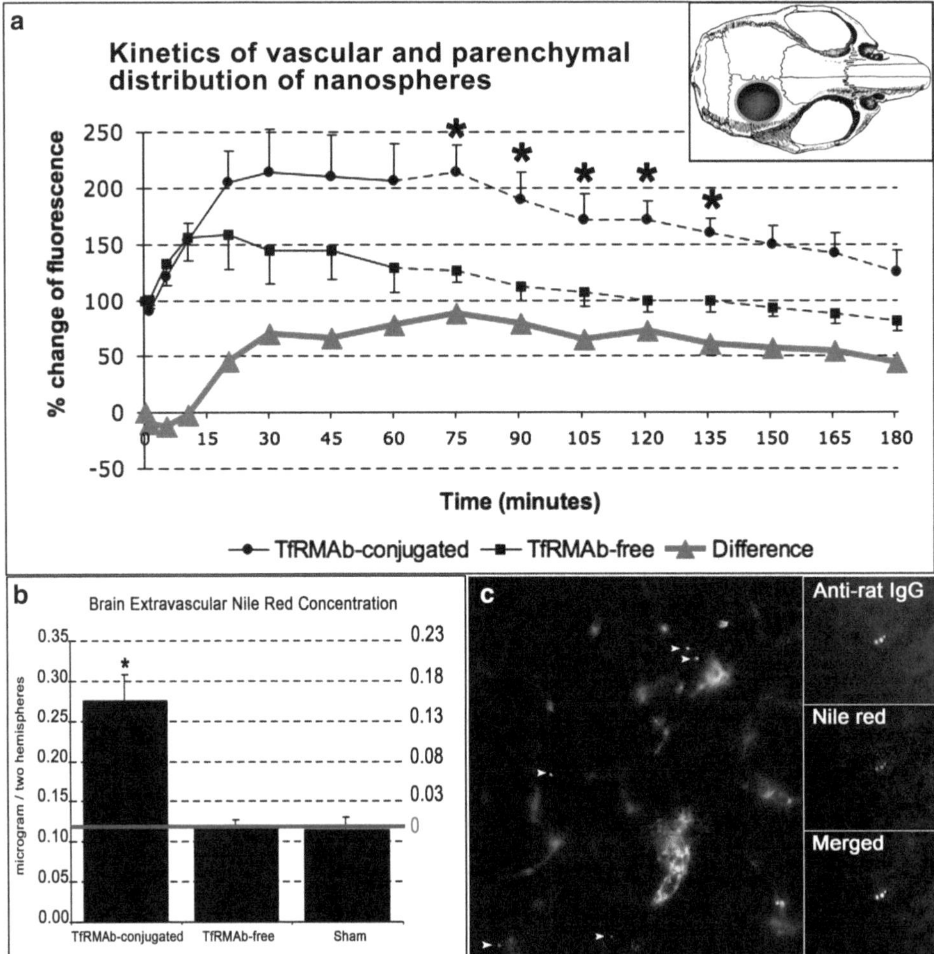

Fig. 5. Nanoparticles are rapidly transported to brain parenchyma after systemic administration. (**a**) The graph illustrates the change in fluorescence recorded from the brain over the course of 3 h after injection of Nile red loaded to TfRMAb-conjugated or unconjugated (TfRMAb-free) nanoparticles. The difference between the two lines (*triangles*) reflects the fluorescence coming from the nanoparticles within the parenchyma and illustrates the time course of nanoparticle penetrance to the brain. (**b**) Nile red concentration in brain postvascular tissue increased only when TfRMAb-conjugated nanoparticles were administered. The graph illustrates spectrophotometric measurements at 549 nm from brain homogenates obtained 1 h after injection of TfRMAb-conjugated or unconjugated nanoparticles or from sham-operated mice. Only values above the horizontal thick line, below which values correspond to the tissue background readings, were taken into consideration. $*p<0.05$ compared with the other groups. (**c**) Confirmation of penetration of the nanoparticles to the parenchyma by fluorescence microscopy on brain sections obtained 1 h after injection. FITC-conjugated anti-rat IgG antibody (*green*) labeled the nanoparticles bearing TfRMAb, clearly demonstrating that the nanoparticles were dispersed within the extracellular space (*arrowheads*) outside the vessel lumens (*longitudinal structures*). Some FITC-conjugated nanoparticles exhibited green as well as red fluorescence because they had not released of all the Nile red loaded within an hour (images on the *right*). Scale bars: (**c**), *left*, 15 μm; (**c**), *right*, 5 μm. (Reprinted from ref. (7) with permission).

4. Capture fluorescent images under a microscope at 100× magnification in a dark room by using appropriate camera and imaging software (see Note 10; Fig. 5).
5. Save images in TIFF format, and calculate the mean fluoroscence intensity of the area imaged with the software.
6. Record sequential images before starting the experiment to obtain baseline. After the systemic (intravenous) injection of Nile red-loaded nanoparticles, record pictures at predetermined time intervals (e.g. 1, 5, 10, 20, and 30 min, and then every 15 min for 3 h) using the same exposure time and gain settings.
7. Determine the changes in fluorescence (red) intensity from baseline after injection of the antibody-conjugated or the unconjugated nanoparticles. Since unconjugated nanoparticles cannot penetrate the brain, they reflect the Nile red fluorescence coming from the nanoparticles within the circulation. Subtract this signal from the signal obtained with antibody-conjugated nanoparticles to obtain the net signal coming from the nanoparticles penetrating the brain.
8. At the end of the experiment, perfuse mice transcardially with saline to flush intravascular content including the nanoparticles and then extract the brain, liver, and spleen. Tissues are immediately frozen and kept at −80°C until the determination of Nile red by UV spectrophotometry, which gives an estimate of the nanoparticles translocated to the parenchyma of the tissues examined (see Fig. 5).

3.5. Histological Detection of Nanoparticles in the Brain

1. Obtain fresh frozen, 20 μm-thick coronal brain sections.
2. Fix the sections with 96% alcohol for 10 min, wash with PBS, and then immunostain with FITC-conjugated goat anti-rat IgG antibody (Sigma, at 1:100 and 1:200 dilutions, in PBS) at room temperature for 60 min to detect the nanoparticles conjugated with TfRMAb, which is a rat IgG2a (see Note 11).
3. Block the sections with 10% normal goat serum for 10 min at room temperature and immunostain with anti-FGF-2/bFGF polyclonal antibody (Millipore, 1:100 in PBS) at +4°C overnight. After washing with PBS for 5 min two times, incubate with cy3-conjugated AffiniPure goat anti-rabbit IgG (Jackson Immunoresearch, 1:200 in PBS) at room temperature for 90 min to immunostain human bFGF-loaded nanoparticles.
4. Finally, coverslip with mounting medium containing Hoechst 33258 to counterstain the nuclei in order to obtain tissue orientation under the microscope.
5. Check co-localization of the IgG immunoreactive spots (green) under a fluorescent microscope with Nile red fluorescence or with immunoreactivity for human bFGF, depending on

whether the nanoparticles have been loaded with Nile red or bFGF (see Fig. 5).

6. We routinely carry out negative controls by omitting the primary antibody.

4. Notes

1. The concentrations of the polymer and cross-linker solutions are very critical; concentration changes may result in undesirable aggregation or precipitation of nanoparticles or may change the particle size. Use the same concentration and timing to obtain uniform nanoparticle batches.
2. If you want to load a new peptide or another hydrophilic substance into the nanoparticles, you may either prepare a solution of substance and add it into the polymer/cross-linker solution or directly dissolve it in the polymer/cross-linker solution. Attention must be paid to the final concentration of the solution if you choose to add the drug solution into the polymer/cross-linker.
3. Also, use a fixed TPP dropping rate into polymer solution to obtain uniform batches. Using a syringe without plunger may provide reproducible drop rates for every trial.
4. Use a vial with a cap to prevent contamination of solutions from environmental pollutants during magnetic stirring.
5. While working with peptides like bFGF or Z-DEVD-FMK, it is a good idea to separate them into small aliquots and store them at −20°C until use. Keep the volume as low as possible because every aliquot can be used only once.
6. During release studies, filter the supernatant through a 13 mm diameter 0.22-μm cellulose acetate syringe filter before determining the concentration.
7. For intravital microscopy experiments, it is essential to maintain the physiological conditions (e.g. normal body temperature, blood pressure, and oxygenation) so as not to cause cerebrovascular dysfunction.
8. Wash the cranium frequently with saline to avoid heating while drilling the cranial window.
9. Drill the whole window area until the bone is thinned to a film. Gently lift and peel the bony film with a pair of forceps without damaging the underlying dura mater and cortex.
10. Make sure that the cranium is well fixed and not moving with animal's respiration, and that the artificial cerebrospinal fluid

level over the window and the lighting conditions are constant while recording sequential fluorescent images.

11. For histological detection of nanoparticles in the brain parenchyma, it is preferable to use two markers fluorescing at different wavelengths (e.g. anti-IgG (green) and Nile red) and co-localize both signals to ensure that the observed spots are not microscopic staining artifacts.

Acknowledgements

We wish to thank all of our colleagues including our collaborators in Spain (Eduardo Fernandez-Megia, Ramon Novoa-Carballal, and Ricardo Riguera from Universidad de Santiago de Compostela) for their important contributions to the development of the nanoparticles used in studies cited in this chapter. Dr. Turgay Dalkara's work is supported by the Turkish Academy of Sciences.

References

1. Abbott, N. J., Ronnback, L., and Hansson, E. (2006) Astrocyte-endothelial interactions at the blood-brain barrier, Nature Reviews Neuroscience 7, 41–53.
2. Pardridge, W. M. (1998) CNS drug design based on principles of blood-brain barrier transport, J Neurochem 70, 1781–92.
3. Bickel, U., Yoshikawa, T., and Pardridge, W. M. (2001) Delivery of peptides and proteins through the blood-brain barrier, Adv Drug Deliv Rev 46, 247–79.
4. Pardridge, W. M. (2002) Drug and gene delivery to the brain: the vascular route, Neuron 36, 555–8.
5. Huwyler, J., and Pardridge, W. M. (1998) Examination of blood-brain barrier transferrin receptor by confocal fluorescent microscopy of unfixed isolated rat brain capillaries, J Neurochem 70, 883–6.
6. Pardridge, W. M., Wu, D., and Sakane, T. (1998) Combined use of carboxyl-directed protein pegylation and vector-mediated blood-brain barrier drug delivery system optimizes brain uptake of brain-derived neurotrophic factor following intravenous administration, Pharm Res 15, 576–82.
7. Karatas, H., Aktas, Y., Gursoy-Ozdemir, Y., Bodur, E., Yemisci, M., Caban, S., Vural, A., Pinarbasli, O., Capan, Y., Fernandez-Megia, E., Novoa-Carballal, R., Riguera, R., Andrieux, K., Couvreur, P., and Dalkara, T. (2009) A nanomedicine transports a peptide caspase-3 inhibitor across the blood-brain barrier and provides neuroprotection, J Neurosci 29, 13761–9.
8. Chang, J., Jallouli, Y., Kroubi, M., Yuan, X. B., Feng, W., Kang, C. S., Pu, P. Y., and Betbeder, D. (2009) Characterization of endocytosis of transferrin-coated PLGA nanoparticles by the blood-brain barrier, Int J Pharm 379, 285–92.
9. Skarlatos, S., Yoshikawa, T., and Pardridge, W. M. (1995) Transport of [125I]transferrin through the rat blood-brain barrier, Brain Res 683, 164–71.
10. Lee, H. J., Engelhardt, B., Lesley, J., Bickel, U., and Pardridge, W. M. (2000) Targeting rat anti-mouse transferrin receptor monoclonal antibodies through blood-brain barrier in mouse, J Pharmacol Exp Ther 292, 1048–52.
11. Jefferies, W. A., Brandon, M. R., Hunt, S. V., Williams, A. F., Gatter, K. C., and Mason, D. Y. (1984) Transferrin receptor on endothelium of brain capillaries, Nature 312, 162–3.
12. Pardridge, W. M., Buciak, J. L., and Friden, P. M. (1991) Selective transport of an anti-transferrin receptor antibody through the blood-brain barrier in vivo, J Pharmacol Exp Ther 259, 66–70.
13. Rawat, M., Singh, D., and Saraf, S. (2006) Nanocarriers: promising vehicle for bioactive drugs, Biol Pharm Bull 29, 1790–8.

14. Peracchia, M. T., Fattal, E., Desmaele, D., Besnard, M., Noel, J. P., Gomis, J. M., Appel, M., d'Angelo, J., and Couvreur, P. (1999). Stealth PEGylated polycyanoacrylate nanoparticles for intravenous administration and splenic targeting. J Control Release 60, 121–8.
15. Prabaharan, M. (2008) Review paper: chitosan derivatives as promising materials for controlled drug delivery, J Biomater Appl 23, 5–36.
16. Fernandez-Urrusuno, R., Calvo, P., Remunan-Lopez, C., Vila-Jato, J. L., and Alonso, M. J. (1999) Enhancement of nasal absorption of insulin using chitosan nanoparticles, Pharm Res 16, 1576–81.
17. Pan, Y., Li, Y. J., Zhao, H. Y., Zheng, J. M., Xu, H., Wei, G., Hao, J. S., and Cui, F. D. (2002) Bioadhesive polysaccharide in protein delivery system: chitosan nanoparticles improve the intestinal absorption of insulin in vivo, Int J Pharm 249, 139–47.
18. Vila, A., Sanchez, A., Tobio, M., Calvo, P., and Alonso, M. J. (2002) Design of biodegradable particles for protein delivery, J Control Release 78, 15–24.
19. Datta, S. R., Brunet, A., and Greenberg, M. E. (1999) Cellular survival: a play in three Akts, Genes Dev 13, 2905–27.
20. Hara, H., Friedlander, R. M., Gagliardini, V., Ayata, C., Fink, K., Huang, Z. H., ShimizuSasamata, M., Yuan, J. Y., and Moskowitz, M. A. (1997) Inhibition of interleukin 1 beta converting enzyme family proteases reduces ischemic and excitotoxic neuronal damage, Proceedings of the National Academy of Sciences of the United States of America 94, 2007–12.
21. Schulz, J. B., Weller, M., and Moskowitz, M. A. (1999) Caspases as treatment targets in stroke and neurodegenerative diseases, Annals of Neurology 45, 421–29.
22. Chen, J., Nagayama, T., Jin, K. L., Stetler, R. A., Zhu, R. L., Graham, S. H., and Simon, R. P. (1998) Induction of caspase-3-like protease may mediate delayed neuronal death in the hippocampus after transient cerebral ischemia, Journal of Neuroscience 18, 4914–28.
23. Thornberry, N. A., and Lazebnik, Y. (1998) Caspases: Enemies within, Science 281, 1312–16.
24. Aktas, Y., Yemisci, M., Andrieux, K., Gursoy, R. N., Alonso, M. J., Fernandez-Megia, E., Novoa-Carballal, R., Quinoa, E., Riguera, R., Sargon, M. F., Celik, H. H., Demir, A. S., Hincal, A. A., Dalkara, T., Capan, Y., and Couvreur, P. (2005) Development and brain delivery of chitosan-PEG nanoparticles functionalized with the monoclonal antibody OX26, Bioconjugate Chemistry 16, 1503–11.
25. Illum, L. (1998) Chitosan and its use as a pharmaceutical excipient, Pharm Res 15, 1326–31.
26. Caban, S., Yemisci, M., Gursoy-Ozdemir, Y., Fernandez-Megia, E., Novoa-Carballal, R., Riguera, R., Andrieux, K., Couvreur, P., Capan, Y., and Dalkara, T. (2010) in "AAPS Annual Meeting and Exposition", New Orleans, Louisiana, USA.
27. Ma, J., Qiu, J., Hirt, L., Dalkara, T., and Moskowitz, M. A. (2001) Synergistic protective effect of caspase inhibitors and bFGF against brain injury induced by transient focal ischaemia, Br J Pharmacol 133, 345–50.
28. Cho, Y., Shi, R., and Ben Borgens, R. (2010) Chitosan nanoparticle-based neuronal membrane sealing and neuroprotection following acrolein-induced cell injury, J Biol Eng 4, 2.

Chapter 28

[^{3}H]Serotonin Release Assay Using Antigen-Stimulated Rat Peritoneal Mast Cells

Stephen D. Skaper and Laura Facci

Abstract

The concentration of nerve growth factor (NGF) is elevated in a number of inflammatory and autoimmune states in conjunction with increased accumulation of mast cells. Mast cells, which are of hematopoietic lineage, and NGF appear to be involved in neuroimmune interactions and tissue inflammation. Mast cells themselves are capable of producing and responding to NGF. Here we describe a protocol for the isolation and culture of peritoneal-derived rat mast cells, together with a [^{3}H]serotonin release assay which is useful in assessing the effects of antigens and neurotrophic factors on mast-cell activation.

Key words: Mast cells, Antigen, Degranulation, Serotonin, Nerve growth factor, Nerve–immune interaction

1. Introduction

Neurotrophic proteins promote neurite outgrowth, neuronal cell differentiation, and survival in vivo and in vitro. Nerve growth factor (NGF) represents the first and best characterized member of the neurotrophin family (1). We now know that NGF displays biological activities in a broad spectrum of cell types outside the nervous system and is produced by a wide range of cell populations not normally considered targets for innervation by NGF-dependent neurons, including cells of the immune–hematopoietic lineage. Mast cells are a heterogeneous immune-effector cell type found in connective tissues throughout the body, occur adjacent to blood and lymphatic vessels, and are concentrated beneath mucosal surfaces (2). Mast cells were the first cells of the immune lineage to be recognized as a target for NGF, both in vitro (3) and in vivo (4). In addition, NGF stimulates the proliferation of both B and T lymphocytes (5, 6), promotes human hematopoietic cell growth

Stephen D. Skaper (ed.), *Neurotrophic Factors: Methods and Protocols*, Methods in Molecular Biology, vol. 846, DOI 10.1007/978-1-61779-536-7_28, © Springer Science+Business Media, LLC 2012

and differentiation (7), and acts as an autocrine survival factor for memory B lymphocytes (8). Moreover, human monocytes (9), activated $CD4^+$ T-cell clones (10, 11), and lymphocytes (12) express the NGF receptor TrkA.

Mast cells occur in many peripheral tissues, in perivascular regions in close apposition to innervating sensory or autonomic nerve fibers, and also within the peripheral and central nervous systems. Mast cells secrete and respond to mature NGF (13–15) and are capable of synthesizing higher molecular weight forms of biologically active NGF and other neurotrophins (16), suggesting the possibility for autocrine/paracrine effects of these neurotrophic factors on mast cells. Given the interconnecting links between inflammation, pain, NGF, and mast cells (17), procedures for their isolation and analysis of NGF effects represent an important element in the study of neurotrophic factors. The following chapter presents a protocol for the isolation and purification of rat peritoneal mast cells, and their application to the measurement of $[^3H]$ serotonin release.

2. Materials

2.1. Equipment and Labware

1. Laminar flow biological safety cabinet (CL2)
2. Humidified, water-jacketed culture incubator at 37°C and 5% CO_2/95% air
3. Water bath set at 37°C
4. Dissecting tools (Fine Science Tools (InterFocus Ltd) are very high quality for the price)
5. Bench centrifuge to accommodate 15- and 50-mL tubes
6. 15- and 50-mL polypropylene plastic centrifuge tubes (sterile)
7. 0.45-μm filters (Millipore)
8. 50-mL syringes
9. Syringe needles (1.5-in.)
10. 0.5-mL Eppendorf® LoBind microcentrifuge tubes
11. Vacuum-driven filter units (0.45-μm pore size, 250- and 500-mL capacity) from Millipore (Stericup®)

2.2. Cell Culture Media, Supplements, and $[^3H]$ Serotonin Release Assay Reagents

1. Bovine serum albumin (BSA), Cohen fraction V (Sigma)
2. RPMI-1640 medium (Invitrogen)
3. L-Glutamine (200 mM, sterile, for cell culture) (Invitrogen)
4. Penicillin/streptomycin (10,000 U/mL penicillin and 10,000 μg/mL streptomycin) (Invitrogen)
5. Trypan blue stain 0.4% (Invitrogen)

6. Glutaraldehyde (25% in water) (Sigma)
7. Safranin O (Sigma)
8. Mouse monoclonal anti-DNP IgE, clone SPE-7 (Sigma)
9. Dinitrophenylated human serum albumin (DNP-HSA) (Sigma)
10. [^{3}H]serotonin (Perkin Elmer)
11. β-NGF (recombinant human) (Peprotech)
12. Brain-derived neurotrophic factor (BDNF, recombinant human) (Peprotech)
13. Neurotrophin-3 (NT-3, recombinant human) (Peprotech)
14. Neurotrophin-4 (NT-4, recombinant human) (Peprotech)
15. Substance P (Sigma)
16. Phosphate-buffered saline (PBS) (Invitrogen)

2.3. Culture Medium and Solutions

1. *PBS+0.6% (w/v) glucose*: Dissolve 0.6 g of D-glucose in 100 mL of PBS and filter-sterilize.
2. *40% (w/v) BSA in PBS/0.6% glucose*: Dissolve 40 g of BSA in 100 mL of PBS and filter-sterilize (see Note 1).
3. *30% (w/v) BSA*: Dilute the 40% glucose solution (75 mL) in PBS/0.6% glucose (25 mL).
4. *Culture medium*: To 100 mL of RPMI-1640 medium, add 1 mL of 200 mM L-glutamine stock solution (2 mM final) and 0.5 mL penicillin/streptomycin 100× stock solution (50 U/mL penicillin and 50 μg/mL streptomycin, final).
5. *PIPES buffer pH 7.10*: 25 mM PIPES (7.56 g/L), 100 mM NaCl (5.84 g/L), 5 mM KCl (0.372 g/L), 1 mM $CaCl_2 \cdot 2H_2O$ (0.147 g/L), $MgCl_2 \cdot 6H_2O$ (0.81 g/L), and 5.6 mM D-glucose (1.00 g/L).

3. Methods

3.1. Isolation of Mast Cells

1. Adult male CD Sprague Dawley rats (200–250 g) are used. The recommended procedure for euthanasia is CO_2 asphyxiation and cervical dislocation. This should be carried out in strict compliance with approved institutional and national guidelines.
2. Using a 50-mL syringe (1.5-in. needle), inject into the peritoneal cavity 40 mL of sterile PBS/0.6% (w/v) glucose.
3. Gently massage the abdomen for about 90 s.
4. Carefully open the peritoneal cavity and collect the fluid containing peritoneal cells. Take care to avoid entering internal organs, which can contaminate the wash solution (see Notes 2 and 3).

5. Transfer the peritoneal collection fluid into 50-mL polypropylene tissue culture centrifuge tubes. Do not place on ice.
6. Centrifuge at 200 × *g* for 5 min at room temperature.
7. Remove the supernatant with a 10-mL pipette, leaving 5 mL of liquid in the bottom of the tube.
8. To a 50-mL polypropylene tissue culture centrifuge tube, add 5 mL of the 40% BSA solution, then carefully layer 5 mL of the 30% BSA solution over this (see Note 4).
9. Layer over this BSA gradient up to 20 mL of cell suspension carried over from the initial centrifugation in step 5. Using volumes greater than 20 mL will decrease the efficiency of the separation process—use multiple gradients for large samples.
10. Centrifuge the BSA gradient tube(s) at 200 × *g* for 5 min at room temperature.
11. Remove the BSA phases with a 10-mL pipette, leaving the bottom 10 mL in each tube.
12. Using a long (9-in.) Pasteur pipette, continue removing liquid from the tubes in step 9, leaving only the last 2 mL in the bottom of each tube.
13. To each tube, now add 10 mL of PBS/0.6% glucose.
14. Centrifuge at 200 × *g* for 5 min, room temperature.
15. Remove the supernatant(s) and resuspend each cell pellet in 2 mL of RPMI-1640 + glutamine + penicillin/streptomycin.
16. To 90 μL of this cell suspension, add 10 μL of 0.4% trypan blue.
17. Count cells (see Chap. 3). Expect a yield of 300,000–400,000 cells per rat.
18. Expect a final purity of mast cells >90%. If you want to check this for yourself, you can perform Alcian blue and Safranin staining (optional) (18). See Subheading 3.2 below.
19. Plate cells according to experimental design, using the above RPMI-1640 culture medium.

3.2. Safranin O Staining of Mast Cells

Mast cells can be divided into two types: connective tissue mast cells and mucosal mast cells. Peritoneal mast cells belong to the first type. Mast-cell phenotypes can be classified in terms of their profiles of proteinases and proteoglycans. In this regard, mucosal mast-cell granules stain with Alcian blue rather than with Safranin, while connective tissue mast cells stain with Safranin but not with Alcian blue:

1. This procedure makes use of the Shandon Cytospin three centrifuge (Block Scientific, Bohemia, NY, USA, www.blockscientificstore.com).

2. A 0.1-mL cytofunnel disposable sample chamber is adequate for this purpose. Pipette into the chamber 0.1 mL of mast-cell suspension, which has been diluted to ~400,000 cells per milliliter in culture medium.
3. Cytospin the mast cells onto a glass slide at 700 rpm for 5 min. The cytofunnel chamber is attached to the glass slide. The manufacturer-supplied directions provide illustrative examples of the technique and should be followed in this regard.
4. Fix the cells with 0.1% glutaraldehyde (diluted 1:250 from a 25% stock solution in water) in PBS for 20 min at room temperature.
5. Rinse three times with PBS.
6. Rinse quickly with 1% acetic acid (1:100 (v/v) of glacial acetic acid, in double-distilled H_2O) for 10 s.
7. Stain in 0.1% Safranin O solution (0.1 g in 100 mL of 0.125 N HCl) for 5 min.
8. Rinse twice with PBS.
9. Keep the slide(s) in double-distilled H_2O or PBS.
10. Observe under a light microscope without phase contrast optics—Safranin-stained mast-cell granules will appear red.

3.3. Measuring [^{3}H] Serotonin Release from Antigen-Stimulated Mast Cells

Mast cells have high-affinity IgE receptors ($Fc_{\varepsilon}RI$) (19) and have been used extensively to study the signaling pathways leading to exocytotic release of inflammatory mediators during antigen stimulation of mast cells (20, 21). In this assay, a mouse monoclonal IgE which is specific for dinitrophenol (DNP) haptens (clone SPE-7) is used. DNP-HSA is employed as the triggering agent in this assay. Cells are preloaded with [^{3}H]serotonin (5-[1,2-^{3}H(N)] hydroxytryptamine binoxalate) which is released upon $Fc_{\varepsilon}RI$ cross-linking.

1. Isolate rat peritoneal mast cells, as described in Subheading 3.1 above.
2. Plate mast cells in a 96-well tissue culture plate (brand is not important) at a density of 50,000 cells per well in RPMI-1640 medium with 2 mM L-glutamine and 50 U/mL penicillin + 50 μg/mL streptomycin + 0.1 μCi of [^{3}H]serotonin. Final volume: 100 μL per well.
3. Incubate cells for 18 h in a 5% CO_2 incubator (37°C).
4. Replace this medium with 100 μL of PIPES buffer containing anti-DNP IgE (0.3 μg/mL) to sensitize the cells.
5. Incubate for 1 h at 37°C.
6. Replace the IgE solution with 100 μL per well of prewarmed PIPES buffer containing 0.1–1.0 μg/mL of DNP-HSA.

The level of conjunction averages 30–40 mol of DNP per mol of albumin.

7. Incubate for 15 min at 37°C.
8. Collect the incubation medium and add 100 μL per well of 1% Triton X-100 to solubilize the cells (see Note 5).
9. To one vial for liquid scintillation counting of radioactivity, transfer 70 μL of incubation medium, and to a second vial, transfer 70 μL of the Triton extract. Add to each vial an appropriate volume of liquid scintillation counting fluid and place in a liquid scintillation analyzer.
10. Calculate the percentage of [^{3}H]serotonin release as [released dpm/(released dpm + cell-associated dpm)] × 100.
11. Always subtract the background (spontaneous release) from the stimulated release value ("net" release).

3.4. Assay Application to Study of NGF Regulation of Mast-Cell Function

NGF is known to induce mediator release from mast cells (22). Effects of NGF on stimulus-evoked mediator release from mast cells can be assessed using the [^{3}H]serotonin release assay described here. A representative experiment is now described, in which rat peritoneal mast cells are incubated with NGF, together with specific antigen (15) or substance P (23) to induce degranulation. The steps below are taken from Subheading 3.3 above:

1. Seed mast cells into a 96-well tissue culture plate, 50,000 cells per milliliter.
2. Load with [^{3}H]serotonin for 18 h.
3. Remove this solution and prime cells with anti-DNP IgE for 1 h at 37°C.
4. Replace the IgE solution with 100 μL of prewarmed PIPES buffer containing DNP-HSA (0.1 μg/mL) or substance P (30 μM).
5. In replicate wells containing DNP-HSA or substance P, add the following neurotrophins to a final concentration of 100 ng/mL: NGF, BDNF, NT-3, or NT-4. The neurotrophin is added to the DNP-HSA or substance P–containing solution from a 25-μg/mL stock solution (i.e., 1:250) (see Note 6).
6. Continue incubation for a further 15 min.
7. Determine net release of [^{3}H]serotonin.
8. Note that a significant increase in secretion of preloaded [^{3}H]serotonin occurred in mast cells treated with specific antigen or substance P, in combination with NGF, but not BDNF, NT-3, or NT-4 (see Table 1).

Table 1
Exogenous NGF regulates mast-cell function

	Net [³H]serotonin release (%)				
Stimulus	**None**	**NGF**	**BDNF**	**NT-3**	**NT-4**
Antigen	9.3 ± 2.1	20.3 ± 4.2[a]	7.3 ± 3.1	6.1 ± 3.7	5.5 ± 1.8
Substance P	6.7 ± 2.9	18.9 ± 5.9[a]	5.9 ± 4.0	7.9 ± 3.5	–

Rat peritoneal mast cells were preloaded with [³H]serotonin and then challenged with either a specific antigen (IgE) or substance P (30 μM), without or with the indicated recombinant neurotrophin (100 ng/mL). Net release of [³H]serotonin was determined, as described in Subheading 3.3. Values are means ± SD (three experiments). [a]$p < 0.05$. NGF alone did not affect basal [³H]serotonin release. Reprinted with permission from Elsevier (17)

4. Notes

1. Dissolving this amount of BSA can be difficult. We find it easiest to add the required amount of PBS to a beaker with stir bar, place on the magnetic stir plate, and add the entire amount of BSA. Turn on the stir plate (low speed) and allow the BSA to slowly dissolve from the interface between powder and liquid. A 100-mL solution at 40% BSA will take several hours to fully dissolve. Do not stir vigorously (avoid foaming) and do not heat. Vacuum-driven filter units (0.45-μm pore size) from Millipore (Stericup® or Steritop®) work well for sterilization.
2. This is a critical step: The peritoneal lavage should be straw-colored and will be lightly opaque. The major contaminants are blood cells and debris from compromised internal organs, intestine in particular. Difficulty may be encountered when withdrawing the lavage using the syringe used for injection of the lavage solution. An alternative method collection is to make a small incision on the side of the injection and enter the peritoneal cavity with a 25-mL pipette. Withdraw the contents of the cavity toward the periphery, again avoiding internal organs (we actually prefer this route for collection).
3. Once proficiency is developed, it is possible to process up to 25 rats in a period of 90 min. However, the technique is not foolproof, and even an experienced researcher should not be surprised if he/she is unable to achieve a success rate better than 80%.
4. Layering of the BSA solutions is best done using a 5-mL pipette and automatic pipettor—the important thing is not to allow

the two gradient solutions to mix, i.e., a steady hand is needed. Doing this under a tissue culture flow cabinet, we usually do this at eye level (which means getting down on your knees) so that you can physically observe the 30% BSA solution flowing on top of (but not into) the 40% solution beneath.

5. Care must be exercised when treating mast cells with compounds where dimethyl sulfoxide or ethanol is used as solvent, as these will decrease mast-cell release of serotonin. Dimethyl sulfoxide or ethanol, if used, should not exceed 0.2% by volume.

6. The neurotrophins are supplied as solids. When reconstituting, it is important to use a solution with carrier protein, for example, PBS with 0.1% BSA. Neurotrophins (NGF, BDNF, NT-3, NT-4) have a very high isoelectric point (which means they are quite basic) and will stick to glass and plastic (especially polystyrene) surfaces. We recommend reconstitution in the container provided by the supplier to a concentration ≥25 μg/mL. Aliquot this stock into 0.5-mL low-protein-binding Eppendorf tubes (10–20 μL/tube) and store at −20°C for up to 6 months. Once thawed, an aliquot can be safely kept in the refrigerator for up to 4 weeks. Do not subject BDNF to repeated freeze–thaw cycles.

Acknowledgments

L. Facci was supported by Fondazione CARIPARO Progetto Dottorati di Ricerca Anno 2009.

References

1. Levi-Montalcini R (1987) The nerve growth factor 35 years later. *Science 237,* 1154–1162
2. Galli SJ, Nakae S, and Tsai M (2005) Mast cells in the development of adaptive immune responses. *Nat Immunol 6,* 135–142
3. Böhm A, Aloe L, and Levi-Montalcini R (1986) Nerve growth factor enhances precocious differentiation and numerical increase in mast cells in cultures of rat splenocytes. *Accademia Nazionale Lincei 80,* 1–6
4. Aloe L, and Levi-Montalcini R (1977) Mast cells increase in tissues of neonatal rats injected with nerve growth factor. *Brain Res 133,* 358–366
5. Thorpe LW, and Perez-Polo JR (1987) The influence of nerve growth factor on the in vitro proliferative response of rat spleen lymphocytes. *J Neurosci Res 18,* 134–139
6. Otten U, Ehrhard P, and Peck R (1989) Nerve growth factor induces growth and differentiation of human B lymphocytes. *Proc Natl Acad Sci USA 86,* 10059–10063
7. Matsuda H, Coughlin MD, Bienenstock J, and Denburg JA (1988) Nerve growth factor promotes human hemopoietic colony growth and differentiation. *Proc Natl Acad Sci USA 85,* 6508–6512
8. Torcia M, Bracci-Laudiero L, Lucibello M, Rubartelli A, Cozzolino F, Aloe L, et al (1996) Nerve growth factor is an autocrine factor for memory B lymphocytes. *Cell 85,* 1–20
9. Ehrhard PB, Ganter U, Bauer J, and Otten U. (1993) Expression of functional *trk* protooncogene in human monocytes. *Proc Natl Acad Sci USA 90,* 5423–5427

10. Ehrhard PB, Erb B, Graumann U, and Otten U (1993) Expression of nerve growth factor and nerve growth factor receptor tyrosine kinase Trk in activated CD4-positive T-cell clones. *Proc Natl Acad Sci USA 90,* 10984–10988
11. Lambiase A, Bracci-Laudiero L, Bonini S, Bonini S, Starace G, D'Elios MM, et al (1997) Human CD4+ T cell clones produce and release nerve growth factor and express high-affinity nerve growth factor receptors. *J Allergy Clin Immunol 100,* 408–414
12. Santambrogio L, Benedetti M, Chao MV, Muzaffar R, Kulig K, Gabellini N, et al (1994) Nerve growth factor production by lymphocytes. *J Immunol 153,* 4488–4495
13. Horigome K, Pryor ED, Bullock EM, and Johnson EM Jr (1993) Mediator release from mast cells by nerve growth factor. Neurotrophin specificity and receptor mediation. *J Biol Chem 268,* 14881–14887
14. Horigome K, Bullock ED, and Johnson, EM Jr (1994) Effects of nerve growth factor on rat peritoneal mast cells. Survival promotion and immediate-early gene induction. *J Biol Chem 269,* 2695–2702
15. Leon A, Buriani A, Dal Toso R, Fabris M, Romanello S, Aloe L, et al (1993) Mast cells synthesize, store, and release nerve growth factor. *Proc Natl Acad Sci USA 91,* 3739–3743
16. Skaper SD, Pollock M, and Facci L (2001) Mast cells differentially express and release active high molecular weight neurotrophins. *Mol Brain Res 97,* 177–185
17. Skaper SD (2001) Nerve growth factor: a neurokine orchestrating neuroimmune-endocrine functions. *Mol Neurobiol 24,* 183–199
18. Mayrhofer G (1980) Fixation and staining of granules in mucosal mast cells and intraepithelial lymphocytes in the rat jejunum, with special reference to the relationship between the acid glycosaminoglycans in the two cell types. *Histochem J 12,* 513–526
19. Metzger H (1992) The receptor with high affinity for IgE. *Immunol Rev 125,* 37–48
20. Benhamou M, Stephan V, Robbins KC, and Siraganian RP (1992) High-affinity IgE receptor-mediated stimulation of rat basophilic leukemia (RBL-2H3) cells induces early and late protein-tyrosine phosphorylations. *J Biol Chem* 267, 7310–7314
21. Beaven MA, and Metzger H (1993) Signal transduction by Fc receptors: the Fc epsilon RI case. *Immunol Today 14,* 222–226
22. Bruni A, Bigon E, Boarato E, Mietto L, Leon A, and Toffano G (1982) Interaction between nerve growth factor and lysophosphatidylserine on rat peritoneal mast cells, *FEBS Lett 138,* 190–192
23. Fewtrell CM, Foreman JC, Jordan CC, Oehme P, Renner H, and Stewart JM (1982) The effect ofsubstancePonhistamineand5-hydroxytryptamine release in the rat, *J Physiol 330,* 393–411

Chapter 29

Rat Hippocampal Slice Culture Models for the Evaluation of Neuroprotective Agents

Elisabetta Gerace, Elisa Landucci, Tania Scartabelli, Flavio Moroni, and Domenico E. Pellegrini-Giampietro

Abstract

Organotypic slices cultured for weeks in vitro represent an extremely valuable strategy for the investigation of the long-term properties of neuronal circuits under physiological and pathological conditions. Here, we describe how to prepare rat organotypic hippocampal slice cultures and how to expose them for appropriate periods of time to excitotoxic agents or to oxygen and glucose deprivation conditions, in order to mimic the pattern of pyramidal cell damage which is observed in vivo and in other in vitro models. This preparation is very useful not only to study synaptic plasticity or the pathways and mechanisms of neurodegeneration but also to evaluate the effects of neuroprotective agents.

Key words: Organotypic hippocampal slices, Neuroprotection, Oxygen-glucose deprivation, *N*-methyl-D-aspartate, (*S*)-α-Amino-3-hydroxy-5-methyl-4-isoxazolepropionic acid, Kainate, Glutamate receptors

1. Introduction

Organotypic slices cultured for weeks in vitro represent an extremely valuable strategy for the investigation of the long-term properties of neuronal circuits under physiological and pathological conditions (1). In particular, organotypic slice cultures from the hippocampus not only retain a tissue organization and a distribution of glutamate receptors that is very similar to that observed in situ (2) but also exhibit synaptic plasticity mechanisms and a responsiveness to pathological insults (e.g., excitotoxicity, hypoxic, or ischemic conditions) that are comparable to what is obtained in vivo and in other in vitro models (such as acute slices or primary neuronal cell cultures). For example, ischemic-like insults produce a selective damage in the CA1 pyramidal cell layer, whereas kainic acid damages predominantly the CA3 region.

Stephen D. Skaper (ed.), *Neurotrophic Factors: Methods and Protocols*, Methods in Molecular Biology, vol. 846, DOI 10.1007/978-1-61779-536-7_29,

In the past few years, we have shown that organotypic hippocampal slices exposed to oxygen-glucose deprivation (OGD) for periods ranging from 15 to 75 min exhibit 24 h later a time-dependent and gradual increase in CA1 injury that builds up following selective apoptotic degeneration of pyramidal cells (3, 4). In our experience, a period of 30 min OGD has proven optimal to detect the effects of drugs that attenuate CA1 injury (4–7) as well as those that produce an aggravation of OGD toxicity (4, 8). However, in some cases, a "sublethal" period of OGD (20 min) appears to be more appropriate to reveal both the toxic properties of pharmacological agents and the concentration dependency of the protective effects of some antagonists (9).

The concentrations of drugs used in organotypic hippocampal slice experiments are somewhat higher than those expected from their Kd values and those used in cell cultures. Higher concentrations of drugs are required to saturate receptors in brain slices, due to the fact that they diffuse slowly through the thickness of brain tissue in vitro. Moreover, in the case of organotypic slices statically cultured on semiporous membrane inserts at the interface between culture medium and gas atmosphere, the concentrations of drugs need to be further increased because only the bottom of the slice is exposed to the bathing medium to which the drug is added and only a fraction of the drug diffuses across the membrane and reaches the tissue (4, 7). The following chapter presents a protocol for the preparation of rat organotypic hippocampal slice cultures and their exposure to excitotoxic agents or to OGD conditions, in order to mimic the pattern of pyramidal cell damage observed in vivo and in other in vitro models.

2. Materials

All solutions are prepared using ultrapure water (prepared by purifying deionized water to attain a sensitivity of 18.2 MΩ-cm at room temperature) and analytical grade reagents. Deionized-purified water, nonsterile glassware and plasticware, and surgical tools and equipment are sterilized by autoclave before use. All procedures are carried out using standard tissue culture sterile techniques.

2.1. Dissecting Medium

This is the medium used for hippocampal dissection and slice preparation. The solution is prepared under a sterile laminar flow hood:

1. Prepare 100 mL of 45% glucose solution by combining 45 g of D-(+)glucose and 80 mL of sterile water in a glass flask. Stir and

heat gently to dissolve. Add water to 100 mL, filter (0.2 μm), and collect into a sterile recipient. Aliquot in Eppendorf sterile tubes and store at −20°C for 1–2 months.

2. Prepare approximately 100 mL of Dissecting Medium in two 50-mL Falcon sterile tubes placed in a test-tube rack under the hood at room temperature: Add 50 mL of Hanks' Balanced Salt Solution (HBSS, Sigma H9269) (see Note 1) to each of the Falcon tubes using a sterile pipette. Then add 0.5 mL of a 45% glucose solution to each of the tubes and 0.75 mL of amphotericin B (see Note 2).
3. Vortex to mix all ingredients and store at 4°C for maximum 1–2 weeks.

2.2. Normal Medium

This is the medium used to grow and maintain the hippocampal slice cultures. Prepare approximately 100 mL of Normal Medium in two 50-mL Falcon sterile tubes placed in a test-tube rack under the hood at room temperature, as follows:

1. Add 25 mL of Eagle's Minimal Essential Medium (MEM, Sigma M2279) (see Note 3) to each of the Falcon tubes using a sterile pipette. Then add 12.5 mL of HBSS to each of the tubes, 12.5 mL of Horse Serum (see Note 4), 0.5 mL of a 45% glucose solution, and 0.75 mL of amphotericin B.
2. Finally, add 0.25 mL of 200 mM glutamine (see Note 5) to each of the tubes.
3. Vortex to mix all ingredients and store at 4°C for maximum 1–2 weeks.
4. Warm up to 37°C in a water bath before use.

2.3. Serum-Free Medium

This is the medium used for excitotoxicity experiments and for the 24-h recovery period following OGD and excitotoxicity experiments. Prepare approximately 100 mL of Serum-Free Medium in two 50-mL Falcon sterile tubes placed in a test-tube rack under the hood at room temperature, as follows:

1. Add 37.5 mL of Eagle's MEM to each of the Falcon tubes using a sterile pipette. Then add 12.5 mL of HBSS to each of the tubes, 0.5 mL of a 45% glucose solution, and 0.75 mL of amphotericin B.
2. Finally, add 0.25 mL of 200 mM glutamine to each of the tubes.
3. Vortex to mix all ingredients and store at 4°C for maximum 1–2 weeks.
4. Warm up to 37°C in a water bath before use.

2.4. Glucose- and Serum-Free Medium

This is the medium used for OGD:

1. We prepare 100 mL of Glucose- and Serum-Free Medium in two 50-mL Falcon sterile tubes placed in a test-tube rack under the hood at room temperature.
2. Add 37.5 mL of Eagle's MEM to each of the Falcon tubes using a sterile pipette. Then add 12.5 mL of HBSS and 0.75 mL of amphotericin B to each of the tubes.
3. Finally, add 0.5 mL of 200 mM glutamine to each of the tubes.
4. Vortex to mix all ingredients and store at 4°C for maximum 1–2 weeks.
5. Warm up to 37°C in a water bath and saturate by bubbling with 95% N_2/5% O_2 for 15 min before use.

2.5. Experimental Reagents

1. Propidium iodide (PI, Sigma P4170, see Note 6): Prepare approximately 10 mL of a 1-mg/mL solution in sterile water, filter (0.2 μm), aliquot in foil-wrapped Eppendorf tubes, and store in a closed box at 4°C. PI is a highly polar compound which is normally excluded from cells with an intact plasma membrane. When the membrane is damaged, PI enters the cells and upon binding to exposed DNA becomes highly fluorescent (the absorption maximum is 535 nm, and the fluorescence emission maximum is 617 nm).
2. Glutamate (Sigma G8415): Prepare a 100-mM stock solution in water. Add a few drops of 0.1 N NaOH to dissolve. Filter (0.2 μm), aliquot in Eppendorf tubes, and store at –20°C.
3. *N*-methyl-D-aspartate (NMDA, Sigma M3262): Prepare a 100-mM stock solution in water. Filter (0.2 μm), aliquot in Eppendorf tubes, and store at –20°C.
4. (*S*)-α-Amino-3-hydroxy-5-methyl-4-isoxazolepropionic acid (AMPA, Ascent Scientific 005): Prepare a 10-mM stock solution in water. Filter (0.2 μm), aliquot in Eppendorf tubes, and store at –20°C.
5. Kainic acid (Tocris Bioscience 0222): Prepare a 20-mM stock solution in water. Filter (0.2 μm), aliquot in Eppendorf tubes, and store at –20°C.

2.6. Special Equipment

1. Laminar flow hood. Both a vertical or horizontal laminar flow hood can be used.
2. Cell culture CO_2 incubator. The conditions are 37°C and 100% humidity in 95% air/5% CO_2.
3. McIlwain tissue chopper (Mickle Laboratory Engineering Co. Ltd., UK). Use a new double-edge blade (Gillette Platinum or from the chopper manufacturer) for every slice preparation session.

4. Millicell-®CM (0.4 μm, 30-mm diameter) semiporous cell culture inserts (Millipore PICM03050).
5. Dissecting stereo microscope. We use an Olympus SZ51 Stereozoom microscope.
6. Fiber optic cold light source. We use a KL200 Schott illuminator for stereo microscopy.
7. Inverted fluorescence microscope. We use an Olympus IX-50 microscope equipped with a Xenon-arc lamp, a low power objective (4×), and a rhodamine filter set (such as Omega Optical filter sets O-5725 and O-5729).
8. Paddle pastettes (Alpha Laboratories LW4295).
9. Modular incubator hypoxia chamber (Billups-Rothenberg MIC-101). Airtight sealed chamber with inlet and outlet valves for fast gas exchange, which allows for rapid and economic creation of a nonfluctuating hypoxic environment.
10. Diazocarb (95% N_2/5% O_2) cylinder (50 L) with flow meters.
11. Low light level CCD camera (we use a Basler SCA640 camera) controlled by software (we use *InCyt Im1*™, Intracellular Imaging Inc.) running on a personal computer.

3. Methods

3.1. Preparation of Hippocampal Slice Cultures

1. Under a sterile laminar flow hood, fill each well of 3 or more 6-well sterile Corning cell culture plates (see Note 7) with 1.2 mL of prewarmed (37°C) sterile Normal Medium. Place one Millicell-®CM insert into each well and keep the plates in an incubator at 37°C and 100% humidity in 95% air/5% CO_2 until use.
2. Place the following items on an aluminum foil sheet under the hood: one Whatman (Grade No. 1) filter paper disc (diameter: 9–10 cm) for each pup, one polyethylene disc (diameter: 2 cm) for each pup (see Note 8), a dissecting scalpel with a new blade, two pairs of scissors (one large and one small), two pairs of small tweezers (one straight and one curved), 3–4 paddle pastettes, and one plastic teaspoon. Turn on the UV light for 15 min and then turn it off before starting.
3. For each pup, fill a 100 × 15-mm sterile Petri dish with 5 mL of ice-cold Dissecting Medium and place a polyethylene disc on top of the cutting table of the chopper (see Note 9). Insert the Whatman filter paper disc inside the inverted lid of the Petri dish and place the two scissors, the curved tweezers, and the teaspoon into a 100-mL beaker (filled to one-third with 70% ethanol). Take the beaker and the Petri dish lid with the filter paper to the animal house.

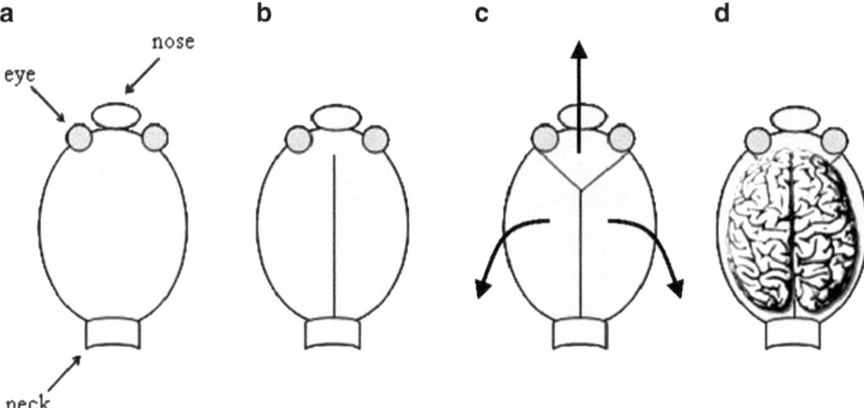

Fig. 1. How to remove the brain from a rat pup. After removing the skin (**a**), cut the skull bone with small scissors along the sagittal middle line from back to front (**b**), ending with two diagonal cuts toward each eye (**c**). Remove the skull bone with curved tweezers and, using a teaspoon, transfer the whole pup brain (**d**) to a filter paper placed on an inverted Petri dish lid.

4. Cut the head of a rat pup (see Note 10) with the large pair of scissors. Use the small scissors to delicately remove the skin, cut the skull bone along the sagittal middle line from back to front, ending with two diagonal cuts toward each eye (see Fig. 1). Remove the skull bone with the curved tweezers, and using the teaspoon, transfer the whole pup brain to the filter paper on the Petri dish lid.
5. Under the sterile hood, place the Petri dish lid with the brain onto a piece of black cardboard and illuminate with the Schott cold light source. Divide the two hemispheres with the scalpel and, with the use of the paddle pastette, remove the cerebellum, brainstem, and midbrain. Cut out the portion of the cortex which is rostral to the hippocampus with the scalpel. Insert the paddle pastette between the remaining cortex and the hippocampus, raise and remove the hippocampus, and place it on the polyethylene disc on top of the cutting table of the chopper (see Note 11).
6. Arrange the two hippocampi on the polyethylene disc, side by side. Start the chopper (set at 420 μm) and cut the whole length of both hippocampi. Remove the polyethylene disc with the straight tweezers, turn it over, and dip it gently into the Petri dish containing the 5 mL of ice-cold Dissecting Medium, thus releasing the slices into the medium.
7. Allow gentle separation of the slices using a paddle pastette. Take out the 6-well plates with the Millicell-®CM insert from the incubator. Select the best slices (see Note 12) using the stereomicroscope and place four of them very gently onto each humidified insert using a paddle pastette. Place the plates in an incubator at 37°C and 100% humidity in 95% air/5% CO_2.

8. Change the Normal Medium every 2–3 days (1 mL per well). After 24 h, the slices will start to adhere to the insert membranes. The slices will be mature and ready for the experiments after 12–14 days in vitro (DIV).

3.2. Excitotoxicity in Organotypic Hippocampal Damage

Hippocampal slices cultured for 14 DIV retain an organotypic organization in which the pyramidal and granule cell layers can be clearly defined when observed under phase-contrast microscopy or following toluidine blue staining (4, 7). Control slices incubated with PI display very low fluorescence levels, but when cultures are exposed to glutamate receptor agonists, PI staining increases dramatically. Maximal damage is achieved in this system by exposing the slices to 10 mM glutamate for 24 h: Virtually all neuronal populations are destroyed, and PI fluorescence in CA1 and CA3 increases severalfold above background levels (see Fig. 2). In order to obtain the layer-selective injury shown in Fig. 2, we use relatively low concentrations of NMDA (10 μM), AMPA (10 μM), or kainate (5 μM) for 24 h. It is also possible to expose the slices to higher concentrations of NMDA (300 μM), AMPA (100 μM), or kainate (100 μM) for a brief period of time (60 min), but in this case, the damage will be more diffuse and less selective (see examples in 8). All slices used in one experimental session are prepared from the same litter, i.e., from neonatal rat pups born on the same day:

1. Before the experiment, perform a "PI viability test" (see Note 13) by adding 6 μL of the 1-mg/mL stock solution of PI into the wells containing the cultures slices (final concentration: 5 μg/mL). Rinse gently and observe under the fluorescence microscope 30 min later. Cultures displaying distinct PI fluorescence in the pyramidal cell layers are excluded. A mild PI uptake is sometimes observed in the dorsal blade of the dentate gyrus, even in healthy slices.
2. Briefly rinse the slices and then incubate with 1.2-mL Serum-Free Medium that has previously been prewarmed to 37°C.
3. Under the hood, add to each well the appropriate microliters from the stock solutions (use intermediate dilutions if needed) in order to reach the desired final concentration of glutamate receptor agonists: 10 mM glutamate, 10 μM NMDA, 10 μM AMPA, and 5 μM kainate. Treat control slices with Serum-Free Medium.
4. Incubate for 24 h by placing the 6-well plates in the incubator at 37°C and 100% humidity in 95% air/5% CO_2 for recovery. Neuronal injury is evaluated 24 h later.

3.3. OGD in Organotypic Hippocampal Slices

To mimic the conditions that occur following cerebral ischemia in vivo, 14 DIV organotypic hippocampal slices are exposed to OGD in a hypoxia chamber for selected periods of time. Exposure

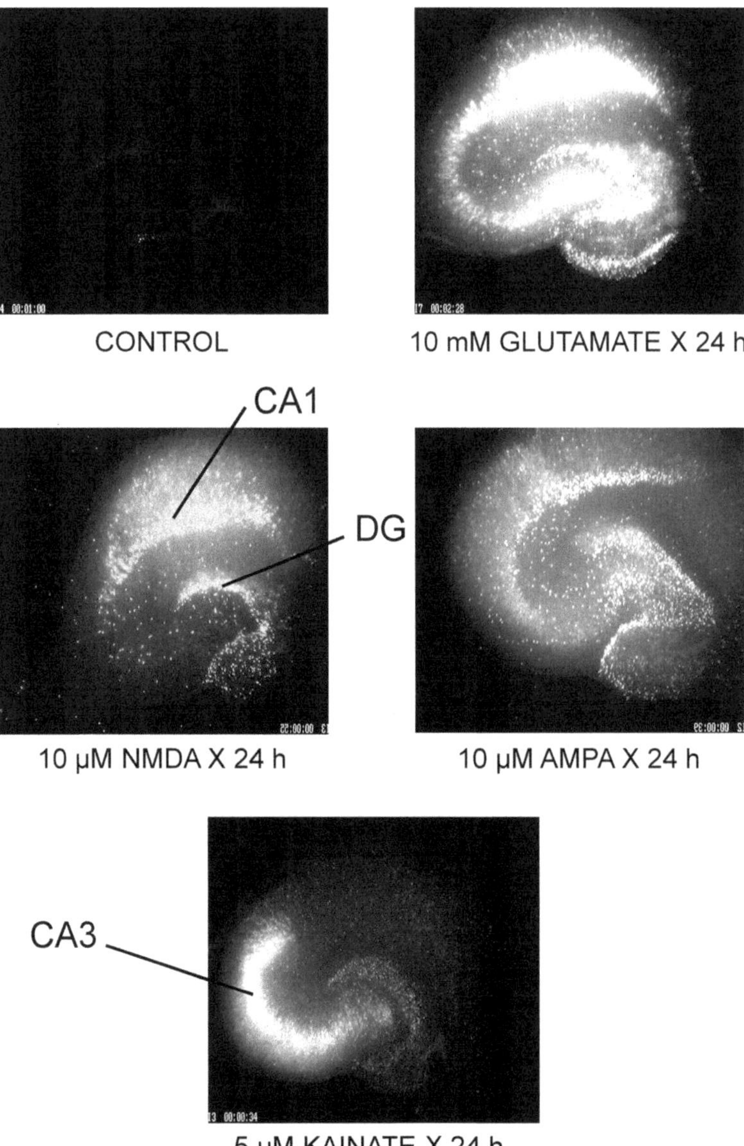

Fig. 2. Neuronal death induced by glutamate receptor agonists in rat organotypic hippocampal slices. Cultured slices are exposed to drugs, incubated with PI (5 μg/mL), and observed under fluorescence optics. Hippocampal slices under normoxic and drug-free (control) conditions display background PI fluorescence. Hippocampal slices observed following 24 h exposure to 10 mM glutamate display a maximal degree of neuronal death in all hippocampal layers. Hippocampal slices observed following 24 h exposure to 10 μM NMDA display more intense fluorescence in the CA1 pyramidal cell layer, whereas following exposure to 10 μM AMPA PI labeling is abundant in all layers. Following 24 h exposure to 5 μM kainate fluorescence is more intense in the CA3 pyramidal cell layer.

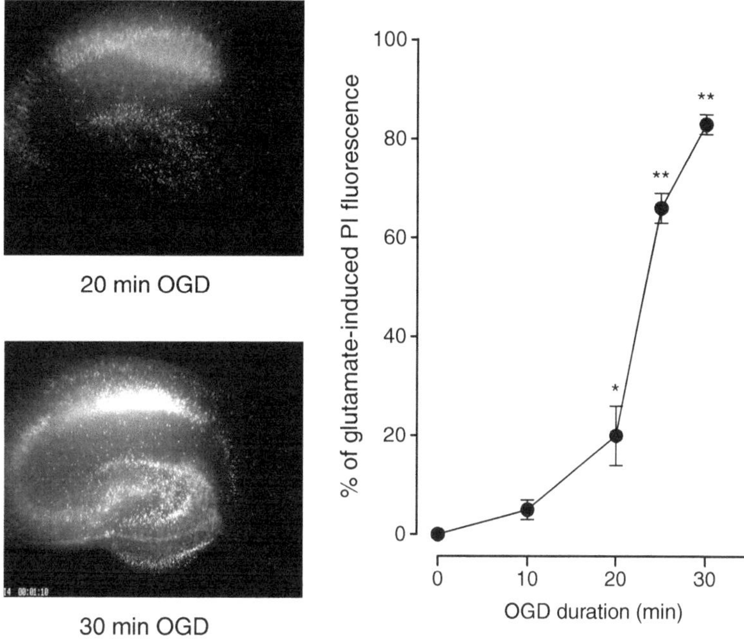

Fig. 3. Neuronal death induced by OGD in rat organotypic hippocampal slices. Cultured slices are exposed to OGD for the indicated period and 24 h later incubated with PI for fluorescence detection of its optical density in the CA1 region. OGD causes increasing levels of CA1 injury when applied for 20–40 min. Data are expressed as percentage of maximal damage produced by 24 h exposure to 10 mM glutamate. Hippocampal slices on the left, photographed under fluorescence optics, display intense PI labeling in the CA1 subregion when exposed to increasing periods of OGD. Values represent the mean ± SEM of at least five experiments performed in triplicate. $^{*}P<0.05$ and $^{**}P<0.01$ vs. basal PI fluorescence (ANOVA + Tukey's *w* test).

to OGD for periods ranging from 20 to 40 min leads to significant increases in the levels of PI fluorescence in the CA1 region, as measured 24 h after the insult (4, 7) (see Fig. 3).

1. Perform the "PI viability test" as indicated in Subheading 3.2, step 1.
2. Briefly rinse the slices and then incubate with 1.2-mL Glucose- and Serum-Free Medium that has previously been prewarmed to 37°C and saturated with 95% N_2/5% O_2.
3. Place one or two multiwell plates (without the closing lid) under the hood in the Billups-Rothenberg hypoxia chamber. Seal the chamber, attach the gas tube to the inlet valve, and flush with 95% N_2/5% O_2 at 2–3 psi for 10 min keeping the outlet valve open. Clamp the outlet valve and flush for another minute, clamp the inlet valve and then place the airtight chamber in the incubator at 37°C for the desired period of time (see Note 14). Treat control slices with glucose-containing and normoxic Serum-Free Medium for the same period of time.

4. At the end of the OGD exposure, remove the Glucose- and Serum-Free Medium and replace with glucose-containing and normoxic prewarmed (37°C) Serum-Free Medium.
5. Return the multiwells to the incubator at 37°C and 100% humidity in 95% air/5% CO_2 for recovery. Neuronal injury is evaluated 24 h later.

3.4. Assessment of Neuronal Damage

Cell injury is assessed in organotypic hippocampal slices by measuring the intensity of PI fluorescence (4, 7). Preliminary experiments in our laboratory, as well as a number of previous studies using organotypic hippocampal slices exposed to ischemia-related insults (10, 11), have shown that there is a linear correlation between relative PI fluorescence and the number of injured cells as detected by morphological criteria:

1. At the end of the 24-h recovery period, add 6 μL of the 1-mg/mL stock solution of PI into the wells containing the cultures slices (final concentration: 5 μg/mL). Rinse gently and observe under the fluorescence microscope 30 min later.
2. Digitize the image using a CCD camera controlled by software and store the images in the hard disc of the personal computer.
3. To quantify cell death, the CA1, CA3, and dentate gyrus hippocampal subfields are identified and encompassed in a frame using the drawing function of an image software (ImageJ; NIH, Bethesda, USA). A PI fluorescence intensity measurement for any given subfield consists of the mean of the fluorescence intensity values of each pixel in the area defined by the frame.
4. Percentage of cell death in each experiment is expressed as % death = $100\,(F_{24} - F_B)/(F_{max} - F_B)$ (see Subheading 4), where F_{24} is subfield fluorescence 24 h after exposure to excitotoxins or OGD, F_B is background PI fluorescence determined in control slices, and F_{max} is maximal fluorescence after complete neuronal death induced by 24 h exposure to 10 mM glutamate.

4. Notes

1. HBSS is shipped at room temperature. Store at 4°C after opening.
2. Amphotericin B (Sigma A2942) is shipped as a 250-μg/mL solution in a 100-mL bottle. Aliquot upon arrival in 1.5-mL sterile Eppendorf tubes and store at −20°C for 2–4 months.

3. Eagle's MEM must always be stored at 4°C.
4. Horse Serum (Gibco 26050) is shipped in dry ice in 500-mL bottles. Thaw at 37°C in a water bath, aliquot in sterile 50-mL Falcon tubes, and store at −20°C until use.
5. Glutamine (Sigma) is shipped as a 200-mM solution in a 100-mL bottle. Aliquot upon arrival in sterile 1.5-mL Eppendorf tubes and store at −20°C for 2–4 months. This ingredient should be added last in all solutions to avoid precipitation. Vortex the Eppendorf tube before use.
6. PI is extremely volatile as a powder and light-sensitive; therefore, extreme care should be exercised when handling. Moreover, wear gloves and a mask because PI is a known mutagen.
7. We usually prepare three 6-well Corning plates for each rat litter (approximately 10 pups).
8. We cut the 2-mm polyethylene discs from the bottom of the wrappers of the Millicell-®CM inserts.
9. It is very important to secure the polyethylene disc under the clips of the cutting table to prevent displacement of the hippocampi while cutting the sections.
10. We use Wistar Hannover rats at postnatal day 8–9 (P8-P9), weighing approximately 15 g.
11. In order to spread the hippocampi straight on the polyethylene disc, we use two paddle pastettes soaked with Dissecting Medium. This is also helpful to avoid damaging the hippocampus.
12. Select slices in which the CA3 and CA1 pyramidal cell layers and the dentate gyrus granule cell layer are both compact and clearly visible. Exclude those with any visible damage and the slices from the dorsal hippocampus (which are smaller and less resistant).
13. If PI is thought to interfere with the experiment, slice cultures can also be observed under phase-contrast optics. Damaged cells will appear brownish, and if they are numerous, the slice should be definitely discarded.
14. To achieve a reliable, intense, and selective damage of the CA1 pyramidal cell layer (see Fig. 3) that can be attenuated by neuroprotective drugs (4, 7), we perform an OGD of 30 min, which is calculated by adding a 10-min flush with the open outlet valve +20 min in which the airtight sealed hypoxia chamber is kept at 37°C in the incubator. To obtain a milder degree of CA1 damage, which is particularly useful to detect exacerbation of OGD induced by pharmacological agents, we expose the slices to 20 (10 + 10) min OGD (9).

Acknowledgements

This work was supported by grants from the University of Florence, the Italian Ministry of University and Research (MIUR, PRIN 2008 project), and the Compagnia di San Paolo (Turin, Italy).

References

1. Gähwiler BH, Capogna M, Debanne D, McKinney RA, Thompson SM (1997) Organotypic slice cultures: a technique has come of age. Trends Neurosci 20: 471–477
2. Bahr BA (1995) Long-term hippocampal slices: a model system for investigating synaptic mechanisms and pathologic processes. J Neurosci Res 42: 294–305
3. Moroni F, Meli E, Peruginelli F, Chiarugi A, Cozzi A, Picca R, et al (2001) Poly(ADP-ribose) polymerase inhibitors attenuate necrotic but not apoptotic neuronal death in experimental models of cerebral ischemia. Cell Death Differ 8: 921–932
4. Pellegrini-Giampietro DE, Cozzi A, Peruginelli F, Leonardi P, Meli E, Pellicciari et al (1999) 1-Aminoindan-1,5-dicarboxylic acid and (S)-(+)-2-(3′-carboxybicyclo(1.1.1)pentyl)-glycine, two mGlu1 receptor-preferring antagonists, reduce neuronal death in in vitro and in vivo models of cerebral ischemia. Eur J Neurosci 11: 3637–3647
5. Cozzi A, Meli E, Carlà V, Pellicciari R, Moroni F, Pellegrini-Giampietro DE (2002) Metabotropic glutamate 1 (mGlu1) receptor antagonists enhance GABAergic neurotransmission: a mechanism for the attenuation of post-ischemic injury and epileptiform activity? Neuropharmacology 43: 119–130
6. Moroni F, Attucci S, Cozzi A, Meli E, Picca R, Scheideler MA, et al (2002) The novel and systemically active metabotropic glutamate 1 (mGlu1) receptor antagonist 3-MATIDA reduces post-ischemic neuronal death. Neuropharmacology 42: 741–751
7. Pellegrini-Giampietro DE, Peruginelli F, Meli E, Cozzi A, Albani-Torregrossa S, Pellicciari R, et al (1999) Protection with metabotropic glutamate 1 receptor antagonists in models of ischemic neuronal death: time-course and mechanisms. Neuropharmacology 38: 1607–1619
8. Werner CG, Scartabelli T, Pancani T, Landucci E, Moroni F, Pellegrini-Giampietro DE (2007) Differential role of mGlu1 and mGlu5 receptors in rat hippocampal slice models of ischemic tolerance. Eur J Neurosci 25: 3597–3604
9. Landucci E, Scartabelli T, Gerace E, Moroni F, Pellegrini-Giampietro DE (2011) CB1 receptors and post-ischemic brain damage: studies on the toxic and neuroprotective effects of cannabinoids in rat organotypic hippocampal slices. Neuropharmacology 60: 674–682
10. Pringle AK, Iannotti F, Wilde GJC, Chad JE, Seeley PJ, Sundstrom LE (1997) Neuroprotection by both NMDA and non-NMDA receptor antagonists in in vitro ischemia. Brain Res 755: 36–46
11. Abdel-Hamid KM and Tymianski M (1997) Mechanisms and effects of intracellular calcium buffering on neuronal survival in organotypic hippocampal cultures exposed to anoxia/aglycemia or to excitotoxins. J. Neurosci 17: 3538–3553

Chapter 30

A 6-Hydroxydopamine In Vivo Model of Parkinson's Disease

Giulia Mercanti, Gianfranco Bazzu, and Pietro Giusti

Abstract

Animal models of Parkinson's disease are essential to explore pathophysiological hypotheses and to test new treatment options, including neurotrophic factors. Catecholaminergic neurotoxins used to generate such models are 6-hydroxydopamine and 1-methyl-4-phenyl-1,2,3,6-tetrahydropyridine. These neurotoxins predominantly kill dopaminergic neurons through oxidative damage and mitochondrial failure, although 1-methyl-4-phenyl-1,2,3,6-tetrahydropyridine fails to induce a significant dopaminergic neurodegeneration in rats. The present chapter describes a protocol for the 6-hydroxydopamine rat model based on stereotaxic injection performed only unilaterally, which mimics an early-to-mid stage of the disease.

Key words: Parkinson's disease, Animal models, Neurotoxins, Dopaminergic neurons, 6-Hydroxydopamine, 1-Methyl-4-phenyl-1,2,3,6-tetrahydropyridine

1. Introduction

Parkinson's disease (PD) is a slowly progressing neurodegenerative disease where the midbrain dopaminergic neurons are lost. The symptoms become evident when about 60–80% of the dopamine (DA) content is lost, which corresponds to about 50–60% loss of DA neurons in the substantia nigra pars compacta (SNpc) (1). PD is characterized by cardinal motor symptoms including rigidity, postural imbalance, bradykinesia, akinesia, and resting tremor (2). Experimental animal models represent a critical component of any cascade directed to the identification and development of new therapeutics for PD, including neurotrophic factors (3–7). The ideal PD model would mimic the human etiopathological condition in both the distribution of damage and its temporal profile. The etiopathology of PD is currently unknown, and different models may therefore mimic only particular aspects of the neurodegenerative process, such as mitochondrial dysfunction, oxidative stress, and misfolded protein aggregation. The most common

Stephen D. Skaper (ed.), *Neurotrophic Factors: Methods and Protocols*, Methods in Molecular Biology, vol. 846, DOI 10.1007/978-1-61779-536-7_30,

model reproduces the DA deficiency syndrome rather than the process of progressive dopaminergic degeneration and involves stereotaxic injection of specific catecholaminergic neurotoxins into different brain regions involved in voluntary motor coordination. Catecholaminergic neurotoxins used to generate such models are 6-hydroxydopamine (6-OHDA) and 1-methyl-4-phenyl-1,2,3,6-tetrahydropyridine (MPTP) (8) and predominantly kill dopaminergic neurons through oxidative damage and mitochondrial failure through inhibition of respiratory chain complexes (9). While they elicit motor deficits in rodents and nonhuman primates, MPTP fails to induce a significant dopaminergic neurodegeneration in rats (10).

Severe lesioning of the nigrostriatal DA system is required for studies of restorative or symptomatic treatments. 6-OHDA is a toxin that enters DA neurons through the high affinity DA transporter and accumulates in the cytosol, where it inhibits mitochondrial respiratory chain complexes I and IV (9). At the same time, free radicals form which finally results in DA degeneration from oxidative stress (11). The toxin does not cross the brain–blood barrier and has to be injected directly into areas containing DA fibers.

The protoxin MPTP is highly lipophilic and, when systemically administered, crosses the brain–blood barrier to enter the brain parenchyma where it is oxidized to 1-methyl-4-phenyl-2,3-dihydropyridinium in cells containing monoamine oxidase B. 1-Methyl-4-phenyl-2,3-dihydropyridinium then undergoes spontaneous oxidation to the active toxic molecule 1-methyl-4-phenylpyridinium (MPP^+) and is released into the extracellular space. MPP^+ enters DA neurons via the DA transporter and is concentrated within the mitochondria, where it impairs oxidative phosphorylation by inhibiting complex I activity (1). Alterations in energy metabolism and generation of free radicals lead to neurodegeneration of DA neurons. MPTP neurotoxicity in nonhuman primates was the first effective primate model of PD. Surprisingly, MPTP, for reasons unknown, is unable to destroy the dopaminergic innervation in other species, except for certain strains of mice (C57 black and Swiss Webster) (12).

The MPTP-treated mouse, although widely used, is not reproducible and robust. Stereotaxic injection of MPP^+ in rats has been used but offers no advantages over the 6-OHDA rat model, and it is not routinely employed (12). For these reasons, this chapter will focus on the 6-OHDA rat model based on stereotaxic injection. The stereotaxic lesion is most commonly performed only unilaterally because animals subject to bilateral destruction of the DA system develop aphagia and adipsia and require extensive monitoring and care to survive (13). The complete lesion model mimics a late stage of the human disease. The unilateral model, on the other hand, mimics an early-to-mid stage of the disease and has the advantage of exhibiting a behavioral impairment on the side contralateral to the lesion, leaving the intact side as an internal control. This model can be well standardized; the lesion obtained is quite stable for 2 weeks following neurotoxin injection and is cost-effective. Different levels of lesion can be produced by injecting the neurotoxin at different levels of the nigrostriatal pathway: a partial lesion, by injecting toxin directly into the striatum (which is taken up by axon terminals and causes retrograde degeneration in the SNpc); a very severe state of DA depletion, corresponding to an advanced stage of PD (14) if the injection site is the medial forebrain bundle (MFB); or a more specific but moderate DA depletion, when the lesion is performed in the SNpc.

2. Materials

2.1. Surgical Apparatus and Tools

1. Kopf 900 stereotaxic instruments supplemented with ear bars (Kopf instruments).
2. 10-μL Hamilton syringe (Hamilton Bonaduz AG, Bonaduz, GR, Switzerland).
3. Microinjection needles (27–30 gauge) connected to the Hamilton syringe with a polyethylene tubing (PE 20, Becton Dickinson) of about 25-cm length.
4. Surgical tools: scissors, sharp forceps, scalpels, sterile cotton pads, syringes and needles, metal clips, suture clips and applying forceps, and ear punch (Fine Science Tools).
5. Electric shaver.
6. Dental cement.
7. Dremel 300 Series drill with engraving cutter 109 (Dremel Italia).
8. Harvard compact infusion syringe pump (American Laboratory Trading, Inc. East Lyme, CT, USA).
9. Heating pad for rodents and small animals (two Biological Instruments, Besozzo, Italy).

2.2. List of Materials

1. Fentanyl citrate salt, 50 μg/mL (Sigma-Aldrich.), + medetomidine hydrochloride (DormitorVet®, Pfizer Animal Health) 50 μg/mL, 20:1 mixture.
2. Atipamezole hydrochloride (Antisedan®, Pfizer Animal Health).
3. Ketamine 80 mg/kg + xylazine 12 mg/kg (ketamine hydrochloride/xylazine hydrochloride solution, Sigma-Aldrich).
4. Chloral hydrate: dissolve 4 mg in 100 mL of sterile 0.9% (w/v) NaCl (Sigma-Aldrich).
5. 6-OHDA-HCl (Sigma-Aldrich). Dissolve in oxygen-free water containing 0.02% ascorbic acid. The solution should be

prepared fresh every 90 min, kept on ice, and protected from exposure to light. Oxidized solutions turn red.

6. Sterile saline solution: 0.9% (w/v) NaCl with 0.02% ascorbic acid (Sigma-Aldrich).
7. Sterile deionized water.
8. Ethanol 70%.

3. Methods

3.1. Surgical Procedures and Administration of 6-OHDA

All the animal procedures are carried out following guidelines governing animal experimentation (European Union decree of 24/11/1986 (86/609/IIC)) for the techniques described.

Before starting with the lesion, prepare all reagents and materials that will be used. Once the animal is under anesthesia, everything must be ready and near the stereotaxic instrument to avoid any distraction from the surgical procedure.

1. Prepare a vial containing the solution of sterile 0.9% NaCl with 0.02% (w/v) ascorbic acid and put on ice.
2. Aliquot small amounts of approximately 1–2 mg of 6-OHDA-HCl into Eppendorf® tubes and cover with aluminum foil to avoid exposure to light. Store the aliquots at −20°C and avoid repeated thawing/freezing. Be careful because 6-OHDA rapidly oxidizes when exposed to light and room temperature and takes on a pinkish color.
3. Set up the surgical work station with all the surgical tools listed above; disinfect with 70% ethanol.
4. Connect the needle with the PE tube to the Hamilton syringe. Test by filling the syringe with the saline–ascorbic acid solution and press it through the syringe and needle. This procedure needs some practice and may seem time-consuming but is critical that there is no leakage from the PE tube or any occlusion as this could prevent the toxin from reaching the target structures at the correct concentration (see Notes 1–3).
5. Place the needle on the stereotaxic frame and make sure that its orientation is perfectly vertical and straight to avoid injection on the wrong site.
6. Turn on the heating pad in such way that it is warm enough when the animal will be placed in the stereotaxic frame.
7. Once everything is ready, weigh and anesthetize the animal. Surgery can be performed using one of the different injectable anesthesias proposed below:
   - Fentanyl citrate salt 50 μg/mL + medetomidine HCl 50 μg/mL, 20:1 mixture, 6.3 mL/kg intraperitoneal. To wake

up the animals faster, use atipamezole hydrochloride (Antisedan®) 1 mg/kg as an antagonist.

- Ketamine 80 mg/kg + xylazine 12 mg/kg intramuscular (ketamine hydrochloride/xylazine hydrochloride solution). Weigh the animal and give 1 μL solution per gram body weight.
- Chloral hydrate, 400 mg/kg intraperitoneal (weigh 4 g of chloral hydrate and dissolve in 100 mL of sterile 0.9% NaCl).

8. Afterwards, leave the animal in a cage with bedding until unconscious (see Note 4).
9. Wipe the fur on the top of the skull with a cotton pad soaked in 70% ethanol and shave the fur with an electric shaver. This facilitates suturing of the skin and decreases the chance of infection at the incision site.
10. Place the animal in the stereotaxic frame on a flat-skull position; enter the ears with the ear bars and fix the skull to avoid left-right movements of the head (see Fig. 1a). Next, fix the teeth of the animal on the tooth bar so as to prevent the head moving up and down (see Fig. 1b). This procedure is delicate and takes time, but it is essential that the head is positioned correctly so that the bregma and lambda are at the same horizontal level.

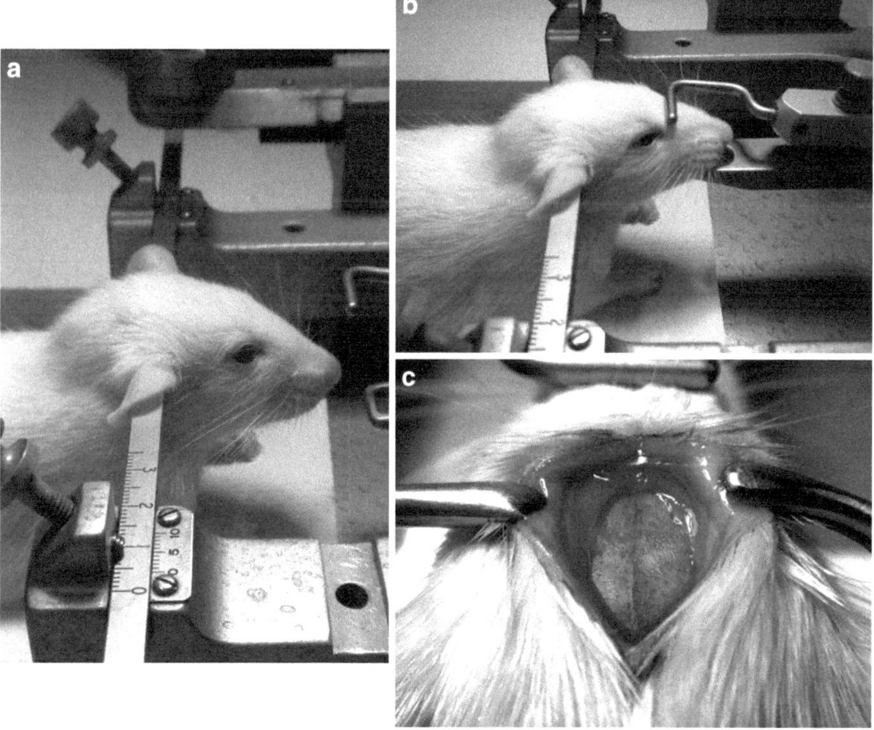

Fig. 1. Stereotaxic surgery. Flat-skull positioning of the animal with ear bars (**a**) and tooth bar (**b**) on the stereotaxic frame before starting surgery. (**c**) Shows the operatory field with skull bone exposure.

Only in this way it is possible to correctly calculate the coordinates to find the right injection site (see Note 5).

11. Perform a midline incision on the skin of about 2 cm in length starting between the eyes and remove the connective tissue above the bone so that the bregma can be easily located. With the two metal clips, keep the skin open (see Fig. 1c).
12. Injection coordinates are expressed in millimeters relative to bregma and the dural surface (15):

    To target the sensorimotor part of the striatum: AP = +0.5; L = −2.5; DV = −7.0 (4 μL).

    To target the SNpc: AP = −5.7; L = −2.0; DV = −8.7 (4 μL).

    To target the MFB: first injection AP = −4.4; L = −1.2; DV = −7.8, tooth bar = −2.3 (2 μL); second injection AP = −4.0; L = −0.8; DV = −8, tooth bar = +3.4 (2 μL).
13. Find the bregma and point the needle exactly on it, read the coordinates anteroposterior (AP) and lateral (L) on the manipulator (*x*, *y*, *z*), then retract the needle and calculate the location of injection subtracting the coordinates for different lesion sites (see Fig. 2). Move the manipulator on the injection site and mark it with a pencil (see Note 6).

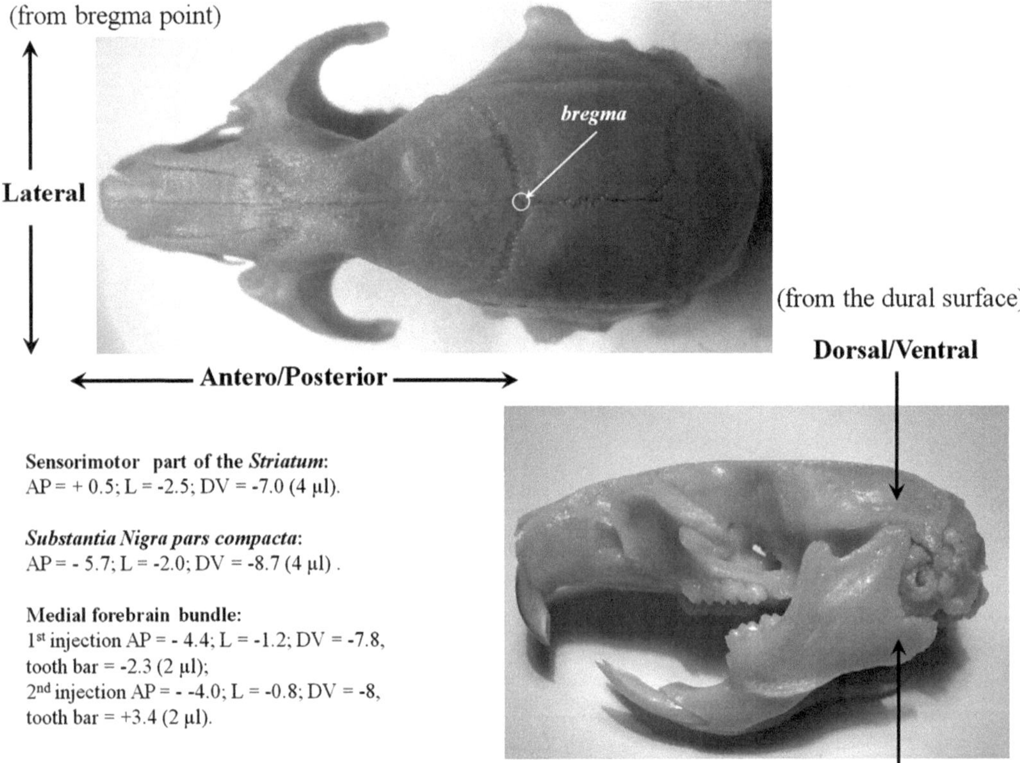

Fig. 2. Stereotaxic coordinates. Dorsal and lateral views of rat skull. Antero/posterior (A/P) and lateral (L) coordinates are referred to bregma point; dorsal/ventral (D/V) coordinates are calculated from the dural surface.

14. Retract the needle and drill a small hole in the skull bone above the injection coordinate. Be careful not to rupture the meninges when drilling (see Note 7).
15. Calculate the dorsal-ventral (DV) coordinate of the injection having the dural surface as a reference plane, then make a small hole in the meninges with either sharp forceps or an injection needle. This will facilitate the injection and avoid damage to the needle tip.
16. Dissolve 4 mg of 6-OHDA-HCl in 1 mL of sterile 0.9% NaCl with 0.02% ascorbic acid. The concentration of 6-OHDA-HCl in this solution should be 3.6 mg/mL (corresponding to 3.0 mg/mL free base 6-OHDA). Always keep the aliquots in the freezer until use and check the color of the solution. Once dissolved, keep on ice and in the dark because 6-OHDA is light and temperature sensitive and rapidly oxidizes. Prepare a fresh solution every 90 min, approximately.
17. Rinse the syringe with saline containing 0.02% ascorbic acid.
18. Fill the syringe with the 6-OHDA solution.
19. Place the Hamilton syringe on the pump, connect the needle with the PE tube to the Hamilton syringe, fill the circuit with the toxin solution, and set the infusion flow rate to 0.5 μL/min.
20. Slowly lower the injection needle down to the desired depth.
21. Start the infusion pump and pressure-eject the desired volume of 6-OHDA solution at a maximum rate of 0.5 μL/min.
22. Leave the needle in place for 5 min before slowly retracting it (see Fig. 3).
23. Empty the syringe and clean with two washes of ethanol and one of saline containing 0.02% ascorbic acid.
24. Rehydrate the wound area with sterile water and close by suturing.
25. Apply analgesic postoperatively and mark the tail or the ear with a progressive identification number.
26. Inject 2 mL of saline subcutaneously to prevent dehydration during recovery from anesthesia. For ethical reasons, analgesic treatment should be administered to the animals before they regain consciousness following anesthesia. The postoperative analgesia consists of:
    - Glucose solution (glucose 5%): 3 mL subcutaneously
    - Steroidal anti-inflammatory (i.e., betamethasone disodium phosphate 1.5 mg/mL, Bentelan®, Glaxo Wellcome): 100 μL intramuscular
    - Nonsteroidal anti-inflammatory (i.e., indomethacin, Liometacen®, Chiesi): 100 μL intramuscular
    - Antibiotic (i.e., ceftriaxone, Rocefin® 1, Roche): 100 μL subcutaneously

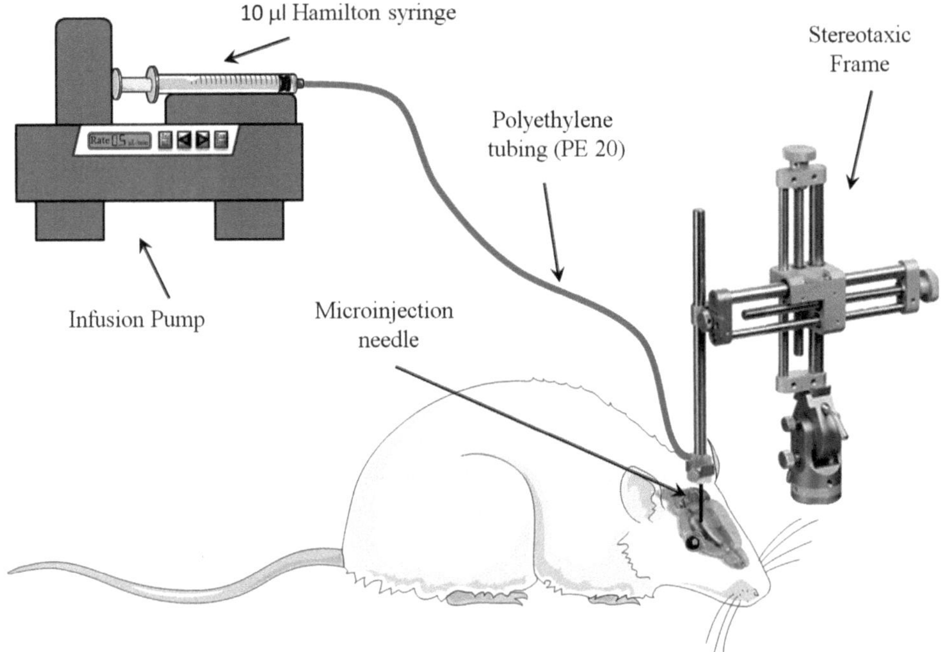

Fig. 3. Schematic representation of lesion procedure. The toxin 6-OHDA-HCl is injected into the selected area by means of a 27/30-G needle connected to a polyethylene tube and a Hamilton syringe. A precision infusion pump has a constant rate of 0.5 μL/min. At the end of toxin infusion, the needle is left in place for additional 5 min, retracting it slowly.

27. Place the animal in a clean cage until it regains consciousness and recovers.
28. Disinfect the surgical tools and the table.

3.2. Postoperative Care

1. To prevent dehydration, administer sterile physiological saline or glucose–saline solution subcutaneously (5 mL) immediately after surgery.
2. During surgery, the body temperature of the rat decreases. To offset this temperature loss, place a wool pad underneath the animal's body. When surgery is finished, place the rat into its cage and cover the body with a cloth and the eyes with paper. The latter is especially important if the rat is put under a heating lamp—this is to prevent the eyes from drying.
3. On the days following surgery, check that the animals resume their normal activity, that the wound heals, and that they do not lose weight. If some animals have problems eating, give them a special diet with food pellets completely soaked in 15% sugar/water solution, which should be put directly into the cage.

3.3. Comments: Choice of Lesion Site

6-OHDA does not cross the blood–brain barrier and therefore is classically injected into the SNpc and/or the MFB and produces a massive, virtually complete lesion of nigral dopaminergic cell bodies; SNpc neurons begin to die within the first 12 h of injection, and marked lesion of striatal dopaminergic terminals, paralleled by dopamine depletion, is established within 2–3 days (16). The injection is commonly carried out unilaterally, with the controlateral hemisphere serving as control. Bilateral injections are generally avoided, due to the high mortality rate associated with this procedure, which induces marked aphagia and adipsia in the operated animals (13) that requires intense nursing care. In the mid-1990s, a variant of the original procedure was proposed, in which 6-OHDA is injected into the striatum, where the terminals of SNpc neurons are located. The intrastriatal injection of 6-OHDA induces prompt damage of striatal terminals, followed by delayed, progressive cell loss of SNpc neurons, which are secondarily affected through a "dying back" mechanism. The degree of SNpc damage obtained with this procedure is less marked, compared with the intra-SNpc injection, remaining confined to 50–70% of the nucleus (17). However, the target site for administration of a neurotoxin in the brain for producing lesions that mimic parkinsonism in animals remains ambiguous (18).

4. Notes

1. An alternative to using a Hamilton microinjection needle is a glass capillary with an outer diameter of approximately 50 μm at the tip (the diameter can be checked on a microscope). The purpose of the glass capillary is to provide an injection cannula of the smallest possible size in order to reduce nonspecific damage to the tissue. The inner diameter of the capillary at the end opposite to the tip should not be smaller than 0.8 mm to fit on the Hamilton syringe needle. Place a small amount of parafilm around the needle of the Hamilton syringe and insert the needle into the glass capillary. Make sure the needle is sealed and does not leak. Test this by filling the syringe with saline containing 0.02% ascorbic acid and pass the solution through the capillary. This procedure needs some practice and may seem time-consuming on the first trial. However, it is critical that there is no leakage from the borders of the capillary, since this could prevent the toxin from reaching the target structure in the brain.
2. Mount the syringe with or without the capillary on the stereotaxic frame and make sure that the needle/capillary assembly is vertically straight in its orientation.
3. Pay particular attention to the cleanliness of the needle to prevent infection and the formation of blood clots which can cause mechanical damage to brain tissue during subsequent injections.
4. Test reflexes by pinching the tail or paws to ensure that the animal is anesthetized.

5. While the animal is fixed to the stereotaxic frame, pay particular attention to keep the head up to avoid that the animal may suffocate.
6. Note the time when the anesthesia is done, the coordinates of the injection site for each animal, and the time when 6-OHDA is dissolved.
7. When drilling the hole in the skull bone, hold the drill in a vertical position and avoid applying too great a pressure on the bone which may cause the drill bit to penetrate the dura. Make small circular and concentric movements with the drill bit and make an opening not too large.

References

1. Dauer W, and Przedborski S (2003) Parkinson's disease: mechanisms and models. Neuron 39:889–909
2. Recchia A, Rota D, Debetto P, Peroni D, Guidolin D, Negro A, Skaper SD and Giusti P (2008) Generation of alpha-synuclein-based rat model of Parkinson's disease. Neurobiol Dis 30:8–18
3. Kordower JH, Emborg ME, Bloch J, Ma SY, Chu Y, Leventhal L et al (2000) Neurodegeneration prevented by lentiviral vector delivery of GDNF in primate models of Parkinson's disease. Science 290:767–73
4. Gash DM, Zhang Z, Ovadia A, Cass WA, Yi A, Simmerman L et al (1996) Functional recovery in parkinsonian monkeys treated with GDNF. Nature 380:252–255
5. Biju K, Zhou Q, Li G, Imam SZ, Roberts JL, Morgan WW et al (2010) Macrophage-mediated GDNF delivery protects against dopaminergic neurodegeneration: a therapeutic strategy for Parkinson's disease. Mol Ther; doi:10.1038/mt.2010.107
6. Yasuda T and Mochizuki H (2010) Use of growth factors for the treatment of Parkinson's disease. Expert Rev Neurother 10:915–924
7. Laganiere J, Kells AP, Lai JT, Guschin D, Paschon DE, Meng X et al (2010) An engineered zinc finger protein activator of the endogenous glial cell line-derived neurotrophic factor gene provides functional neuroprotection in a rat model of Parkinson's disease. J Neurosci 30:16469–16474
8. Lambert CE, and Bondy SC (1989) Effects of MPTP, MPP^+ and paraquat on mitochondrial potential and oxidative stress. Life Sci 44:1277–1284
9. Glinka Y, Gassen M, and Youdim MB (1997) Mechanism of 6-hydroxydopamine neurotoxicity. J Neural Transm Suppl 50:55–66
10. Przedborski S, Jackson-Lewis V, Naini A, Jakowec M, Petzinger G, Miller R, Akram M (2001) The parkinsonian toxin 1-methyl-4-phenyl-1,2,3,6-tetrahydropyridine (MPTP): a technical review of its utility and safety. J Neurochem 76:1265–1274
11. Olanow CW (1993) A radical hypothesis for neurodegeneration. Trends Neurosci 16:439–444
12. Jenner P (2008) Functional models of Parkinson's disease: a valuable tool in the development of novel therapies. Ann Neurol 64(Suppl 2):S16–S29
13. Ungerstedt U (1971) Adipsia and aphagia after 6-hydroxydopamine induced degeneration of the nigro-striatal dopaminergic system. Acta Physiol Scand Suppl 367:95–122
14. Ungerstedt U (1968) 6-Hydroxy-dopamine induced degeneration of central monoamine neurons. Eur J Pharmacol 5:107–110
15. Paxinos G, Watson C (1986) Rat Brain in Stereotaxic Coordinates San Diego, Academic Press
16. Deumens R, Björklund A and Prickaerts J (2002) Modeling Parkinson's disease in rats: an evaluation of 6-OHDA lesions of the nigrostriatal pathway. Exp Neurol 175:303–317
17. Lee CS, Sauer H and Bjorklund A (1996) Dopaminergic neuronal degeneration and motor impairments following axon terminal lesion by intrastriatal 6-hydroxydopamine in the rat, Neuroscience 72:641–653
18. Sindhu KM, Banerjee R, Senthilkumar KS, Saravanan KS, Raju BC, Rao JM et al (2006) Rats with unilateral median forebrain bundle, but not striatal or nigral, lesions by the neurotoxins MPP^+ or rotenone display differential sensitivity to amphetamine and apomorphine. Pharmacol Biochem Behav 84:321–329

Chapter 31

Brain Microdialysis in Freely Moving Animals

Gianfranco Bazzu, Alice Biosa, Donatella Farina, Ylenia Spissu, Giammario Calia, Sonia Dedola, Gaia Rocchitta, Rossana Migheli, Pier Andrea Serra, and Maria Speranza Desole

Abstract

Brain microdialysis is an analytical technique used for the dynamic monitoring of brain neurochemistry in awake, freely moving animals. This technique requires the insertion of a small dialysis catheter, called a microdialysis probe, into a specific brain region, and its perfusion with an artificial extracellular fluid. The microdialysate samples, obtained from the probe outlet, can be analysed using high-performance liquid chromatography with electrochemical detection for the quantification of oxidizable molecules recovered from the extracellular space. In this chapter, we describe a protocol for performing a microdialysis setup and experiment in freely moving rats and mice. Furthermore, the high-performance liquid chromatographic determination of ascorbic acid, uric acid, catecholamines, indolamines and derivatives is described in detail.

Key words: Ascorbic acid, Amperometric detection, Dopamine, 3,4-Dihydroxyphenylacetic acid, L-3,4-Dihydroxyphenylalanine, High-performance liquid chromatography, Homovanillic acid, 5-Hydroxyindoleacetic acid, In vivo microdialysis, 3-Methoxytyramine, Noradrenaline, Serotonin, Stereotaxic surgery, Uric acid

1. Introduction

Microdialysis is an analytical technique used for continuous measurement of the concentrations of low molecular weight dialysates in the extracellular fluid of virtually any tissue. The microdialysis principle was first employed in the 1960s when rudimentary dialysis systems were implanted into animal tissues, mainly into rat brain, to study the neurochemistry in a dynamic manner (1). The modern brain microdialysis technique was used for the first time in 1972 (2), and in 1974, it was employed for monitoring dopamine release (3). Today, studied compounds include endogenous molecules such as neurotransmitters (to determine their variations) or exogenously administered drugs (to determine their pharmacokinetic).

Stephen D. Skaper (ed.), *Neurotrophic Factors: Methods and Protocols*, Methods in Molecular Biology, vol. 846, DOI 10.1007/978-1-61779-536-7_31, © Springer Science+Business Media, LLC 2012

The brain microdialysis technique requires the insertion of a small microdialysis probe into a discrete brain region to mimic a small blood capillary having a diameter around 200 μm. The microdialysis probe encloses a cylindrical semipermeable membrane (hollow fibre) perfused with an isosmolar liquid similar, in composition, to the extracellular fluid. The membrane is connected to a perfusion circuit by means of a piece of polyethylene tubing (inlet), while a second piece of tubing (outlet) allows the recovery of the microdialysate sample: the perfusion fluid enriched with the unbound dialysates present in the extracellular space passively diffused through their concentration gradient. In a similar manner, drugs can be added to the perfusion fluid and locally administered (intracerebral retrodialysis). The low perfusion flow rate, ranging from 0.1 to 5 μL/min (1), allows for maximum efficiency of recovery. The microdialysis technique has some unique advantages including the continuous monitoring of the extracellular space of (virtually) any tissue, high spatial resolution, limited local traumatisation and lack of tissue proteins (including enzymes) in the microdialysate. The principal disadvantage is the poor temporal resolution (minutes) compared to other techniques (4). Microdialysis is mainly used for studying the changes in extracellular concentrations of analytes. It is difficult to calculate the real analyte concentrations in the extracellular space due to the irregular geometry of this compartment; moreover, a total equilibrium cannot be established because of the constant perfusion of the probe with fresh perfusate (1). Microdialysis probes can be calibrated by either measuring the loss of a compound using an analyte-containing perfusion fluid or the gain of compound using an analyte-containing calibration solution. The most frequently used calibration methods are no-net-flux methods (5, 6) and the retrodialysis method (7).

Today, brain microdialysis is used not only in awake, freely moving animals (mice, rats, primates etc.) but also in humans (8, 9). In this chapter, we describe the materials and the methods necessary for performing a microdialysis setup and experiment in freely moving rats and mice. We have selected the striatum as the most targeted region for microdialysis probe implantation and the analysis of ascorbic acid, uric acid, catecholamines, indolamines and derivatives by high-performance liquid chromatography (HPLC) as one of the most used techniques for analysing microdialysate samples.

2. Materials

2.1. Chemicals and Solutions

1. Artificial extracellular fluid (aECF): dissolve 13.32 mg of $CaCl_2$ in 100 mL of double-distilled water and stir the solution for 10 min until completely dissolved. Add 20.14 mg of KCl, 858.48 mg of NaCl and 8.1 mg of $MgCl_2$ and stir briefly.

The final solution will have the following concentration: 147 mM NaCl, 1.2 mM $CaCl_2$, 2.7 mM KCl and 0.85 mM $MgCl_2$. Filter the solution using 0.22-μm (pore diameter) syringe filters (Millex-HP, Millipore, Billerica, USA). Divide the solution in aliquots of 5 mL each and store at −20°C. The solution is stable for several weeks.

2. Standard solution for microdialysis probe calibration:
   (a) Ascorbic acid (AA): dissolve 1.76 mg in 1 mL of mH_3PO_4 1%.
   (b) Uric acid (UA): dissolve 1.68 mg in 1 mL of mH_3PO_4 1%.
   (c) Norepinephrine (NE): dissolve 1.62 mg in 1 mL of mH_3PO_4 1%, then dilute by adding 10 μL of this solution to 1 mL of mH_3PO_4.
   (d) L-Dihydroxyphenylalanine (L-DOPA): dissolve 1.97 mg in 1 mL of mH_3PO_4 1%, then dilute by adding 10 μL of this solution to 1 mL of mH_3PO_4.
   (e) Dihydroxyphenylacetic acid (DOPAC): dissolve 1.68 mg in 2 mL of mH_3PO_4 1%.
   (f) Dopamine (DA): dissolve 1.90 mg in 1 mL of mH_3PO_4 1%, then dilute by adding 10 μL of this solution to 1 mL of mH_3PO_4.
   (g) 5-Hydroxyindolacetic acid (5HIAA): dissolve 1.91 mg in 2 mL of mH_3PO_4 1%.
   (h) Homovanillic acid (HVA): dissolve 1.82 mg in 2 mL of mH_3PO_4 1%.
   (i) 3-Methoxytyramine (3MT): dissolve 2.04 mg in 1 mL of mH_3PO_4 1%, then dilute by adding 10 μL of this solution to 1 mL of mH_3PO_4.
   (j) 5-Hydroxytryptamine (5HT): dissolve 4.05 mg in 1 mL of mH_3PO_4 1%, then dilute by adding 10 μL of this solution to 1 mL of mH_3PO_4.

   The final solution is prepared by adding different aliquots of solutions a–j above to 2,796 μL of aECF: 30 μL of AA, 15 μL of UA, 30 μL of NE, 30 μL of L-DOPA, 3 μL of DOPAC, 30 μL of DA, 3 μL of 5HIAA, 3 μL of HVA, 30 μL of 3MT and 30 μL of 5HT. The final concentrations will be AA 100 μM, UA 50 μM, NE 1 μM, L-DOPA 1 μM, DA 1 μM, DOPAC 5 μM, 5HIIA 5 μM, HVA 5 μM, 3MT 1 μM and 5HT 1 μM. Filter the solution using 0.22-μm (pore diameter) syringe filters (Millex-HP, Millipore). Divide the solution in aliquots of 500 μL each (using 1.5-mL Eppendorf vials) and store at −80°C. The solution will be stable for few weeks.stock solutions of neurochemicals are prepared in 1% mH_3PO_4, obtained by dissolving 1 g of mH_3PO_4 in 100 mL of distilled water, corresponding to a concentration of 102.04 mM.

3. Anaesthetic and euthanasia solutions: prepare a 4% chloral hydrate solution for anaesthesia by dissolving 4 g of solid in 100 mL of sterile saline (0.9% NaCl) and store at 4°C. Prepare an 8% solution for euthanasia dissolving 8 g of chloral hydrate in 100 mL of sterile saline and store at 4°C. Before use, filter the solutions using 0.22-μm syringe filters (Millex-HP, Millipore). These solutions can be used for a week after preparation.
4. D-Amphetamine solutions for pharmacological treatments (10): for subcutaneous (s.c.) administration, dissolve 1.2 mg of D-amphetamine hydrochloride in 2.0 mL of sterile saline. For retrodialysis administration, prepare a 200 μM solution by dissolving 1.35 mg of powder in 1 mL of sterile water, then dilute 40 μL of the former solution into 1,960 μL of aECF solution. These solutions must be prepared immediately before use and filtered using 0.22-μm syringe filters (Millex-HP, Millipore).
5. HPLC mobile phase: prepare 1 L of mobile phase by dissolving 84.5 g of KH_2PO_4 and 2.4 g of K_2HPO_4 in 850 mL of Milli-Q water. Add 292.24 mg of EDTA and stir for few minutes until completely dissolved. Add 150 mL of HPLC grade methanol and at the end, add 80 mg of 1-octanesulfonic acid sodium salt. Adjust to pH = 2.7 with concentrated H_3PO_4. Degas the mobile phase solution using an ultrasonic bath for at least 30 min.
6. Standard solution for HPLC calibration:

   stock solutions of neurochemicals are prepared in 1% mH_3PO_4 as described in Subheading 2.1, step 1.

   (a) AA: dissolve 1.76 mg of powder in 1 mL of mH_3PO_4 1%.

   (b) UA: dissolve 1.68 mg of powder in 1 mL of mH_3PO_4 1% and dilute by adding 10 μL of this solution to 1 mL of mH_3PO_4.

   (c) NE: dissolve 1.62 mg of powder in 1 mL of mH_3PO_4 1% and dilute by adding 1 μL of this solution to 1 mL of mH_3PO_4.

   (d) L-DOPA: dissolve 1.97 mg in 1 mL of mH_3PO_4 1% and dilute by adding 1 μL of this solution to 2 mL of mH_3PO_4.

   (e) DOPAC: dissolve 1.68 mg in 1 mL of mH_3PO_4 1% and dilute by adding 10 μL of this solution to 1 mL of mH_3PO_4.

   (f) DA: dissolve 1.90 mg in 1 mL of mH_3PO_4 1% and dilute by adding 1 μL of this solution to 2 mL of mH_3PO_4.

   (g) 5HIAA: dissolve 1.91 mg in 1 mL of mH_3PO_4 1% and dilute by adding 10 μL of this solution to 1 mL of mH_3PO_4.

(h) HVA: dissolve 1.82 mg in 1 mL of mH_3PO_4 1% and dilute by adding 10 μL of this solution to 1 mL of mH_3PO_4.

(i) 3MT: dissolve 2.04 mg in 1 mL of mH_3PO_4 1% and dilute by adding 1 μL of this solution to 2 mL of mH_3PO_4.

(j) 5HT: dissolve 4.05 mg in 1 mL of mH_3PO_4 1% and dilute by adding 1 μL of this solution to 2 mL of mH_3PO_4.

The final solution is prepared by adding different aliquots of solutions a–j above to 2,851 μL of aECF: 2.3 μL of AA, 30 μL of UA, 9 μL of NE, 7.6 μL of L-DOPA, 18 μL of DOPAC, 4.2 μL of DA, 15 μL of 5HIAA, 15 μL of HVA, 30 μL of 3MT and 18 μL of 5HT. The final concentrations will be AA 7.5 μM, UA 1.0 μM, NE 15 nM, L-DOPA 12.7 nM, DA 7.0 nM, DOPAC 0.6 μM, 5HIIA 0.5 μM, HVA 0.5 μM, 3MT 50 nM and 5HT 30 nM. Filter the solution using 0.22-μm (pore diameter) syringe filters (Millex-HP, Millipore). Divide the solution in aliquots of 50 μL each (using 0.5-mL Eppendorf vials) and store at –80°C. The solution is stable for several weeks.

2.2. Animals and Stereotaxic Surgery

1. *Animals*: male Wistar rats weighing 300 g or 10-week-old C57BL/6 mice (weighing 30 g) from Charles River (Ballina, Italy) can be used for in vivo microdialysis experiments. The animals must be maintained under standard animal care conditions (12:12-h light/dark cycle, lights coming on at 7 a.m., room temperature 21°C), with food and water *ad libitum*. Prior to the start of any experiment, the health of the rodents has to be assessed according to published guidelines (11). All procedures need to be specifically licenced under the European Community directive 86/609 included in national legislatures (i.e. Decreto No. 116/1992 of the Ministry of Public Health in Italy).
2. *Brain atlas for stereotaxic coordinates*: the George Paxinos' volumes entitled "The Rat Brain in Stereotaxic Coordinates" (12) and "The Mouse Brain in Stereotaxic Coordinates" (13) are the reference books for setting up a microdialysis experiment in rats and mice, respectively.
3. *Stereotaxic apparatus*: stereotaxic frame for small animals (Model 900, David Kopf Instruments, UK) for rats equipped with a mouse adaptor for mice (Kopf Mouse Adaptor from 2Biol, Varese, Italy). We recommend placing the stereotaxic frame on a vibration-free marble table to increase precision during probe implantation and to reduce brain damage.
4. *Neurosurgery materials*: alcohol-based permanent marker (OHPen 96 fine, Stabilo, Germany), Betadine disinfectant (Meda Pharma, Italy), sterile cotton wool, anatomical and surgical tweezers, scalpel and scalpel blades, 27-G needles, 1- and 2.5-mL disposable syringes (for intraperitoneal (i.p.) injection

of the anaesthetic solution), scissors, crocodile clips, Dremel drill and tips (small tip for screw hole and trephine tip), small screws and dental cement (Paladur, Hereus Kulzer GmbH, Hanau, Germany), 5/0 suture needles and needle holder.

2.3. Microdialysis Probe and Circuits

1. *Microdialysis probe construction*: razor blades, scalpel, scalpel blades, alcohol-based permanent marker, 100 mm of microdialysis hollow fibre (AN69 Hospal Industrie, France) obtained from a blood dialysis cartridge, epoxy resin (Pattex, Henkel Italia, Milano, Italy), low-magnification (2–20×) stereo microscope, 100 mm of PE50 tubing (0.58 I.D. polyethylene tubing, Intramedic BD Becton Dickinson Italia S.p.A., Milano, Italy), 75 mm of silica capillary (Polymicro Technologies, Phoenix, AZ, USA), 23-Gauge (23 G) and 27-G needles (hypodermic needle, Artsana S.p.A.,Casnate, como Italy), diamond wheel (Dremel® 545 Diamond Wheel, Dremel, Milano, Italy), drill (Dremel® 200 Series, Dremel) and needle holder
2. *Microdialysis probe in vitro calibration*: 300 mm of PE50 tubing, 1-mL Hamilton syringe, 250-μL Eppendorf vials, 27-G needles and microinfusion pump (KDS-101-CE, KD Scientific, Halliston, MA)
3. In vivo microdialysis: 800 mm of PE50 tubing, 1-mL Hamilton syringe, 250-μL Eppendorf vials, 27-G needles, swivel (375/22 Single channel swivel, 22 gauge, Instech Laboratories, Inc., Plymouth Meeting, PA USA) and microinfusion pump (KDS-101-CE)

2.4. HPLC

1. Alltech 426 HPLC pump (Alltech Srl, Milan, Italy)
2. Rheodyne injection valves (Alltech Srl) equipped with a 25-μL loop
3. 50-μL Hamilton syringes (Alltech Srl)
4. Alltech Adsorbosphere™ C18 Column (100 × 4.6 mm; Alltech Srl)
5. Antec EC controlled equipped with VT03 cell having a glassy carbon working electrode, a Ag/AgCl reference electrode and a stainless steel counter electrode (Alfatech Srl, Milan, Italy)
6. HPLC polyetheretherketone (PEEK) tubing and accessories for 25-μL loop and connections (Alfatech Srl)
7. Varian data acquisition software (Varian Star v. 5, Varian Inc., USA) and hardware (Varian 800 MIB, Varian)
8. Personal computer (1 GB RAM and 250 GB hard disk) equipped with an Ethernet board (operating system: Windows XP/Vista/Windows 7)

2.5. Statistical Analysis

Software for data storage, scientific graphing and statistical analysis: Microsoft Excel 2007 (Microsoft Inc., USA), GraphPad Prism 5.0 (GraphPad Software, La Jolla, CA, USA), KaleidaGraph 4.0 (Synergy software, Reading, PA, USA)

3. Methods

3.1. Microdialysis Probe Construction

1. Prepare three short microdialysis hollow fibres (25 mm in length) cutting a straight microdialysis fibre with a sharp razor blade (see Fig. 1a).

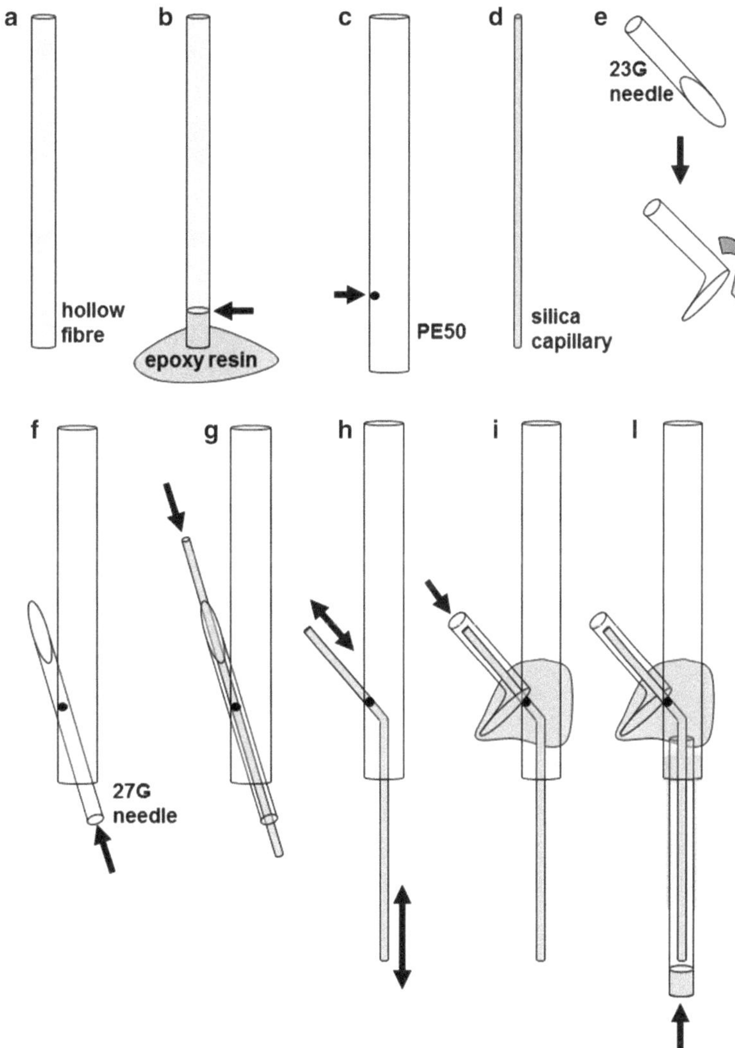

Fig. 1. Microdialysis probe construction procedure. See text for a detailed explanation of the steps necessary for obtaining a working probe implantable in rat or mice striatum.

2. Prepare the epoxy resin (see Note 1) on a small piece of polycarbonate (usually a piece of transparent film measuring 5 × 5 cm) and seal one tip of the hollow fibre immersing it for 1 mm in the epoxy resin. The glue will rise in the hollow fibre by capillary action, forming a cylinder with a meniscus in the upper part (see Fig. 1b, arrow). Perform this step under a microscope. Repeat this for each of the three fibres and leave them to dry for at least 30 min at room temperature.
3. Cut one piece (30 mm in length) of PE50 tubing using the razor blade and mark it with the permanent marker at 7 mm from one end (see Fig. 1c, arrow).
4. Cut one silica capillary (25 mm in length) using the scalpel blade (see Fig. 1d).
5. Cut a 23-G needle (10 mm in length) using the drill mounting a diamond wheel. Gently curve the 23-G needle tip to obtain an L-shape using a needle holder (see Fig. 1e).
6. Insert a 27-G needle in the PE50 and perforate the tube wall in the direction of the permanent mark (see Fig. 1f); leave the needle in this position.
7. Insert the silica capillary inside the 27-G needle up to two-thirds of its length (see Fig. 1g) and gently remove the 27-G needle while maintaining the position of the silica capillary. At the end of this operation, the silica capillary will be inside the PE50 tubing as illustrated in Fig. 1h.
8. The silica capillary has to be repositioned until it comes out 10 mm from the PE50 outlet (see Fig. 1h, arrows).
9. Insert the tip of the silica capillary (at the opposite side of the outlet) inside the 23-G needle (prepared in the step 5) as illustrated in Fig. 1i.
10. Prepare the epoxy resin and glue the silica capillary, the PE50 tubing and the 23-G needle at the level of the permanent mark (see Fig. 1i). At this point, the probe skeleton is made, and it needs to dry for 30 min at room temperature.
11. Cut the previously prepared hollow fibres (14 mm in length starting from the epoxy meniscus) with a sharp razor blade.
12. Under microscope, insert the silica capillary inside the straight hollow fibre up to 0.5 mm from the epoxy meniscus (see Fig. 1l).
13. Prepare the epoxy resin and glue the hollow fibre to the PE50 tubing (see Fig. 1l; Note 2). Leave the microdialysis probe to dry for at least 30 min at room temperature.
14. Prepare the epoxy resin and 'paint' the external surface of the hollow fibre using a 27-G needle (previously immersed in the glue) for 6 mm (rats) or 8 mm (mice) starting from the PE50

tubing (see Fig. 11). Gently remove the excess epoxy using a clean 27-G needle. During this step, the membrane will be shielded which allows the dialysis process only in the probe's active area (around 4 mm for rats and 2 mm for mice). Leave the microdialysis probe to dry for at least 30 min at room temperature.

15. Store the microdialysis probe in a fresh and dark environment, far from dust and moisture, and keep it in place for at least 24 h before use to allow complete polymerization of the epoxy resin.

3.2. Microdialysis Probe In Vitro Calibration

3.2.1. Circuit for In Vitro Probe Calibration (see Fig. 2)

(a) Cut one piece (200 mm in length) of PE50 tubing using a razor blade.

(b) Cut a second piece of PE50 tubing (50 mm in length).

(c) Cut a 23-G needle (10 mm in length) using the drill mounting a diamond wheel.

(d) Insert the 23-G needle (for 4 mm) inside the probe outlet.

(e) Fill the Hamilton syringe (1 mL) with the aECF (see Note 3).

(f) Connect one end of the 200 mm PE50 tubing to the Hamilton syringe needle and the other end to the probe inlet.

(g) Connect one end of the 50 mm PE50 tubing to the 23-G needle previously inserted in the probe outlet.

(h) Place the Hamilton syringe on the infusion pump and fill the circuit using the maximum flow rate (see Note 4).

3.2.2. In Vitro Probe Calibration (see Fig. 2)

(a) Place the microdialysis probe on the stereotaxic frame holder and set the infusion pump flow rate to 1.5 μL/min.

(b) Unfreeze one Eppendorf vial, containing the calibration solution, to room temperature.

(c) Insert the probe membrane (hollow fibre) in the calibration solution and wait 10 min to allow for stabilisation of the microdialysis gradient.

(d) Insert the free end of the 50 mm PE50 tubing in a 250-μL Eppendorf vial and wait 20 min.

(e) Remove the Eppendorf vial (containing 30 μL of dialysate) and store at −80°C until HPLC analysis (see Subheading 3.5). At the same time, take a 30 μL sample of the calibration solution (in which the probe is immersed) and store at −80°C. Repeat this step two more times.

(f) After HPLC analysis, calculate the percent recovery of the probe for every neurochemical component in the calibration solution, using the following formula:

$$\text{Recovery}\ (\%) = [\text{microdialysate}]/[\text{calibration solution}] \times 10(\quad (1)$$

and determine the mean ± standard deviation (see Note 5).

3.3. Neurosurgery

1. Anaesthetize the animal using the previously prepared 4% chloral hydrate solution (400 mg/kg intraperitoneal) and switch on the heating mat (37°C).
2. Shave the animal's head and disinfect the skin over the skull using a polyvinylpyrrolidone-based disinfectant (Betadine, Meda Pharma, Italy).
3. Fix the animal's head on the stereotaxic frame (rat) or in the mouse adapter (mouse) (see Fig. 3a) and be sure that the body is in contact with the heating mat.
4. Cut the skin over the skull in the middle line using a sharp scalpel blade, following the rostro-caudal direction, starting from the frontal region up to the occipital bone (see Fig. 3b).
5. Fix the skin borders using small surgical crocodile clips, remove the periosteum from the skull using the scalpel blade (see Note 6) and mark the bregma with the permanent marker (see Fig. 3c).
6. Calculate the coordinates (rostro-caudal and middle lateral) of the right striatum using the Paxinos atlas of rat (or mouse) brain mounting a 27-G needle on the stereotaxic frame holder. Mark the corresponding point on the skull using the permanent marker (see Figs. 1 and 3d).
7. Make the screw hole on the right parietal bone (see Figs. 2 and 3d) using the Dremel drill mounting a small tip and cut a small disk of bone centred on the marked point (see Figs. 1 and 3e) using the trephine tip (cylindrical blade) exposing the *dura mater*. Insert the screw in the parietal hole (see Figs. 2 and 3e).
8. Place the microdialysis probe on the stereotaxic frame holder and recalculate the rostro-caudal and middle-lateral coordinates from the bregma. Rest the tip of the probe on the *dura mater* and calculate the dorso-ventral coordinate. Cut the *dura mater* with a 27-G needle (see Note 7) and gently insert the probe in the brain (see Fig. 3f).
9. Fix the microdialysis probe on the skull using dental cement and anchor it to the screw previously inserted in the right parietal bone (see Fig. 3g). Leave the dental cement to dry for at least 10 min.
10. Remove the probe from the stereotaxic holder and close the skin using a small semicircular suture needle and non-absorbable material (5/0 fine silk filament).
11. Disinfect the skin and remove the animal from stereotaxic frame.
12. Place the animal in a warm and ventilated cage until complete recovery from anaesthesia (see Note 8).

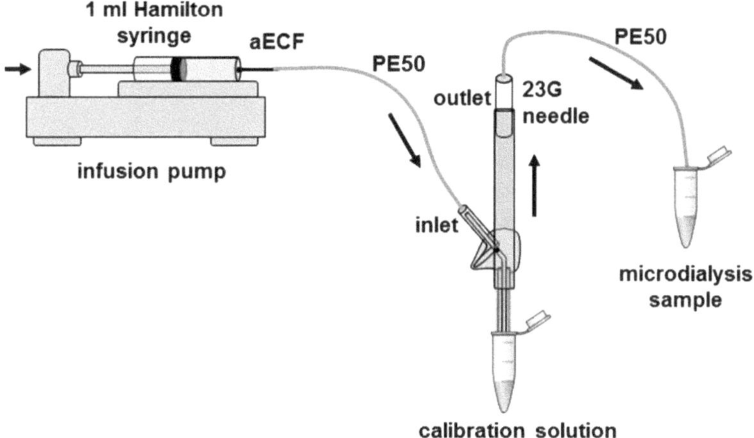

Fig. 2. Circuit for the in vitro calibration of the microdialysis probe before in vivo use. The ratio between the concentrations of neurochemicals, determined in the microdialysate and in the calibration solution, represents the probe recovery.

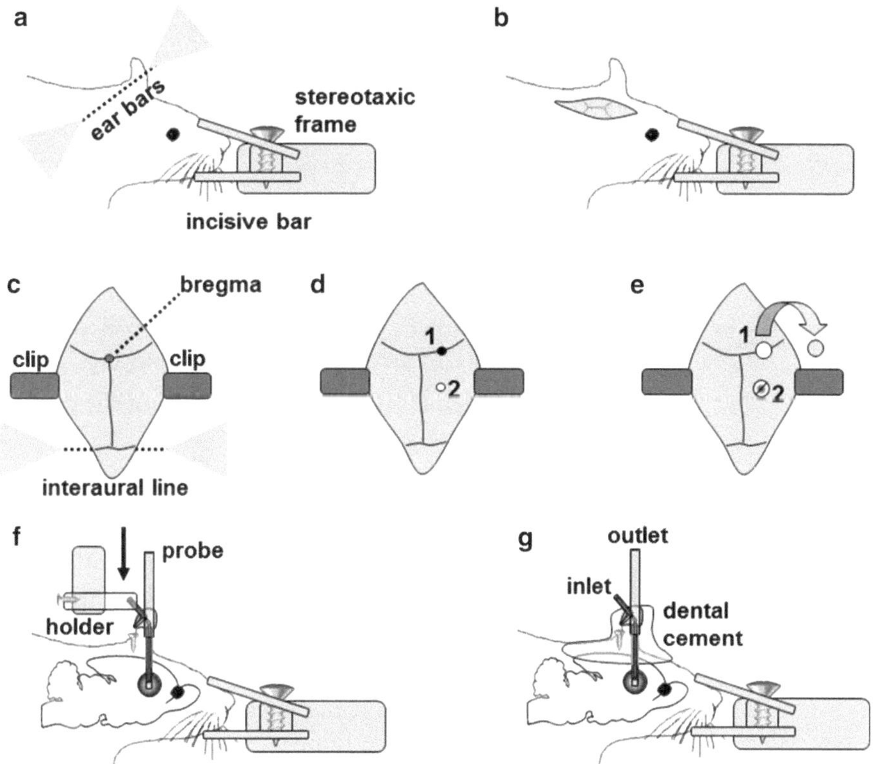

Fig. 3. Neurosurgery procedure used for implantation of the microdialysis probe in the striatum of anaesthetized rats or mice. A stereotaxic frame for small animals is used for rats while it needs to be equipped with a mouse adaptor for mice. See text for a detailed explanation of the steps illustrated in the figure.

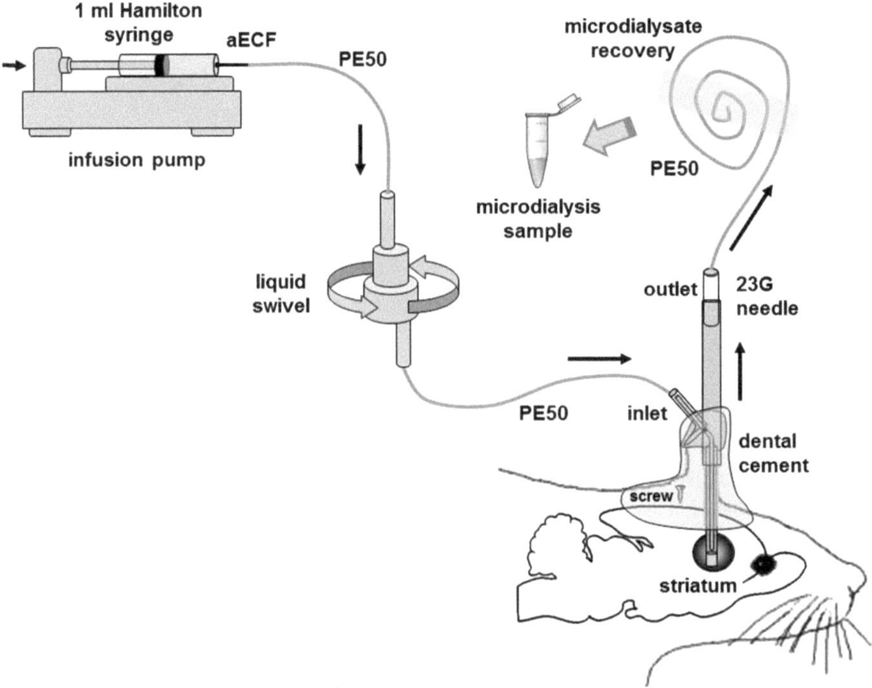

Fig. 4. Circuit for in vivo microdialysis in freely moving animals. The introduction of the liquid swivel prevents the formation of sharp bends and folding of the polythene tubing during the animal's rotations.

3.4. In Vivo Microdialysis

3.4.1. Circuit for In Vivo Microdialysis Experiment (see Fig. 4)

(a) Cut two pieces (300 mm in length each) of PE50 tubing using a razor blade.

(b) Cut a third piece of PE50 tubing (150 mm in length) and roll up it to obtain three or four coils. Fix the coiled tubing with a small piece of adhesive tape as illustrated in Fig. 4.

(c) Cut a 23-G needle (10 mm in length) using the drill mounting a diamond wheel.

(d) Insert the 23-G needle (for 4 mm) inside the probe outlet.

(e) Fill the Hamilton syringe (1 mL) with the aECF (see Note 3).

(f) Connect one end of the first 300 mm PE50 tubing to the Hamilton syringe needle and the other end to the liquid swivel inlet.

(g) Connect one end of the second 300 mm PE50 tubing to the liquid swivel outlet and the other end to the probe inlet (the probe is inserted in the brain).

(h) Place the Hamilton syringe on the infusion pump and fill the circuit using the maximum flow rate (see Note 4).

3.4.2. In Vivo Experiment with Freely Moving Animals (see Fig. 4)

(a) Set the infusion pump flow rate to 1.5 μL/min and wait 60 min to allow for stabilisation of the microdialysis gradient.

(b) Connect one end of the 150 mm PE50 tubing to the 23-G needle previously inserted in the probe outlet and wait 20 min.

(c) Remove the 150 mm PE50 tubing from the animal head disconnecting it from the 23-G needle attached to the probe outlet.

(d) Aspirate the microdialysate present inside the 150 mm PE50 tubing using a 50-μL Hamilton syringe, transfer it (around 30 μL) to a 250-μL Eppendorf vial and store at −80°C (see Note 9) until HPLC analysis (Subheading 3.5). Repeat steps b–d for every microdialysis sample recovered during the in vivo experiment.

(e) Usually three consecutive microdialysis samples are sufficient for defining the baseline of the examined neurochemicals (see Note 10).

(f) After baseline sample recovery and if desired, a pharmacological treatment can be performed via systemic or intracerebral drug administration. D-Amphetamine prepared as previously described (Subheading 2.1, step 4) can be injected subcutaneously (2 mg/kg corresponding to 1.0 mL in rats and 100 μL in mice) or by intracerebral retrodialysis. In the second case, the aECF containing 2 mM D-amphetamine needs to replace the perfusion fluid present in the microdialysis circuit (see Note 11) during the entire period of treatment (usually comprising between 20 and 60 min, corresponding to 1–3 microdialysis samples). Immediately after the systemic drug administration or during its intracerebral retrodialysis, the microdialysis samples are recovered as illustrated previously (steps b–d).

(g) After the end of the intracerebral drug administration, the solution containing only aECF has to be restored (see Note 11).

(h) After the pharmacological treatments, microdialysates can be recovered for a period varying from 120 to 240 min (corresponding to 3–6 samples).

(i) At the end of the experiment, disconnect the animal from the microdialysis circuit and sacrifice it using the euthanasia solution (described in Subheading 2.1.3) injected *intraperitoneum* (2.0 mL in rats or 200 μL in mice). At this point, you can verify the correct positioning of the microdialysis probe as described elsewhere (10).

3.5. HPLC Setup and Sample Analysis

1. As illustrated in Fig. 5, the mobile phase is moved by an isocratic pump, through PE tubes, within the Rheodyne injector equipped with a 25-μL loop. Through the injector, 30 μL sample is loaded and channelled into the C18 column where the substances are separated, revealed by electrochemical cell and quantified by the proper software (14, 15).
2. One hour prior to starting the experiment, check the electrochemical cell for applied potential (+780 mV vs. Ag/AgCl reference electrode), then increase gradually the mobile phase

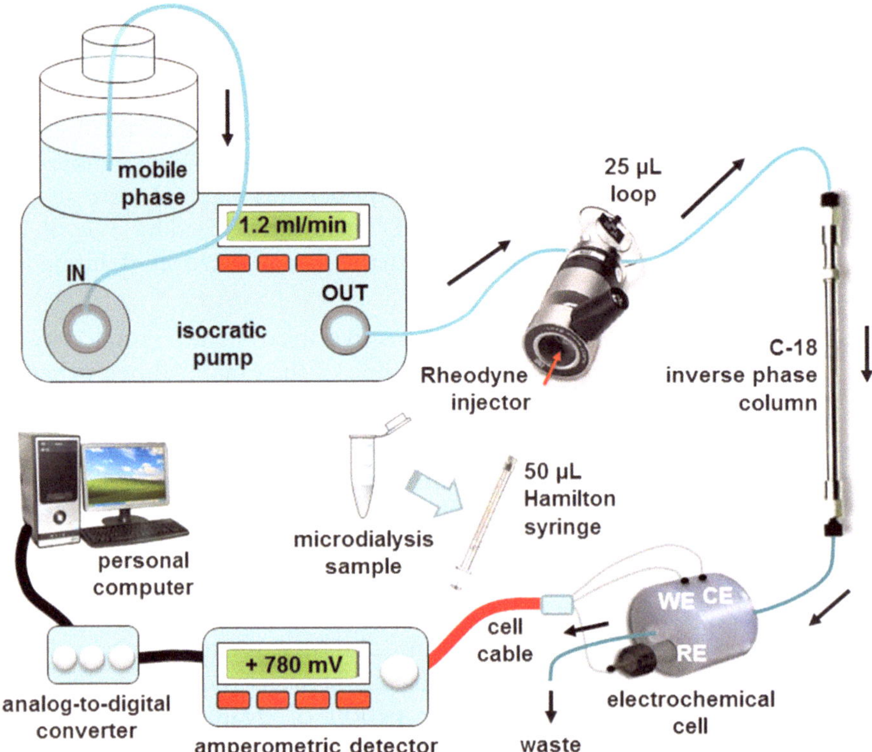

Fig. 5. HPLC setup for ascorbic acid, uric acid, catecholamine and indolamine quantification in microdialysate samples from the striatum of freely moving animals. The electrochemical detection (EC) allows the amperometric oxidation of the abovementioned neurochemicals using a +780 mV potential vs. Ag/AgCl reference electrode. The circuit is completely metal-free for preventing sample degradation during the analysis. *WE* glassy carbon working electrode; *RE* Ag/AgCl reference electrode; *CE* stainless steel counter electrode.

flux from 0.3 mL/min (not-operating flux) to 1.2 mL/min and let stabilise for 45 min.

3. After stabilisation, inject 30 µL of the same aECF solution used for the microdialysis experiment (see Note 12).
4. Thaw and inject 30 µL of standard solution and with the proper software, calculate each peak area. After each injection, carefully wash the syringe with ultrapure water.
5. Thaw (if necessary) and inject 30 µL of microdialysate sample.
6. By means of the provided software, calculate sample area peaks and the concentration of each compound using the following equation:

$$[\text{sample}] = (\text{area of microdialysis sample} / \text{area of standard solution}) \times [\text{standard solution}]. \quad (2)$$

7. Repeat steps 5 and 6 for each microdialysate sample.
8. Report microdialysis sample concentrations (for every neurochemical component analysed) in a spreadsheet and perform statistical analysis.

3.6. Microdialysis Data Analysis

The concentrations in the dialysate will be expressed in nM (DA, L-DOPA, NE, 3HT and 3MT) or μM (AA, UA, DOPAC, HVA and 5-HIAA) and given as mean ± SEM. D-Amphetamine effects on neurochemicals can be evaluated as absolute or percent changes from the preceding baseline (for this purpose data need to be expressed as percentage of baseline). Statistical significance is assessed using analysis of variance (ANOVA) for differences over time. Differences within groups are determined by paired *t* tests, with the Newman-Keuls multiple comparison adjustment. Pearson's correlation coefficient between neurochemicals (i.e. DA vs. 3-MT, AA vs. DA or DA vs. DOPAC + HVA) could be calculated in each animal after drug administration.

4. Notes

1. The preparation of the epoxy resin is a key point of the entire process. It is necessary to mix the two epoxy components (resin and hardener) in equal amounts for 2–3 min, and wait for obtaining the desired consistency before use. Although the epoxy resin becomes solid in few minutes, it needs at least 24 h for complete polymerization.
2. The epoxy resin used in this step has to be less consistent: mix 70% of resin and 30% of hardener. Apply a small amount of epoxy resin and pay attention to the capillary effect (suction effect) inside the PE50 tubing.
3. Prevent the formation of air bubbles inside the syringe by gently aspiring the aECF.
4. Prevent the formation of air bubbles inside the entire circuit (and in particular in the dialyzing part of the probe) filling the tubing by gently pushing the syringe piston.
5. The percent recovery of the probe is an important parameter for comparing microdialysis performance among probes made by the same operator or different operators. Differences in the construction process (mainly related to probe active surface area) can influence this parameter. It is necessary to minimise the standard deviation around the mean for each neurochemical to be sure that the calibration results are consistent.
6. Periosteum is rich in nociceptors while they are not present in the brain; it is thus necessary to remove this membrane to prevent pain in the postoperative period and during the microdialysis experiment.
7. Dural venous sinuses (also called dural sinuses) are venous channels found between layers of *dura mater* and are particularly present under the cranial sutures. A lesion at this level can induce copious bleeding.

8. It could be useful to monitor body temperature, breathing and animal reflexes at regular intervals. A tardive side effect of chloral hydrate is the production of pulmonary secretions treatable with subcutaneous adrenaline.
9. Alternatively, you can directly analyse the microdialysis sample by introducing it in the Rheodyne injector as illustrated in Fig. 5.
10. Sometimes, baseline levels of neurochemicals (DA in particular) are not so stable. In this condition, wait until the percent variation of three consecutive samples is lower than 10%.
11. For facilitating and speeding up the solution exchange procedure, you can disconnect the fluidic circuit upstream of the microdialysis probe and reconnect it filled with the new solution (see also Notes 3 and 4).
12. No chromatographic peaks are visible in normal conditions. If oxidizable interferents are visible after aECF injection, a contaminant is present in the aECF solution. Replace the solution.

Acknowledgments

The authors would like to thank Mr. Gianni Esposito, HPLC expert and brilliant laboratory technician, for his past work and future suggestions. This work was supported by University of Sassari (ex 60% fund), Fondazione Banco di Sardegna and Regione Autonoma della Sardegna.

References

1. Chaurasia CS, Müller M, Bashaw ED, Benfeldt E, Bolinder J, Bullock R, et al (2007) AAPS-FDA Workshop White Paper: Microdialysis Principles, Application and Regulatory Perspectives *Pharm Res* 24, 1014–1025
2. Delgado JMR, DeFeudis FV, Roth RH, Ryugo DK, and Mitruka BM (1972) Dialytrode for long-term intracerebral perfusion in awake monkeys. *Arch Int Pharmacodyn Ther* 198, 7–21
3. Ungerstedt U, and Pycock C (1974) Functional correlates of dopamine neurotransmission. *Bull Schweiz Akad Med Wiss* 30, 44–55
4. O'Neill RD, Rocchitta G, McMahon CP, Serra PA, and Lowry JP (2008) Designing sensitive and selective polymer/enzyme composite biosensors for brain monitoring in vivo. *Trends Anal Chem* 27, 78–88
5. Lönnroth P, Jansson PA, and Smith U (1987) A microdialysis method allowing characterization of intracellular water space in humans. *Am J Physiol* 253, 228–231
6. Olson RJ, and Justice JB, Jr (1993) Quantitative microdialysis under transient conditions. *Anal Chem* 65, 1017–1022
7. Wang P, Wong S, and Sawchuk RJ (1993) Microdialysis calibration using retrodialysis and zero-net flux: application to a study of the distribution of zidovudine to rabbit cerebrospinal fluid and thalamus. *Pharm Res* 10, 1411–1419
8. Charalambides C, Sgouros S, and Sakas D (2010) Intracerebral microdialysis in children. *Childs Nerv Syst* 26, 215–220
9. Nordström CH (2010) Cerebral energy metabolism and microdialysis in neurocritical care *Childs Nerv Syst* 26, 465–472
10. Miele M, Mura MA, Enrico P, Esposito G, Serra PA, Migheli R, et al (2000) On the mechanism of d-amphetamine-induced changes

in glutamate, ascorbic acid and uric acid release in the striatum of freely moving rats. *Br J Pharmacol* 129, 582–588

11. Morton DB, and Griffiths PHM (1985) Guidelines on the recognition of pain, distress and discomfort in experimental animals and a hypothesis for assessment. *Vet Rec* 116, 431–436
12. Paxinos G, and Watson C (1986) The Rat Brain in Stereotaxic Coordinates. Academic Press, San Diego
13. Paxinos G, and Franklin KBJ (2001) The Mouse Brain in Stereotaxic Coordinates Second Edition. Academic Press, San Diego
14. Rocchitta G, Migheli R, Esposito G, Marchetti B, Desole MS, Miele E, et al (2006) Endogenous melatonin protects L-DOPA from autoxidation in the striatal extracellular compartment of the freely moving rat: potential implication for long-term L-DOPA therapy in Parkinson's disease. *J Pineal Res* 40, 204–213
15. Rocchitta G, Migheli R, Mura MP, Grella G, Esposito G, Marchetti B, et al (2005) Signaling pathways in the nitric oxide and iron-induced dopamine release in the striatum of freely moving rats: role of extracellular Ca^{2+} and L-type Ca^{2+} channels. *Brain Res* 1047, 18–29

Chapter 32

Evaluating Motor Neuron Death in Neonatal Rats Subjected to Sciatic Nerve Lesion

Andre Schwambach Vieira, Alexandre Cesar Santos de Rezende, and Fabio Rogerio

Abstract

Neonatal sciatic nerve lesion is a useful experimental model for the study of neuronal cell death. Sciatic nerve transection or crush is the most frequently used approach to evaluate motoneuron loss in the lumbar enlargement of the spinal cord. Here we describe and illustrate the surgical procedures performed in our laboratory to assess motoneuron cell death and the related cellular mechanisms.

Key words: Neonatal rat, Sciatic nerve, Transection, Crush, Motoneuron, Neuronal cell death

1. Introduction

During neurogenesis, programmed cell death is considered to play a role in the establishment of neural pathways. Specifically, neurons produced in excess would be excluded through cellular mechanisms dependent on the availability of trophic factors (1, 2). Since the original studies conducted by Rita Levi-Montalcini (3) on the nature and characterization of neurotrophic factors, it has become clear that distinct populations of neurons are sensitive to specific trophic molecules. Innervated organs and cells of the central and peripheral nervous systems are recognized as sources of such molecules. In addition to the evidence obtained during embryogenesis, the dependence of neuronal cells on trophic factors synthesized by their targets can be experimentally highlighted after transectioning or crushing the fibers of a peripheral nerve (1, 4–10).

A commonly used model to study the effects of the disruption of peripheral trophic support on neurons is transectioning the sciatic nerve of rodents within the first postnatal week. In this approach, the nerve is completely sectioned and separated in two stumps,

Stephen D. Skaper (ed.), *Neurotrophic Factors: Methods and Protocols*, Methods in Molecular Biology, vol. 846, DOI 10.1007/978-1-61779-536-7_32,

proximal and distal, which are kept distant one from the other. Therefore, since physical contact between the stumps is not allowed, axonal regeneration is impaired (11). When performed at the first postnatal day (P0), such type of injury determines the loss of approximately 100% of the lesioned motoneurons in the lumbar enlargement of the spinal cord. Such extensive neuronal cell loss is not achieved when the lesion is performed in older animals, particularly adults. Besides disruption of trophic support from the target, it is hypothesized that immature nonneuronal cells present in the peripheral nerves, especially Schwann cells, would not be able to provide the lesioned motoneurons with trophic factors. The opposite would be true for the sciatic nerve of adult animals (5, 12–14).

Another experimental model used to evaluate motoneuron damage is sciatic nerve crush. Conversely to the nerve fiber transection model, the axons are not totally sectioned. In fact, as neural fibers are disrupted by compression, the nerve stumps are thought to remain connected, which would allow for neuronal regeneration. Specifically, cellular recovery would be favored by a microenvironment established at the injury site, in which growth factors and chemokines released by cells and molecules of the extracellular matrix would allow remodeling of the nerve stumps, axonal recovery, and neuronal regeneration, at least in part, after the lesion. However, it is important to emphasize that Schwann cells of neonatal rats have reduced competence to support regenerating motoneurons when compared to adult Schwann cells. One hypothesis that may be put forward is that immature Schwann cells have reduced production of trophic factors (1, 5, 15, 16).

Motoneuron lesion induced by sciatic nerve transection or crush in neonatal rats is an experimental model that can be used in many research fields of neuroscience, such as neuronal cell death pathways, peripheral nerve regeneration, and functional recovery and evaluation of trophic factors and neuroprotective agents. In our laboratory, we have focused on the use of exogenous growth factors and neuroprotective substances on motoneuron survival after sciatic transection in neonatal rats. In particular, we have been investigating neuronal cell survival and the possible related cellular pathways through motoneuron counting and evaluating protein and gene expression (9, 10, 17, 18). Here we describe the surgical procedures currently performed in our laboratory to induce motoneuron injury in the lumbar spinal cord of neonatal rats by either transectioning or crushing the sciatic nerve.

2. Materials

2.1. Surgical Procedure

1. Microscissors
2. Microforceps
3. Inverted microforceps

4. Microneedle holder
5. Scalpel #11
6. 8-0 suture silk
7. Glass Petri dish
8. Cotton swabs
9. 100-W incandescent lamp
10. Surgical microscope

2.2. Transcardiac Perfusion

1. Ketamine (75 mg/kg)/xylazine (15 mg/kg)
2. 1-mL plastic syringe
3. Surgical table
4. Surgical scissors
5. Forceps
6. Hemostatic forceps
7. 28-G needle attached to flexible tube
8. Peristaltic pump
9. Microscissors
10. 0.9% NaCl solution containing 400 U/L of heparin
11. 4% Buffered formaldehyde solution

2.3. Spinal Cord Dissection

1. Surgical scissors
2. Forceps
3. Scalpel #11
4. Microscissors
5. Microforceps
6. Surgical table
7. Surgical microscope

3. Methods

All the experimental procedures described herein were approved by the Committee on Animal Care of the State University of Campinas (protocol # 509-1).

3.1. Surgical Procedures

1. Separate rat pups from their mother in a small plastic box and maintain them under a 100-W incandescent lamp until the experiment is complete (see Note 1).
2. Immerse one rat in crushed ice for 4 min to induce anesthesia by hypothermia (see Note 2).

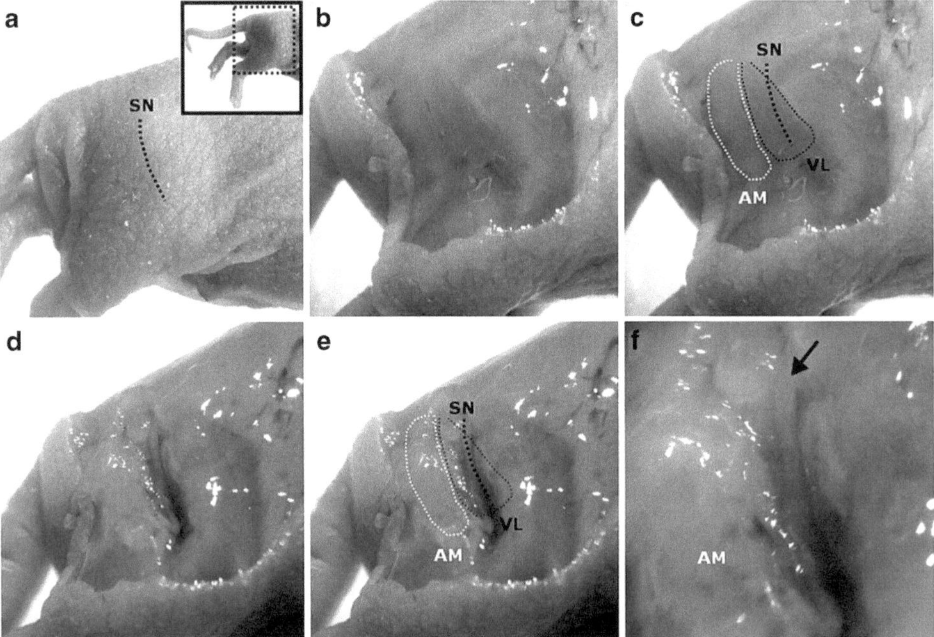

Fig. 1. Anatomical location of the neonatal rat sciatic nerve as seen under the surgical microscope. (**a**) The approximate location of the sciatic nerve (SN) is shown as a *dotted line* on the skin. The inset shows the region magnified in (**a**). (**b**, **c**) The SN runs under the *vastus lateralis* (VL) and the *adductor magnus* (AM) muscles. (**d**, **e**) The VL and AM muscles were removed to expose the SN. (**f**) Higher magnification of the SN. The *arrow* indicates the proximal region of the nerve. (**c**, **d**) The nerve and the muscles are indicated by *dotted lines*.

3. Place the anesthetized pup on a Petri dish. From this moment on, perform all procedures under the surgical microscope.
4. Make a 1-mm incision on the skin of the midthigh 2 mm posterior to the upper half of the femoral bone (see Figs. 1 and 2).
5. Introduce the microforceps between the muscles *vastus lateralis* and *adductor magnus* and separate them by repeatedly opening and closing the forceps (see Figs. 1 and 2).
6. The sciatic nerve runs under the muscles mentioned above (item 5). Using the microscissors and the microforceps, dissect the nerve until the site where it crosses the *incisura ischiadica*.
7. For sciatic nerve *transection*, hold the nerve with the microforceps and cut it using the microscissors, 3 mm proximal and distal to the region being held. Remove the nerve segment which was isolated (see Note 3; Fig. 2).
8. For sciatic nerve *crushing*, press the nerve using the inverted microforceps with fine tips. Keep such pressure for 30 s (see Note 4).
9. Reapproximate the *vastus lateralis* and *adductor magnus* muscles and suture the skin using the 8-0 silk.

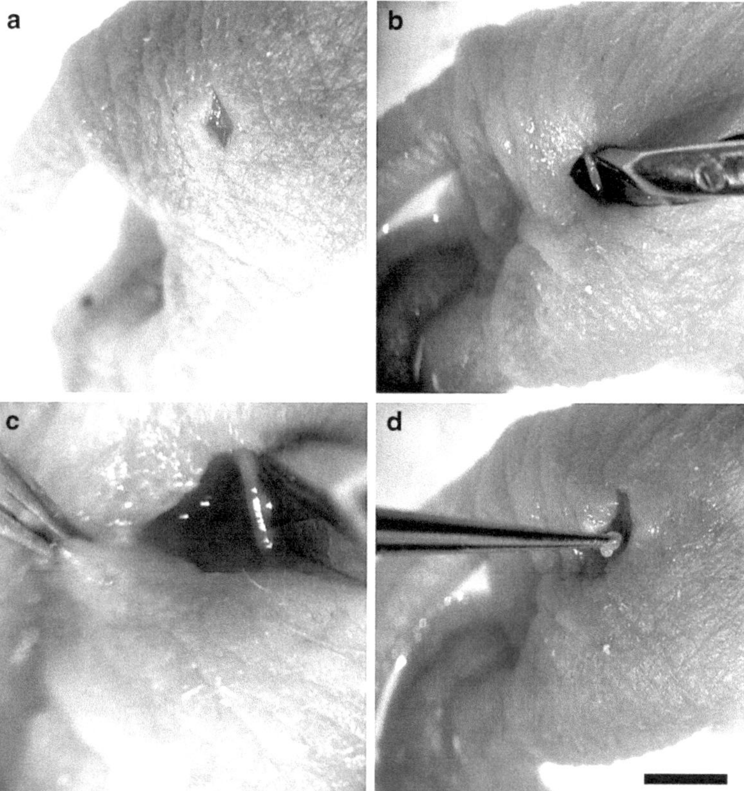

Fig. 2. Anatomical location of the neonatal rat sciatic nerve as seen under the surgical microscope. (**a**) Incision on the midthigh skin. (**b**) Exposed sciatic nerve after separation of the *vastus lateralis* and the *adductor magnus* muscles. (**c**) Higher magnification of the region shown in (**b**). (**d**) The segment of nerve removed after transection is shown between the tips of the microscissors. Bar: 3.0 mm (**a**, **b**, **d**), 1.2 mm (**c**).

10. Place the neonate rat under the 100-W incandescent lamp until recovery is observed (see Note 5).
11. Return the pups to their mother (see Note 5).

3.2. Transcardiac Perfusion

1. Separate rat pups from their mother in a small plastic box and maintain them under a 100-W incandescent lamp (see Note 1).
2. For anesthesia, administer the solution of ketamine (75 mg/kg)/ xylazine (15 mg/kg) intraperitoneally.
3. When the rat does not display any pedal reflex, fixate it on the surgical table.
4. Make a skin incision through the midline, extending from the sternal angle to the pubic region.
5. In the abdominal region, make a bilateral incision parallel to the subcostal region and a midline linear incision. In this procedure, the muscular and peritoneal membrane will be sectioned, exposing the abdominal cavity.

6. Use small scissors to make an incision in the diaphragm in order to access the thoracic cavity.
7. Make bilateral sections along each lateral part of the thoracic cage. Lift the released ribs using the hemostatic forceps in order to expose the heart.
8. Gently introduce the needle in the left ventricle and cut the right atrium with microscissors (see Note 6).
9. Use a peristaltic pump to perfuse the rat with saline solution, followed by a 4% buffered formaldehyde solution (see Note 7).

3.3. Spinal Cord Dissection

1. After perfusion, fixate the rat in the prone position (*decubitus ventralis*) on a surgical table.
2. Make an extensive dorsal skin incision through the midline to expose the paravertebral musculature. From this moment on, perform all procedures under the surgical microscope.
3. Gently remove the paravertebral musculature using a scalpel and the vertebrae with the microforceps. Then, section the meningeal covering to expose the spinal cord (see Note 8).
4. For motoneuron investigation, isolate the lumbar enlargement from the remaining spinal cord using a scalpel. Remove the lumbar enlargement from the spinal canal, sectioning the nerve roots and meningeal membrane using microscissors.
5. The isolated lumbar enlargement may be processed using standard histological protocols for paraffin embedding or frozen processing (see Note 9).

4. Notes

1. We routinely use an incandescent lamp (100 W) under which the pups are kept warm throughout the experiment, as they are isolated from the mother and placed in a plastic cage. A distance of approximately 20 cm between the lamp and the rat is usually sufficient. For anesthetic recovery, the same approach is performed.
2. Deep hypothermia in young rodents is an efficient technique to induce anesthesia (19). Since the pups are not pharmacologically treated, there is no risk of death associated with overdose. After a short period of being unexposed to the incandescent lamp, the rat must be totally immersed in crushed ice until limb movements completely stop. A period of 4–5 min of hypothermia (counting from the immersion) is sufficient to elicit anesthesia for a surgical procedure of about 10 min.
3. It is crucial to keep the nerve stumps at a distance to avoid regeneration. For this goal, an additional useful procedure is to

retract the nerve endings in opposite directions using the microforceps before approximating the muscles for wound closure.

4. The inverted microforceps correspond to custom-made microforceps with a 1-mm tip that opens when pressed and closes when released, that is, the opposite to a standard microforceps. In our experience, such inverted microforceps are extremely helpful to ensure a constant and reproducible pressure applied to the sciatic nerve in crushing experiments performed in different occasions.
5. After the surgical procedure, dry the animals and place them under the incandescent lamp until indirect signals suggestive of anesthetic recovery are observed. We consider active respiratory and upper limb movements as reliable observations that the pup has recovered from anesthesia and is ready to be returned to its mother. Before returning the pups to their mother, make sure that the surgical wound is completely free of blood clots. The presence of blood may interfere with the acceptance of the mother to its pups and eventually with breastfeeding.
6. During transcardiac perfusion, take extreme care to avoid the rupture of the cardiac septum. A useful procedure is to attach a plastic catheter to the 28-G needle. The catheter should be 2 mm shorter than the needle in order to allow the introduction of only 2 mm of the needle inside the heart. The rupture of the septum may be inferred by increase in lung volume and elimination of liquids through the upper airways.
7. In our experience, the ideal perfusion rate is 3.0 mL/min. At this rate, the perfusion with saline takes approximately 5 min, and with formaldehyde, approximately 15 min. After the perfusion, we usually keep the whole rat immersed in a buffered formaldehyde solution for 12–24 h. The goal of this step is to avoid the occurrence of histological artifacts (see Note 8).
8. Extreme care should be taken when dissecting the spinal cord. Excessive manipulation of inadequately perfused specimen or after insufficient period of fixation may induce the occurrence of the so-called dark neurons. In hematoxylin- and eosin-stained sections, dark neurons are shrunken, irregular, and densely basophilic cells. This particular histological artifact may interfere with the morphological evaluation of motoneurons, mainly being confused with dying neurons (20).
9. Morphological evaluation of spinal motoneuron loss induced by sciatic lesion in neonatal rats may be performed through different approaches. Since the sciatic nerve is constituted of motoneurons whose bodies are in the ventrolateral region of the lumbar enlargement (L4 and L5 segments mainly), a peripheral insult allows one to carry out evaluation by counting the remaining cellular bodies. A possible semiquantitative evaluation consists

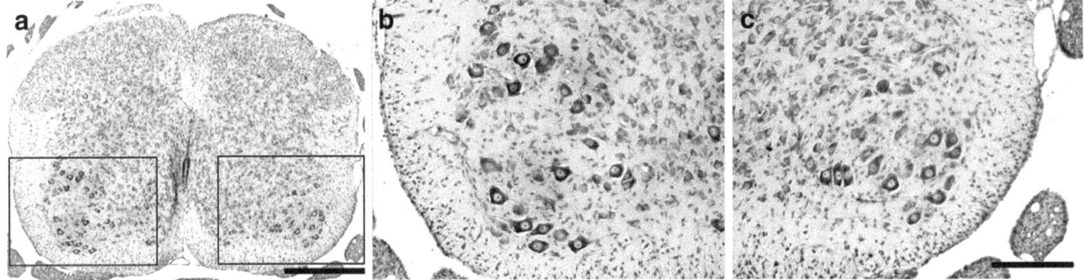

Fig. 3. Transverse sections of the lumbar enlargement (L4) of rats 5 days after sciatic transection performed at an age of 2 days postnatally. (**a**) Reduction of the number of motor neurons is observed in the right ventrolateral region (shown in (**c**)) as compared to the contralateral unlesioned side (shown in (**b**)). Paraffin-embedded sections stained with cresyl violet. Bar: 500 μm (**a**), 100 μm (**b**, **c**).

of counting the nucleoli of motoneurons ipsilateral and contralateral to the axonal injury. In general, we count nucleoli in 20 serial paraffin sections (5 μm) per animal. Specifically, the first section of every four is considered for counting. Then, a ratio of the total number of nucleoli in the lesioned side to that in the unlesioned side is calculated for each animal and used for further statistical analyses (17, 18) (see Fig. 3). Alternatively, the spinal cord may be subjected to stereological techniques for a quantitative approach, such as the "optical fractionator" (21). For this goal, we usually freeze the lumbar enlargement and obtain 40-μm serial sections. Subsequently, a random systematic uniform sampling is performed for the selection of sections used for counting and for the placement of a grid of known size in each sample. Then, an optical dissector, a three-dimensional counting probe, is placed in each point of intersection of the grid that falls in the region of interest. The total number of cells is given by the formula $C = \Sigma Q^{-} \,.\, t/h \,.\, 1/asf \,.\, 1/ssf$, where C is the total number cells, ΣQ^{-} is the sum of cells counted in the optical dissector, t is the section thickness, h is the dissector height, asf is the ratio between counting frame and grid areas, and ssf is the ratio between sections used for counting and total number of sections. For details on the fractionator and other stereological tools, the reader is referred to an introductory review by Gundersen et al. (22).

References

1. Kuno M (1990) Target dependence of motoneural survival: the current status. *J Neurosci Res* 9, 155–172
2. Oppenheim RW (1991) Cell death during development of the nervous system. *Ann Rev Neurosci* 14, 453–501
3. Levi-Montalcini, R (1964) The nerve growth factor. *Ann NY Acad Sci* 118, 149–170
4. Sendtner M, Kreutzberg GW, and Thoenen H (1990) Ciliary neurotrophic factor prevents the degeneration of motor neurons after axotomy. *Nature* 345, 440–441
5. Lowrie MB, and Vrbova G (1992) Dependence of postnatal motoneurones on their targets: review and hypothesis. *Trends Neurosci* 15, 80–84
6. Sendtner M, Holtmann B, Kolbeck R, Thoenen H, and Barde YA (1992) Brain-derived neurotrophic factor prevents the death of motoneurons in newborn rats after nerve section. *Nature* 360, 757–759

7. Greensmith L, and Vrbova G (1996) Motoneurone survival: a functional approach. *Trends Neurosci* 19, 450–455
8. Oliveira AL, Risling M, Negro A, Langone F, and Cullheim S (2002) Apoptosis of spinal interneurons induced by sciatic nerve axotomy in the neonatal rat is counteracted by nerve growth factor and ciliary neurotrophic factor. *J Comp Neurol* 447, 381–393
9. Rezende ACS, Vieira AS, Rogerio F, Rezende LF, Boschero AC, Negro A, et al. (2008) Effects of systemic administration of ciliary neurotrophic factor on Bax and Bcl-2 proteins in lumbar spinal cord of neonatal rats after sciatic nerve transection. *Braz J Med Biol Res* 41, 1024–1028
10. Rezende ACS, Peroni D, Vieira AS, Rogerio F, Talaisys RL, Costa FTM, et al. (2009) Ciliary neurotrophic factor fused to a protein transduction domain retains full neuroprotective activity in the absence of cytokine-like side effects. *J Neurochem* 109, 1680–1690
11. Vejsada R, Sagot Y, and Kato AC (1995) Quantitative comparison of the transient rescue effects of neurotrophic factors on axotomized motoneurons in vivo. *Eur J Neurosci* 7, 108–115
12. Schmalbruch H (1984) Motoneuron death after sciatic nerve section in newborn rats. *J Comp Neurol* 224, 252–258
13. Schmalbruch H (1987) Loss of sensory neurons after sciatic nerve section in the rat. *Anat Rec* 219, 323–329
14. Lowrie MB, Lavalette D, and Davies CE (1994) Time course of motoneurone death after neonatal sciatic nerve crush in the rat. *Dev Neurosci* 16, 279–284
15. Sendtner M, Stöckli KA, and Thoenen H (1992) Synthesis and localization of ciliary neurotrophic factor in the sciatic nerve of the adult rat after lesion and during regeneration. *J Cell Biol* 118, 139–148
16. Magill CK, Moore AM, Yan Y, Tong AY, MacEwan MR, Yee A, et al. (2010) The differential effects of pathway- versus target-derived glial cell line-derived neurotrophic factor on peripheral nerve regeneration. *J Neurosurg* 113, 102–109
17. Rogerio F, De Souza Queiroz L, Teixeira SA, Oliveira AL, De Nucci G, and Langone F (2002) Neuroprotective action of melatonin on neonatal rat motoneurons after sciatic nerve transection. *Brain Res* 926, 33–41
18. Rogerio F, Teixeira SA, Rezende ACS, Cofino R, Queiroz LS, De Nucci G, et al. (2005) Superoxide dismutase isoforms 1 and 2 in lumbar spinal cord of neonatal rats after sciatic nerve transection and melatonin treatment. *Dev Brain Res* 154, 217–225
19. Phifer CB, and Terry LM (1986) Use of hypothermia for general anesthesia in preweanling rodents. *Physiol Behav* 38, 887–890
20. Cammermeyer J (1961) The importance of avoiding "dark" neurons in experimental neuropathology. *Acta Neuropathol* 1, 245–270
21. West MJ, Slomianka L, and Gundersen HJG (1991) Unbiased stereological estimation of the total number of neurons in the subdivisions of the rat hippocampus using the optical fractionator. *Anat Rec* 231, 482–497
22. Gundersen HJ, Bagger P, Bendtsen TF, Evans SM, Korbo L, Marcussen N, et al. (1988) The new stereological tools: dissector, fractionator, nucleator and point sampled intercepts and their use in pathological research and diagnosis. *APMIS* 96, 857–881

Chapter 33

Rodent Spinal Cord Injury Model and Application of Neurotrophic Factors for Neuroprotection

Hari Shanker Sharma and Aruna Sharma

Abstract

Spinal cord injury (SCI) is a serious clinical problem that causes lifetime disabilities to victims and inflicting huge social burden on our society. One of the main lacunae in developing potential therapeutic measures in SCI is a lack of suitable animal models that could be comparable to clinical situations. Thus, development of new animal models of SCI is highly needed to expand our knowledge on cell injury and repair process in order to reduce cord pathology, and in translating advanced therapies in patients of SCI to improve therapeutic strategies. Keeping these views in mind, a suitable animal model is developed in our laboratory that can be used to explore new therapeutic tools in SCI. The details of our methods used to induce SCI in rodents and neuroprotection achieved by use of selected neurotrophic factors are described in this chapter.

Key words: Spinal cord injury, New model, Neurotrophic factors, Neuroprotection, Cell injury, Neurorepair

1. Introduction

Spinal cord injury (SCI) is a devastating disease causing severe disability, e.g., paralysis and paraplegia, in human populations across the globe (1, 2). Thus, victims of SCI require lifetime rehabilitation and support causing serious burden on our society (3). Unfortunately, the treatment strategies for SCI if delayed beyond 3 h may not achieve significant neuroprotection (4, 5). Thus, efforts should be made to treat trauma patients as soon as possible following primary insult to the central nervous system.

Available evidence suggests that most of the tissue destructive changes in the spinal cord occur within the first hour after the primary insult and progress with time (3, 6). The early consequences of SCI include breakdown of the blood–spinal cord barrier (BSCB), reduction in local blood flow, edema formation,

Stephen D. Skaper (ed.), *Neurotrophic Factors: Methods and Protocols*, Methods in Molecular Biology, vol. 846, DOI 10.1007/978-1-61779-536-7_33,

and neuronal, glial, and myelin damage (1, 3–5). These changes will continue further with advancement of time beyond the injury point across the spinal cord in rostrocaudal directions leading to permanent disability (1, 4, 5, 7). Thus, a detailed knowledge on early pathological events is necessary in exploring suitable therapeutic strategies to treat SCI victims.

Recent studies show that treatment schedule initiated after 3–6 h of primary insult is largely ineffective on the spinal cord pathology (1, 4, 5). Thus, studies using early therapeutic interventions to reduce spinal cord cell and tissue damage are needed. Obviously, when spinal cord neurons along with nonneuronal cells, e.g., glial cells, endothelial cells, and myelin, are protected after injury, then only neurorecovery and/or neuroprotection could be achieved (1, 8–12). Restoration of spinal cord cell and tissue function after trauma is crucial to reduce sensory motor disturbances (1, 2, 13–17). Accordingly, neuroregeneration and neurorepair mechanisms appear to be the most important factors for any future spinal cord therapeutic strategies after injury.

Our studies have shown that topical application of neurotrophins, including nerve growth factor (NGF), brain derived neurotrophic factor (BDNF), or other growth factors, e.g., glial-derived neurotrophic factor (GDNF) and ciliary neurotrophic factor (CTNF), reduces spinal cord pathology and functional disabilities in a rat model of SCI when administered repeatedly within 10–30 min after trauma (5, 6, 9–12). This suggests that neurotrophins could be the potential and ideal candidates for effective therapeutic strategies in SCI. However, further studies are needed to understand the potential beneficial effects of various neurotrophins or other compounds in combination to effectively reduce the pathological consequences of SCI. For this purpose, an effective animal model of SCI is needed to explore neuroprotective efficacy of various compounds in the vicinity of the lesion site as well in the ipsilateral or contralateral side of the cord (13, 14, 16, 17). However, to date, there are no models of SCI in which this ipsi- or contralateral aspects of spinal cord pathology can be examined in detail. This is largely due to the fact that most models include compression, weight drop, or transaction, leaving no scope to study the effects of lesion or trauma around the injury site (1, 3–5, 15). Thus, effects of drugs in the ipsilateral or contralateral cord are still lacking in other models of SCI.

In this chapter, we describe a new model of SCI in rats and mice in which both ipsilateral and contralateral cell changes can be examined within the lesion site as well as in the rostrocaudal directions of the cord after trauma. These features make this model unique in expanding our understanding of the pathophysiological consequences of trauma within the spinal cord in a very precise manner.

2. Materials

2.1. Chemicals and Other Reagents

1. Urethane.
2. Pentobarbital.
3. Ketamine.
4. Xylazine.
5. Equithesin (Swedish Pharmacy, Umeå, Sweden): chloral hydrate 4.25 g, pentobarbital sodium 0.97 g, magnesium sulfate 2.1 g, propylene glycol 2.1 g, ethanol anhydrous 42.8 g, and sterile water 9 g (100 mL solution). The anesthetic solution has a short expiry date of 1 month (see ref. (18, 19)).
6. BDNF, recombinant human (Sigma-Aldrich).
7. GDNF, recombinant human (Sigma-Aldrich).
8. Insulin-like growth factor-I, recombinant human (IGF-1) (Sigma-Aldrich).
9. CNTF, recombinant rat (Sigma-Aldrich).
10. NGF-β from rat (Sigma-Aldrich).
11. Neurotrophin-3, recombinant human (NT-3) (Sigma-Aldrich).
12. Evans blue (Merck).
13. [131]-Iodine (Nordion, Belgium).
14. Paraformaldehyde.
15. Glutaraldehyde.
16. Lanthanum chloride.
17. Somogyi fixative (4% neutral paraformaldehyde, 0.05% glutaraldehyde, and 0.25% picric acid).
18. Osmium tetroxide.
19. Araldite.
20. Electron microscope grade Epon 812.
21. Toluidine blue.
22. Uranyl acetate.
23. Lead citrate.

2.2. Neurotrophin Stock Solutions and Storage

1. Dissolve neurotrophic factors in phosphate-buffered saline to a concentration of 100 μg/mL and store at –20°C in a desiccator. At the time of experiments, dilute the neurotrophic factors in solution to a concentration of 1 μg/10 μL so that you could adjust their dose by adjusting the volume (20, 21).
2. The diluted neurotrophic agent solutions can be safely stored at 4°C in a desiccator for 3–7 days. If you need to use it longer than a week, you could store it under desiccation at –20°C,

but avoid freeze–thaw cycles. CNTF, although stable at room temperature for 3 weeks, is preferably stored at −20°C in a desiccator.

2.3. Lab Apparatus and Other Supplies

1. Liston bone cutter, straight: 14 cm/5.5 in. (Harvard Surgical Instruments)
2. Friedman Rongeur, curved: 14 cm/5.5 in. (Harvard Surgical Instruments)
3. Dissecting stereomicroscope (Carl Zeiss, 20–30×)
4. Magnifying glass lamp (*Magnifying Lamp* 900-061).
5. Circular 22 W fluorescent light (Chicago electric)
6. Dual Lab standard stereotaxic frame for rats or mice, with 45° ear bars (Harvard Apparatus)
7. EdgeAhead crescent knife, 1.75 mm edge length, 60°, bevel up
8. Gelco sponge (London)
9. PE 10 cannula
10. 21- and 36-gauge needles, butterfly needles
11. 50-μL-capacity Hamilton microliter syringes (HAMILTON Bonaduz AG, Bonaduz, Switzerland)
12. Precision microinfusion pump (Razel Scientific Instrument Inc, Stamford, CT, USA)
13. 3-in. Gamma counter (Packard)
14. Precision electronic balance (e.g., METTLER TOLEDO)
15. Oven capable of reaching 90°C (e.g., Class 100 Cleanroom Ovens, Terra Universal)
16. Light microscope (Zeiss or similar type)
17. Ultramicrotome with diamond knife (LKB or comparable supplier)
18. Phillips 400 transmission electron microscope (or comparable model)

3. Methods

All the existing animal models have essentially some advantages and drawbacks (1–5). The basic differences between the existing animal models and human SCI are summarized in Table 1. Human SCI results from four main vectors of forces, e.g., flexion, extension, rotation, and compression, that are working simultaneously or in combination (1–3, 5, 6). Fracture of vertebral bodies or luxation of vertebrae leads to concussion, contusion, or laceration of the cord due to these multiple sheer forces. Thus, human SCI

Table 1
Animal vs. human SCI factors, variability, and differences

Clinical vs. experimental spinal cord injury (SCI)

Clinical	Experimental
Multifactorial Flexion, extension, rotation, compression concussion, contusion, laceration	Simple Weight drop, incision,[a] compression, transaction, hemisection, electrolytic lesion, contusion, clip compression
Complex forces are active at the time of injury	Only one injury factor is present
Closed vertebral system injury	Open vertebral system injury After laminectomy
Usually anterior compression	Mainly posterior compression
Mainly occurs during state	Produced under anesthesia conscious
Unlimited or uncontrolled time interval between injury and first hand	Carefully controlled time schedule hospital examination

Human vs. rat spinal cord[b]

Parameters	Human	Rat
Number of neurons	10^9	0.3^6
Length (cm)	43–45	8–10
Weight (g)	34	0.7
Proportion to brain (volume %)	2	35
Glial/neuron ratio	10–15	9–12[c]
Endothelial cell/neuron ratio	12–16[c]	10–14[c]
Weight-drop injury (g)	2,428 971 485	50[b] 20[b] 10[b]

Compiled from various sources (for details see ref. (1–8))
Data modified after Sharma (3–7) (for details see ref. (4, 5))
For details see: http://faculty.washington.edu/chudler/facts.html#spinal visited on Nov 25, 2010
[a]Authors own investigation
[b]Weight of rat spinal cord is about 50 times less than human spinal cord. Normally, 10–50 g weight is dropped on the rat spinal cord for making injury (values represent static weight)
[c]Rough estimate based on various sources on assumption based on 1 or 2 endothelial cells per capillary (minimum)

includes various kinds of physical and mechanical damage that finally determine the nature and extent of the cord pathology and varies in each individual case (1). Furthermore, human SCI results from the anterior cord compression or damage in conscious state on the closed vertebral system, whereas, all animal models represent posterior cord damage under anesthesia after laminectomy (1–3, 8). In addition, the animal models of SCI involve one force vector at one time.

To this end, we developed a new model of SCI in rats and mice to study cell changes in the ipsilateral or contralateral cord in rostrocaudal direction (see Fig. 1). The model consists of a longitudinal incision to the right dorsal horn of the T10–11 segments. The deepest part of the lesion is limited to the Rexed laminae VIII (13–15). The model is quite reproducible, and the extent of lesion varied only within a narrow range (3–7). The primary injury, i.e., the knife wound, is limited to the dorsal horn gray matter, leaving white matter largely intact (16, 17), whereas in other injury models using impact, compression, hemisection, or transection, direct damage to white matter predominantly accounts for loss of sensory motor functions in the ascending and descending spinal tracts (see Table 2).

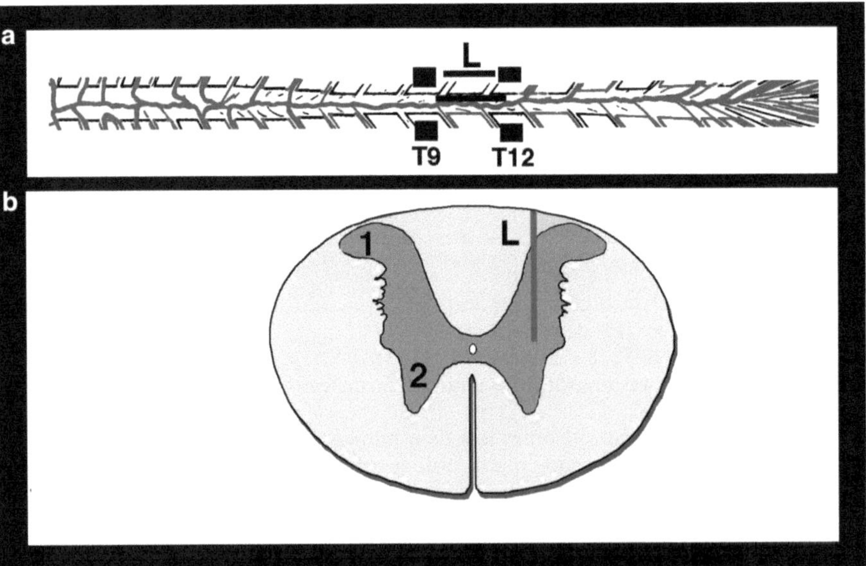

Fig. 1. Diagrammatic representation of spinal cord injury (SCI) in the rat and sampling of spinal cord tissues T9 or T12 for morphological or biochemical analyses (**a**). (**b**) Depth of lesion (L) in the right dorsal horn and selection of tissue areas from the ipsi- or contralateral cord in dorsal horn (1) or ventral horn (2) is clearly identified. Bar = 5 mm. Data modified after Sharma (4, 5, 9, 20).

Table 2
Advantages and shortcomings of various experimental models of spinal cord injury

Type of injury	Advantages	Shortcomings
Weight-drop technique (on exposed spinal cord)	Since second century ad revised by Allen (32)	
	Widely used today in many species	
	Can be graded	
	Reproducible	Results are variable
	Closely simulates some of the biomechanics of human SCI	Differs from clinical conditions because of posterior compression injury
	Histopathology mimics human SCI, e.g., hemorrhage, necrosis, cyst formation	Variable results
	Revised with electromechanical impact device by Bresnahan et al. (33)	
	Better reproducibility	Variable results
Balloon inflation in the spinal extradural space		
Photochemical infarction of vascular endothelium	No laminectomy required	Differs from clinical conditions
Incision[a]	Longitudinal by Sharma and Dey (34)	
	Modified by Sharma and Olsson (13)	
	Reproducible	Differs from clinical conditions
	Can be graded	
	Reproducible results	
	Transverse or longitudinal	
	White matter mainly intact (17)	Does not mimic clinical injury
	Minimal and focal injury	
	Changes can be seen ipsilateral/contralateral	
	Spread of cell injury can be examined in both rostrocaudal and ipsi-/contralateral directions	
	A good model for studying secondary injury mechanisms	Involves simple "cut" factor Hemorrhage due to direct blood vessel damage

Information compiled from various sources (for details see text). Data modified after Sharma and Westman (3); Winkler et al. (7)
[a]Authors' own investigation

3.1. Animal Handling

1. Use rats or mice of specific age, sex, and strain according to protocol. We used male Sprague–Dawley rats (180–250 g) or C57 mice (30–35 g) from Alab, Stockholm, Sweden.
2. Keep the animals under controlled laboratory room temperature (21 ± 1°C) with 12-h light and 12-h dark schedule, with food pellets and tap water supplied ad libitum up to 6–7 days before experiments.

3. Take individual rats/mice out from the cage by gently handling the animal. Avoid lifting animals by the tail since this procedure will induce severe stress, e.g., increase in heart rate and respiration. Use good-quality sterile hand gloves (18, 19, 22).
4. Place the animals on a weighing pan large enough for easy movement of animals. Do not use preweighed restraint boxes. This will create unnecessary stress in animals. Record the body weight, age, sex, and source of the animals. Date and time of the experiments should also be recorded.
5. All experimental conditions for SCI and related techniques should be performed following National Institute of Health Guidelines for the care of laboratory animals and duly approved by Local Institutional Ethics Committee.

3.2. Choice of Anesthetics

Several anesthetics can be used in SCI experiments depending on the local laboratory guidelines and permission from the respective institutional authorities. A brief description of advantages and disadvantages regarding use of particular anesthetics in SCI is described below (see ref. (18, 19, 22)).

3.2.1. Urethane

1. Urethane (ethyl carbamate) is a colorless crystalline substance readily soluble in water (≈1 g in 0.5 mL water at room temperature) with a neutral pH. This is a very safe anesthetic as it induces only mild changes in the cardiovascular and respiratory system without influencing the sympathetic nervous system activity. For neurophysiological investigations, urethane is a good choice for anesthesia because it largely acts on the cortical level and does not depress the respiratory or cardiovascular centers in the brain stem (19).
2. The long-lasting effects of urethane (>12 h) at the level of surgical anesthesia can be achieved using a single dose ranging between 1.5 and 1.8 g/kg, intraperitoneally, in rodents (22) (see Note 1).

3.2.2. Pentobarbital

1. Surgical grade of anesthesia in rats can be achieved by sodium pentobarbital in dose of 40–60 mg/kg, intraperitoneally. However, this anesthesia induces profound bronchial secretions that may be sometimes fatal.
2. To avoid bronchial secretions and poor respiration during the experiment, pretreat with atropine (0.1 mg/kg, subcutaneously, 20–30 min before induction of anesthesia), if this drug is not interfering with the experimental design and protocol.
3. The effects of anesthesia are short lasting (i.e., 60–80 min) and require administration of maintenance doses at regular interval of 45–60 min. The safety margin of pentobarbital anesthesia is also quite narrow.

3.2.3. Combination Anesthesia

1. A combination of ketamine and a muscle relaxant like xylazine is a good choice of anesthesia in various laboratories. However, under this anesthesia, several reflexes, e.g., blinking of eyes, swallowing, and movements of whiskers, or vibrissae, are present although the animal does not respond to pain stimuli.
2. The duration of anesthesia is short lasting and depends on the dose used. The safety margin is considerably higher than sodium pentobarbital alone.

3.2.4. Mixture of Anesthesia

1. In some laboratories, a mixture of various anesthetics is used to produce a safe anesthesia in animals of varying duration. One such anesthetic that is widely used in small animals is Equithesin.
2. Equithesin is a mixture of sodium pentobarbital, ketamine, and muscle relaxant (see below). This anesthesia is very safe in a dose of 3 mL/kg, intraperitoneal, for rats and mice. Respiratory depression is seldom seen. However, the main drawback for this anesthesia is its short duration, i.e., 30–40 min. Thus, repeated maintenance doses are necessary to maintain a certain level of anesthesia that is often difficult.

3.3. Administration of Anesthesia

1 Hold rats/mice firmly from the back skin using a towel, and take care that the animal is not able to move its head otherwise you may get bitten. While holding the animal, firmly turn it around to expose its belly toward you.

2 Use an already prepared anesthetic in a syringe (for dose and choice of anesthesia, see below). Gently administer the anesthetic into the peritoneal space that can be reached from the lower groin with a needle angle of 30° and pushing about 0.5–1 cm inside using a 36-G stainless needle. Just before injection, pull the plunger toward you to see that the needle is inside the peritoneal cavity. In such case, a vacuum is created (18, 19) (see Note 2).

3 If the syringe appears tight and pulling the plunger results in a vacuum, then push the correct dose of anesthesia slowly within 20–30 s.

4 After completion of the injection, gently return the rat to its cage and watch for the effects of anesthetics. The rat will show signs of anesthesia and will gradually fall down on the cage floor and go to sleep. Use a mild tail pinch or ear pinch to check the stage of anesthesia. If the animal does not respond to these stimuli, it is in the surgical grade of anesthesia (Grade IV) (18).

3.4. Exposing the Spinal Column

1. After surgical grade anesthesia is obtained, shave hairs on the back skin gently and expose the skin area covering the spinal column over the thoracic region. Wash the skin surface with 70% alcohol to sterilize the skin surface.

2. Make a longitudinal incision of the skin over the thoracic vertebra pertaining to T8 to T12 (about 2.5–3 cm long). Wipe out any blood coming from the underlying muscle. Identify the thoracic T10 and 11 vertebras by its stout build just rostral to the lumbar enlargement. Normally, this can be achieved by counting vertebral spine from thoracic segments 1–12 (see Note 3).
3. After identifying the T10–11 vertebra, make small incisions around the vertebral muscle to clean the bone surface extending between T9 and T12 vertebrae. Clean any oozing blood using sterile cotton with some mild pressure over it (13, 14) (see Note 4).
4. After exposing and cleaning the T10–11 vertebrae from blood, apply cotton soaked with 0.9% saline over the exposed surface and leave it for 5–10 min to stabilize the spinal column. This will also alleviate stress reaction over the spinal cord due to any pressure applied over the spinal column (13–15) (see Note 5).

3.5. Exposing the Spinal Cord (Laminectomy)

1. After exposing the T10–11 spines and articular processes, insert one tip of a very-fine-pointed side-cutting bone cutter carefully between T10 and T11 vertebral space.
2. Carefully remove the vertebral bone using space between the two adjacent vertebrae. When you are able to perform on the right side, use the same procedure to remove a part of the vertebra on the left side as well. After that, use a precision bone rongeur to widen the space by removing pieces of bones slowly (see Note 6).
3. During the bone removal process, apply cold saline (4°C) over the bone and spinal cord surface frequently to keep the tissue wet all the time (16, 17).
4. Use a dissecting stereomicroscope to remove bone and to avoid direct damage to the spinal cord, if necessary. However, after some experience, a good magnifying glass lamp with a circular 22 W fluorescent light will be enough to perform laminectomy.

3.6. Making the Lesion

1. After laminectomy, stabilize the spinal column of the rat or mice under a stereotaxic apparatus. Use constant cold saline application over the exposed spinal cord.
2. Using one of the vertical axis manipulators of the stereotaxic apparatus, fix the microknife with a straight bevel positioned at 90° over one end of the exposed spinal cord (see Note 7).
3. Adjust the coordinates to fit the knifepoint over the right dorsal horn (about 0.8 mm right to the dorsal spinal artery) of the T10 segment. Carefully lower the tip of microknife to

touch the dura over the right dorsal horn without piercing it. Any accidental piercing will result in leakage of spinal fluid.

4. After positioning the knife over the right dorsal horn, insert the tip of the knife within the spinal cord up to 1.5 mm deep to reach around the Rexed laminae VIII to IX ((13–15, 17); see Fig. 1).
5. After reaching deeper into the desired area, gradually move the knife using stereotaxic apparatus longitudinally backward up to 5 mm (13–17). After the lesion, remove the knife from the spinal cord immediately and cover the wound with cotton soaked in saline to prevent drying of the cord tissue. If there is any bleeding, gently soak it with Gelco sponge (see Fig. 1; Note 8).

3.6.1. Alternative Methods to Induce Spinal Cord Injury

1. An experienced worker could also make a lesion using a #11 sterile carbon steel surgical blade by hand under a magnifying fluorescent lamp in rats or in mice (13).
2. For making lesion by hand, the knife blade may be sealed with araldite leaving only 1.5 mm tip open to avoid accidental injury due to spinal reflex while making incision. The knife preparation can be done one night before the experiment (13, 16) (see Note 9).

3.7. Postoperative Care

1. After the lesion, remove the animals from the stereotaxic platform and place them in a secure plastic cage on good surgical bedding. Do not put spinal cord injured animals in the cage with wooden scrap as small pieces of wood may choke these anesthetized rats/mice causing to death.
2. Place each rat/mice in individual cages to avoid any possible attack by other animal or laceration of the wound in case animals need to be revived for behavioral investigations. However, for the study of spinal cord pathophysiology, it is not necessary to awaken them. In such cases, maintain the anesthetic dose for the entire survival period as needed by the experimental protocol (19).

3.8. Application of Neurotrophic Factors After Spinal Cord Injury

To investigate the role of neurotrophins alone or in combination in inducing neuroprotection after SCI in the above model, we use the following protocol for treatment (20, 21, 23–30).

3.8.1. Topical Application of Neurotrophins over the Spinal Cord

1. Prepare the desired dilution of each neurotrophic factors in separate 50-μL-capacity Hamilton microliter syringes. Connect the syringe to a precision microinfusion pump for slow infusion. Calibrate the speed of injection to deliver 10 μL/min to allow 0.25 μL over the spinal cord within 15 s (20, 21, 23, 24). Should you choose for coapplication, replace with another neurotrophin-filled syringe after 15 s. Normally, a combination of 2–3 neurotrophic factors may be completed within less than a minute.

2. Alternatively, use microsyringes manually to administer neurotrophic factors over the spinal cord by putting the tip over the cord and rotate the plunger inward slowly to deliver 0.25 μL over 5–10 s. Adjust the concentration of neurotrophins from the stock solution or dilution to achieve the desired doses of neurotrophins.
3. When using an infusion pump, attach a polythene cannula (PE 10) to the syringe and place the other end of the cannula over the spinal cord either around the lesion site or over the normal cord in the control group for spinal drug delivery.
4. In case of repeated topical application of neurotrophins alone or in combination, use either manual or microinfusion pump delivery according to the experimental protocol (see Fig. 1). The same system can be adapted to mice model as well, if needed (see Note 10).

3.8.2. Selection of Dosage and Combination of Neurotrophic Factors in SCI

1. In one group of experiments, apply BDNF or IGF-1 in separate group of rats (0.1 μg/10 μL in phosphate-buffered saline) repeatedly, starting from 30 min before injury followed by 10, 30, and 60 min after injury. This application may continue then every hour until 240 min after injury (27–30).
2. In other group of animals, apply various combinations of BDNF with GDNF, NGF, NT-3 or IGF-1, or GDNF starting from 30 to 90 min after SCI (or according to your plan) and allow the animals to survive the desired time after injury. You can control the total dose of the neurotrophins adjusted (0.5 μg each) in 10–30 μL solution (0.5 μg of BDNF + 0.5 μg of GDNF, or IGF-1, NGF or NT-3) and apply topically over the injured spinal cord within 10 s (20, 21, 23, 24). Control group may receive 10–30 μL of 0.9% saline instead of neurotrophins (25–28).

3.9. Parameters for Assessing Neuroprotection

Spinal cord injury induces rapid leakage of proteins across the BSCB and will cause edema and cell injury. These cell and tissue injuries progress over time across the longitudinal and transverse axis of the spinal cord, leading to functional disability. It is likely that neurotrophic factors if applied soon after SCI will reduce BSCB disturbances, edema formation, and cell injury (20, 29, 30). The following subsections describe parameters, which may be used to assess the neuroprotective effects of neurotrophins on SCI.

3.9.1. BSCB Permeability

To measure BSCB breakdown in spinal cord injury, we use Evans blue and radioiodine ([131]-Iodine) as tracers. These tracers when introduced into the circulation bind to serum proteins. Thus, their leakage within the cord represents tracer–protein complex indicating vasogenic edema formation (18, 19).

1. Anesthetize the rats or mice with suitable anesthesia at the end of the experiment and expose the right femoral vein by making an incision on the skin of the hind leg.
2. Use a hypodermic 1-mL plastic syringe filled with these tracers (Evans blue 2% solution, 3 mL/kg and radioiodine, 100 μCi/kg) connected to a fine-tip 36-G needle. Using needle puncture, with care slowly inject the tracers into the femoral vein (19).
3. Alternatively, insert a PE-10 cannula in the right femoral vein after anesthesia to inject the tracer. However, in mice, right jugular vein cannulation may be used to administer the tracer (6).
4. After 5 min of tracer injection, open the chest rapidly and expose the heart. Insert a perfusion cannula (butterfly needle, 21 G) connected to either a perfusion apparatus or a peristaltic pump attached to 0.9% sterile saline solution (250 mL) into the left ventricle of the heart. After this procedure, cut the right auricle and start saline perfusion at a pressure of 90 mmHg for 45 s to wash out the remaining tracer within the blood vessels (see Note 11).
5. After perfusion with saline (approximately 50 mL for rats and 20 mL for mice), dissect the spinal cord and identify several regions around the rostral and caudal portion of the cord (see Table 3). Analyze the leakage of Evans blue into the spinal cord visually under a magnifying lens.
6. Leakage of Evans blue or radioiodine can be determined in the blue areas by counting the radioactivity in a 3-in. Gamma counter. The Evans blue could be measured using colorimetric methods after detection of radioactivity biochemically as described in details earlier (18, 19, 22). Leakage of Evans blue or radioactivity is normally expressed as percentage increase over the whole blood concentration (22).

3.9.2. Spinal Cord Edema Formation

Edema in the spinal cord could be determined by measuring water content of different spinal cord segments.

1. At the end of the experiment, remove the spinal cord and divide it quickly into several segments such as T9, T10, T11, and T12 and weigh them accurately using a precision electronic balance (see Note 12).
2. Immediately after taking wet weight of the samples, place them in an oven maintained at 90°C for 72 h. At this time, total water may be evaporated and you could take the dry weight of the sample (see Note 13).

3.9.3. Spinal Cord Morphology

1. Perfuse the rats or mice through the heart with either cold 0.9% sterile saline (as mentioned above) or using cold phosphate-buffered saline (0.1 M, pH 7.0) to washout blood inside the microvessels (about 100–150 mL for rat and 60–80 mL for mice).

Table 3
Postinjury treatment with neurotrophins either alone or in combination on spinal cord injury (SCI) induced blood–spinal cord barrier (BSCB) permeability, spinal cord edema formation, and cell injury in the T9 segment in rats. The SCI was performed by making a longitudinal incision into the right dorsal horn of the T10–11 segments, and the animals were allowed to survive 5 h after trauma [for details see text]

Parameters measured	Control	5 h SCI	Neurotrophins treatment alone (1 μg)[a]			Neurotrophins combination (1 μg)[a]			
			BDNF	GDNF	NT-3	BDNF + GDNF	BDNF + GDNF	BDNF + NT-3	BDNF + NT-3
			+30 min	+30 min	+30 min	+60 min	+90 min	60 min	90 min
BSCB permeability									
Evans blue (mg %)	0.24 ± 0.04	1.65 ± 0.12**	0.87 ± 0.32[b]	0.79 ± 0.22[c]	1.67 ± 0.13[ns.]	0.72 ± 0.18[c]	0.89 ± 0.23[b]	1.68 ± 0.34**	1.73 ± 0.14**
[131]Iodine (%)	0.35 ± 0.06	1.96 ± 0.14**	0.94 ± 0.12[c]	0.89 ± 0.12[c]	1.88 ± 0.26[ns]	0.81 ± 0.16[c]	1.07 ± 0.24[b]	1.98 ± 0.44**	2.08 ± 0.51**
Edema formation									
Water content (%)	66.12 ± 0.18	69.34 ± 0.23**	67.67 ± 0.21[b]	67.23 ± 0.18[b]	68.86 ± 0.34[ns]	67.16 ± 0.12[c]	68.21 ± 0.12[b]	69.84 ± 0.13**	69.98 ± 0.43**
Cell injury									
Neuronal damage	Nil	4	2 ± 1[d]	2 ± 1[d]	4 ± 1[ns]	2 ± 2[d]	2 ± 2[d]	4 ± 1	4 ± 1
Endothelium damage (La3+)	Nil	4	2 ± 1[d]	2 ± 2[d]	4 ± 1	2 ± 1[d]	3 ± 1[d]	3 ± 1[d]	4 ± 1

Data modified after Sharma (5–8) (see text)
Values are mean ± SD of 5–6 rats in each group
*$P < 0.05$; **$P < 0.01$ (compared from control)
[a]BDNF, GDNF, or NT-3 (Total amount, 1 μg in 10 μL) was applied topically in separate group of animals after SCI. In combination, BDNF, GDNF, or NT-3 (0.5 μg each, total dose 1 μg) was used in identical manner
[b]$P < 0.05$
[c]$P < 0.01$ (compared from 5 h SCI), ANOVA followed by Dunnett's test from one control
[d]$P < 0.05$, Chi-square test from 5 h SCI group, *ns* not significant (from 5 h SCI group)

2. After the blood washout, perfuse with cold 0.4% buffered paraformaldehyde or Simonyi fixative through the heart (about 250–300 mL for rat and 150–200 mL for mice) (see Note 14).
3. Cut small pieces of spinal cord tissues from T9, T10–11, or T12 segments and embed them in paraffin using standard procedures (14, 16).
4. Cut 3-μm thick paraffin sections from each block and stain the sections using hematoxylin–eosin or Nissl staining to study neuronal damage (15).
5. Examine these sections with a light microscope and save the digital images at ×10 and ×20 magnifications using commercial software for further analyses (see Note 15).

3.9.4. Ultrastructural Changes in the Spinal Cord

1. Perfuse with Somogyi fixative for better preservation of cell membrane and nuclei (22).
2. To study BSCB breakdown at the ultrastructural level, add lanthanum chloride (2.5%, w/v) into the fixative before perfusion. Lanthanum is an ion that can be seen at the ultrastructural level across the microvessels without any special processing of the tissue (25–28).
3. After perfusion, cut small tissue pieces from the desired spinal cord segments and postfix them in osmium tetroxide and then embed in araldite or electron microscope grade Epon 812.
4. Cut semithin sections (1 μm thick) from each block on a microtome using a glass knife and stain them with toluidine blue. Analyze these sections under a microscope and identify special areas of the cord for ultrastructural observations (14, 15) (also see Fig. 2).
5. After identifying the desired areas of the spinal cord, trim the block and cut ultrathin sections on an ultramicrotome using diamond knife. Collect these ultrathin sections on a one whole copper grid (or mesh grid) and counterstain them with uranyl acetate and lead citrate before viewing them under a Phillips 400 transmission electron microscope.
6. Save the images using a digital camera using a basic magnification of 4,000–6,000 of the grid for comparison (see ref. (4, 5)). Lanthanum can be seen as crystallized dark black structures around the microvessels and in the neuropil if BSCB breakdown occurred (4, 5) (see Fig. 3).

3.9.5. Pathological Findings After Neurotrophin Treatment in SCI

1. Topical application of BDNF or IGF-1 in high quantity reduced the trauma-induced BSCB disruption to Evans blue albumin and [131]-I (radioiodine) in the T9 and the T12 segments. In these neurotrophin-treated animals, trauma-induced edema formation and cell injuries are considerably attenuated (see Fig. 3).

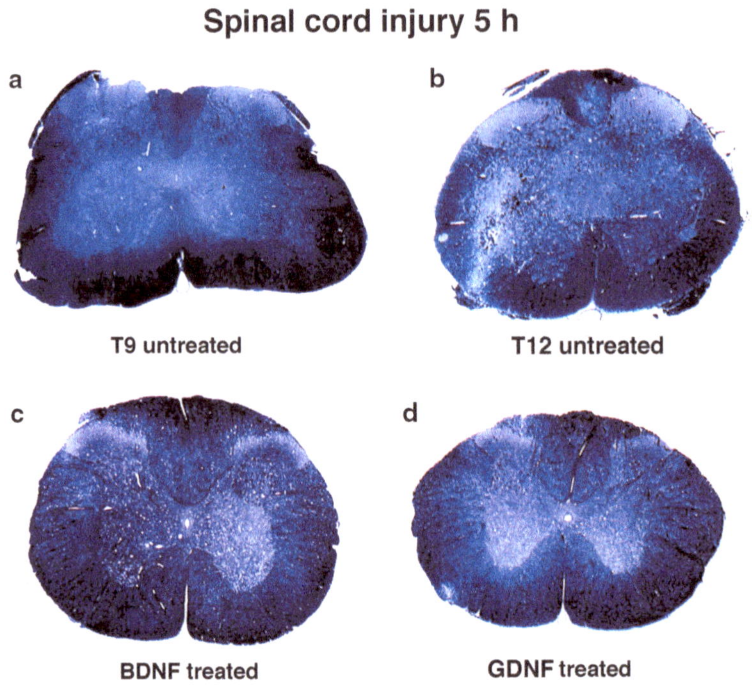

Spinal cord injury 5 h

a T9 untreated

b T12 untreated

c BDNF treated

d GDNF treated

Fig. 2. Epon-embedded toluidine blue–stained high-resolution (1 μm) thick sections from spinal cord under light microscopy show profound expansion and edema in the T9 and T12 segment of an untreated rat (**a**, **b**). Treatment with BDNF (**c**) or GDNF (**d**) alone remarkably attenuated spinal cord general expansion and edema after injury. Bar = 3 mm. Data modified after Sharma (4, 5, 9).

2. Morphological examination revealed less distortion of nerve cells, glial cells, and myelin vesiculation in the BDNF-treated or IGF-1-treated injured rats (see Fig. 3). At the ultrastructural level, the neuropil is quite preserved, and vacuolation, perivascular edema, and myelin damage were minimal in the neurotrophin-treated animals (see Fig. 3). Infiltration of lanthanum across the spinal cord endothelial cell membrane and in the basal lamina is also reduced in these neurotrophin-treated rats (see Fig. 3).
3. These beneficial effects of neurotrophins were most pronounced when the BDNF or IGF-1 was administered within 10 min of spinal trauma. Later application of neurotrophins was less effective in reducing spinal cord pathology.

3.10. Neurotrophin Combination Is More Effective in Inducing Neuroprotection

1. Application of BDNF in combination with NT-3 or NGF (0.5 μg each) 30 min after SCI markedly improved spinal cord morphology (see Table 3). This effect was markedly absent when neurotrophins were applied either 60 or 90 min after SCI. Interestingly, even a triple combination of these neurotrophins (BDNF + NT-3 + NGF, 0.5 μg each) did not induce

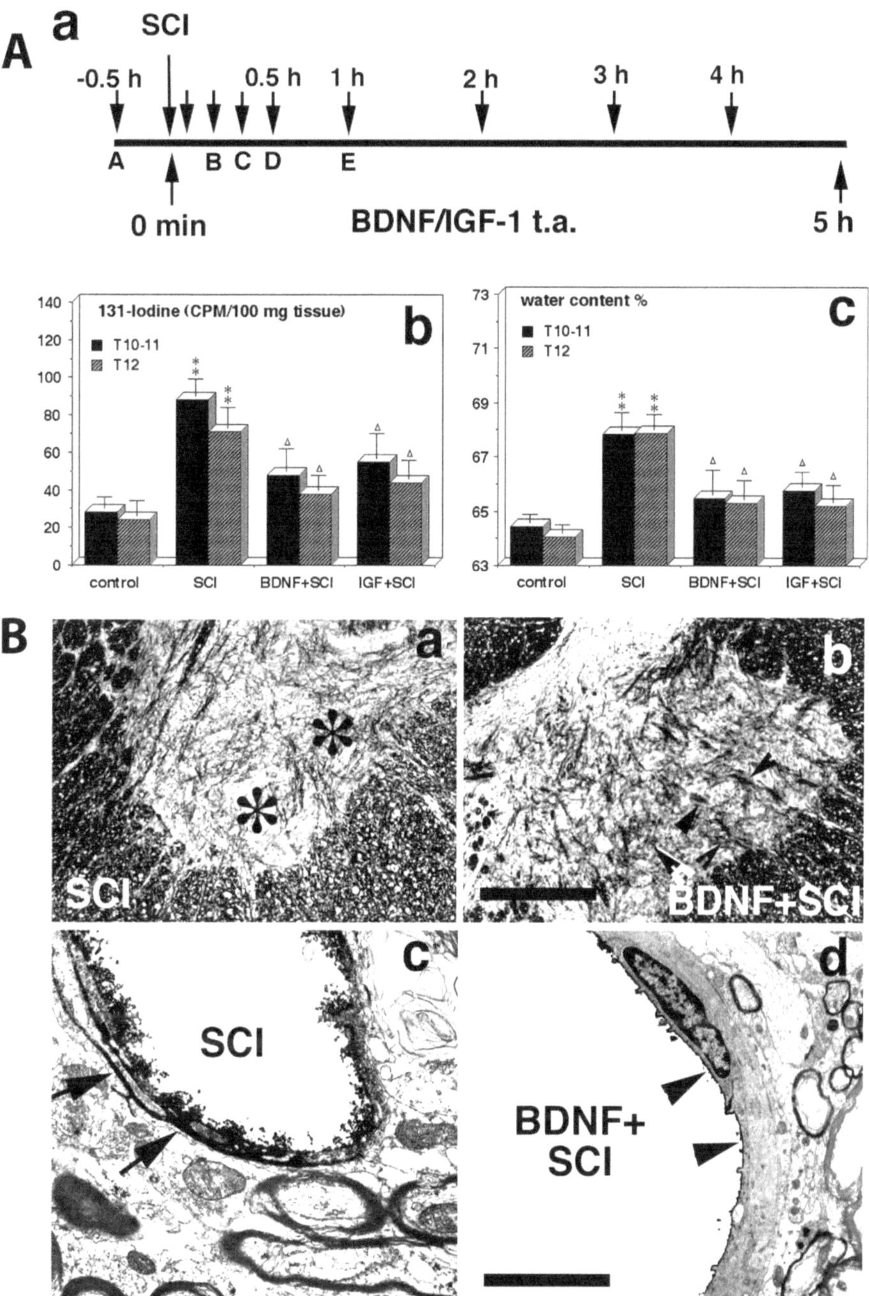

Fig. 3. Repeated application of BDNF or IGF-1 over spinal cord following injury induces neuroprotection. (**A**a) BDNF or IGF-1 was applied (1 μg in 10 μL) over the spinal cord starting from 30 min before injury followed by 5, 10, 15, and 30 min after trauma, and thereafter, 60, 120, 180, and 240 min after injury. The animals were allowed to survive 5 h (a). BDNF or IGF-1 reduced BSCB breakdown to radiotracer in different segments (b) that coincided well with reduction in edema formation (c). Morphological investigations showed preservation of myelin as seen using myelin basic protein (MBP) immunoreactivity (**B**b) as compared to untreated spinal cord MBP (**B**a). Ultrastructural studies showed exudation of lanthanum into the spinal cord neuropil and across the microvessels (*arrows* **B**c) that was significantly reduced by BDNF treatment (**B**d). Bar: **B**:a,b = 50 μm, **B**:c.d: 1 μm. Data modified after Sharma (4, 5, 9).

significant improvement in motor function beyond 60 min SCI (results not shown). On the other hand, combining BDNF with IGF-1 and GDNF (0.5 μg each) improved motor function even after 90 min of SCI (see Table 3).

2. Profound reduction in the BSCB permeability to Evans blue and radioiodine was also noted in traumatized animals that received BDNF in combination with NT-3 or NGF 30 min after SCI (see Table 3). No apparent reduction in BSCB permeability was seen even when these neurotrophins were applied in a triple combination (BDNF + NT-3 + NGF, 0.5 μg each) beyond 60 min SCI (Sharma HS, unpublished observations). However, a triple combination of BDNF with IGF-1 and GDNF (0.5 μg each) significantly reduced the BSCB leakage even after 90 min of SCI (see Table 3).
3. This combination of BDNF with NT-3 and/or NGF was also effective in reducing edema formation if applied at 30 min but not after 60 or 90 min SCI (see Table 3). On the other hand, a combination of BDNF with IGF-1 and GDNF remarkably reduced spinal cord edema formation when applied even 60–90 min after injury (see Table 3).
4. Cell changes in the spinal cord were markedly reduced by the triple combination of BDNF, IGF-1, and GDNF treatment that was effective when these neurotrophins were administered 60–90 min after SCI (see Table 3, Fig. 4). One the other hand, a combination of BDNF with NT-3 and/or NGF did not reduce cell changes in the spinal cord if applied 60 min after SCI (see Table 3).

4. Notes

1. Urethane has some carcinogenic effects, and thus, care should be taken for its use in the laboratory and exposure to humans. Local safety regulations for the use and disposal of urethane should be followed strictly (see ref. (18)).
2. If the needle pierces any other organ, pulling the plunger may draw blood into the syringe. If this happens, then withdraw anesthesia and try again after some time.
3. In rats, the cervical vertebra largely represents the similar area of the spinal cord unlike humans where the vertebrae do not normally represent the same level.
4. Sometimes, topical application of sterile saline (0.9%) over the muscle will help reduce bleeding. Expose the vertebra slowly and make it dry with often application of cold saline over the vertebral column so that spinal cord is not damaged by direct exposure to air.

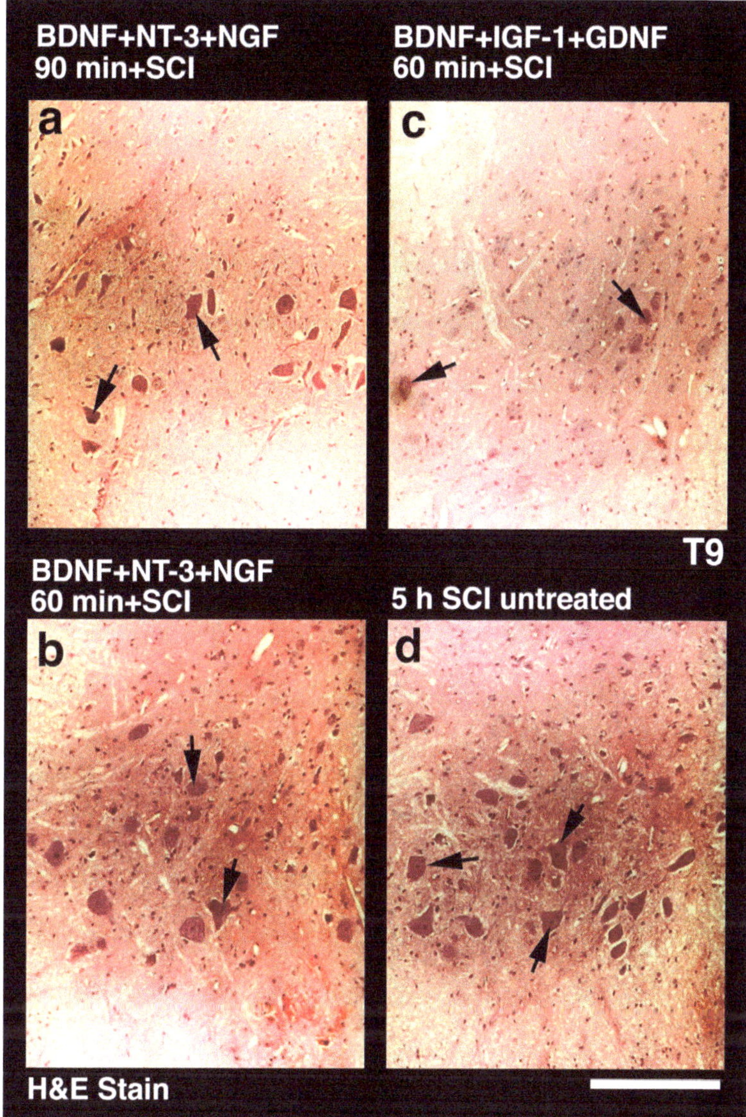

Fig. 4. A suitable combination of neurotrophic factors is more effective than their treatment alone. Thus, treatment with BDNF, IGF-1, and GDNF in combination reduces motoneuron damage after trauma markedly (**c**) as compared to untreated injured group (**d**). Whereas, a combination of BDNF, NGF, and NT-3 was not that effective in reducing cell damage after SCI either given 60 or 90 min after primary insult (**a**, **b**). Bar = 40 μm. Data modified after Sharma (4, 5, 9, 20).

5. Watch carefully the respiration and also maintain animal body temperature using a heating pad or any other device able to control body temperature. Do not allow rectal temperature to fall more than −0.5°C. Avoid direct heating of the spinal cord or spinal column over the muscle or skin using infrared lamp. This will increase the tissue temperature of the cord and spinal column, leading to disturbances in respiration and other vital organ functions.

6. Always use the bone rongeur moving outward to cut the bone in order to avoid direct contact of metal parts with the spinal cord. Try to make space over the dorsal spinal cord (about 5–6 mm), keeping the dorsal spinal artery in the middle. Avoid any deep lateral opening over the spinal cord or spinal roots. Also, make special care not to damage the dura matter while removing the vertebral bone over the cord (6).
7. The Crescent microsurgery knife can easily penetrate tissue without making injuries, and a precision lesion could be produced without any laceration of the cord.
8. Take special care about the possible spinal reflex when any metal touches the spinal cord. For this purpose, stabilization of rat spinal cord is necessary by suitable clamps or using a firm hold by hand over the back of the rat/mouse as and when appropriate to avoid spinal reflex while making the injury (1, 4, 5). Normally, under surgical grade of anesthesia, spinal reflex will be of minor degree during the lesion procedure. Any unwanted movement of spinal cord due to spinal reflex will damage spinal cord to different extents and result in variability of data collection. If this happens, discard those animals from data analyses.
9. Our laboratory compared results of spinal cord edema formation in rats and mice after microknife incision using stereotaxic apparatus and free-hand lesion using scalpel blade (16). No significant differences between handmade lesion and stereotaxic apparatus-induced SCI could be noticed with regard to edema formation or cell changes in the cord (16–19).
10. Repeated topical application of neurotrophins (0.1 μg–0.25 μg /10 μL in phosphate-buffered saline) could gain rapid access within the normal or injured cord tissue. Select the timing of topical application starting from 10 to 30 min before injury or to selected intervals after trauma ranging from 5 to 120 min after the initial insult. An early manipulation of the spinal cord after injury could result in thwarting spinal cord cell and tissue injury and limit autodestructive changes in the cord.
11. Immediately before saline perfusion, take about 1 mL whole blood sample after cardiac puncture and store it for analyses of blood radioactivity or Evans blue concentration at the time of killing.
12. The average wet weight of spinal cord segments could vary from 60 to 90 mg in normal rats. Trauma to the cord may increase the wet weight of the spinal cord samples from the identical segments. Before weighing spinal cord injury samples, remove large blood vessels or blood clots or hemorrhagic spots over the injury site, as their presence will increase the wet weight erroneously (19).

13. If in doubt, take dry weight at least three times after an interval of 4–6 h. When the dry weight in the last three measurements became stable, use this weight to calculate the water based on the differences between wet and dry weight of these samples (for details see ref. (14–16)).
14. For good fixation of spinal cord tissues, wrap the perfused animals separately in aluminum foil and keep it at 4°C for overnight. On the next day, remove the spinal cord and place the samples in the same fixative at 4°C for 3–4 days.
15. If the images require processing using Photoshop or any other image analysis program, always use identical filter or color processing on the computer for control and experimental samples (31).

Acknowledgments

Research work reported in this chapter is supported by grants from Swedish Medical Research Council (2710 HSS), Göran Gustafsson Foundation, Stockholm, Sweden; Alexander von Humboldt Foundation, Bonn, Germany; and Ministry of Science & Technology, Government of India, New Delhi. We very much appreciate the technical assistance of Kärstin Flink (Uppsala) and Franziska Drum (Berlin) in these investigations. We thank Tomas Winkler (Uppsala), Conrad Johanson (Providence, RI, USA), Rajendra D Badgaiyan (Boston, MA, USA), and Syed F Ali (NCTR, US-FDA, Jefferson, AR, USA) for critically reading the manuscript and providing important inputs for improvement.

References

1. Schwab ME, Bartholdi D (1996) Degeneration and regeneration of axons in the lesioned spinal cord. Phys Rev 76, 319–370
2. Stålberg E, Sharma HS, Olsson Y (1998) Spinal Cord Monitoring. Basic Principles, Regeneration, Pathophysiology and Clinical Aspects, Springer, Wien, New York, pp 1–527
3. Sharma HS, Westman J (2004) The Blood-Spinal Cord and Brain Barriers in Health and Disease, Academic Press, San Diego, pp. 1–617 (Release date: Nov. 9, 2003)
4. Sharma HS (2011) Early microvascular reactions and blood-spinal cord barrier disruption are instrumental in pathophysiology of spinal cord injury and repair: novel therapeutic strategies including nanowired drug delivery to enhance neuroprotection. J Neural Transm 118, 155–176
5. Sharma HS (2008) New perspectives for the treatment options in spinal cord injury. Expert Opin Pharmacother 16, 2773–2800
6. Sharma HS (2004) Pathophysiology of the blood-spinal cord barrier in traumatic injury. In: HS Sharma and J Westman (Eds) The Blood-Spinal Cord and Brain Barriers in Health and Disease, Elsevier Academic Press, San Diego, pp. 437–518
7. Winkler T, Sharma HS, Stålberg E, Westman (1998) Spinal cord bioelectrical activity, edema and cell injury following a focal trauma to the spinal cord. An experimental study using pharmacological and morphological approaches. In: Spinal Cord Monitoring. Basic Principles, Regeneration, Pathophysiology and Clinical Aspects, Eds: Stålberg E, Sharma H S, Olsson Y, Springer Wien New York, pp. 283–363

8. Fehlings MG, Tator CH (1988) A review of experimental models of acute spinal cord injury, In: L Illis (Ed), Spinal Cord Dysfunction: Assessment, Oxford University Press, Oxford, pp, 3–33
9. Sharma HS (2007) Neurotrophic factors in combination: a possible new therapeutic strategy to influence pathophysiology of spinal cord injury and repair mechanisms. Curr Pharm Des 13, 1841–1874
10. Sharma HS, Nyberg F, Gordh T, Alm P, Westman, J (1998) Neurotrophic factors attenuate neuronal nitric oxide synthase upregulation, microvascular permeability disturbances, edema formation and cell injury in the spinal cord following trauma. In: Spinal Cord Monitoring. Basic Principles, Regeneration, Pathophysiology and Clinical Aspects, Eds: Stålberg E, Sharma HS, Olsson Y, Springer Wien New York, pp. 181–210
11. Sharma HS (2005) Pathophysiology of blood-spinal cord barrier in traumatic injury and repair. Curr Pharm Des 11, 1353–1389
12. Sharma HS (2003) Neurotrophic factors attenuate microvascular permeability disturbances and axonal injury following trauma to the rat spinal cord. Acta Neurochir Suppl 86, 383–388
13. Sharma HS, Olsson Y. (1990) Edema formation and cellular alterations following spinal cord injury in the rat and their modification with p-chlorophenylalanine. Acta Neuropathol 79, 604–610
14. Olsson Y, Sharma HS, Pettersson CA (1990) Effects of p-chlorophenylalanine on microvascular permeability changes in spinal cord trauma. An experimental study in the rat using ^{131}I-sodium and lanthanum tracers. Acta Neuropathol 79, 595–603
15. Sharma HS (2009) Blood–Central Nervous System Barriers: The Gateway to Neurodegeneration, Neuroprotection and Neuroregeneration, In: A Lajtha, N Banik, SK Ray (Eds) Handbook of Neurochemistry and Molecular Neurobiology: Brain and Spinal Cord Trauma. Springer Verlag, Berlin, Heidelberg, New York, pp. 363–457
16. Winkler T, Sharma HS, Stålberg E, Olsson Y (1993) Indomethacin, an inhibitor of prostaglandin synthesis attenuates alteration in spinal cord evoked potentials and edema formation after trauma to the spinal cord: an experimental study in the rat. Neuroscience 52, 1057–1067
17. Sharma HS, Winkler T, Stålberg E, Olsson Y, Dey PK (1991) Evaluation of traumatic spinal cord edema using evoked potentials recorded from the spinal epidural space. An experimental study in the rat. J Neurol Sci 102, 150–162
18. Sharma HS (2007) Methods to produce hyperthermia-induced brain dysfunction. Prog Brain Res 162, 173–199
19. Sharma HS (2005) Methods to produce brain hyperthermia. Current Protocols in Toxicology 23, UNIT 11.14, pp 1–26, DOI: 10.1002/0471140856.tx1114s23
20. Sharma HS (1999) Pathophysiology of blood–brain barrier, brain edema and cell injury following hyperthermia: New role of heat shock protein, nitric oxide and carbon monoxide. An experimental study in the rat using light and electron microscopy, Acta Universitatis Upsaliensis 830, 1–94
21. Sharma HS (2010) Selected combination of neurotrophins potentiate neuroprotection and functional recovery following spinal cord injury in the rat. Acta Neurochir Suppl 106, 295–300
22. Sharma HS (2007) A select combination of neurotrophins enhances neuroprotection and functional recovery following spinal cord injury. Ann NY Acad Sci 1122, 95–111
23. Sharma HS (2006) Post-traumatic application of brain-derived neurotrophic factor and glia-derived neurotrophic factor on the rat spinal cord enhances neuroprotection and improves motor function. Acta Neurochir Suppl 96, 329–334
24. Sharma HS, Nyberg F, Gordh T, Alm P (2006) Topical application of dynorphin A (1–17) antibodies attenuates neuronal nitric oxide synthase up-regulation, edema formation, and cell injury following focal trauma to the rat spinal cord. Acta Neurochir 96, 309–315
25. Sharma HS (2005) Neuroprotective effects of neurotrophins and melanocortins in spinal cord injury: an experimental study in the rat using pharmacological and morphological approaches. Ann NY Acad Sci 1053, 407–421
26. Sharma HS, Westman J, Gordh T, Alm P (2000) Topical application of brain derived neurotrophic factor influences upregulation of constitutive isoform of heme oxygenase in the spinal cord following trauma an experimental study using immunohistochemistry in the rat. Acta Neurochir 76, 365–369
27. Winkler T, Sharma HS, Stålberg E, Badgaiyan RD (2000) Neurotrophic factors attenuate alterations in spinal cord evoked potentials and edema formation following trauma to the rat spinal cord. Acta Neurochir 76, 291–296
28. Sharma HS, Nyberg F, Gordh T, Alm P, Westman J (2000) Neurotrophic factors influence upregulation of constitutive isoform of heme oxygenase and cellular stress response in the spinal cord following trauma. An experimental study using immunohistochemistry in the rat. Amino Acids 19, 351–361

29. Sharma HS, Nyberg F, Westman J, Alm P, Gordh T, Lindholm D (1998) Brain-derived neurotrophic factor and insulin like growth factor-1 attenuate upregulation of nitric oxide synthase and cell injury following trauma to the spinal cord. An immunohistochemical study in the rat. Amino Acids 14, 121–129
30. Sharma HS, Nyberg F, Gordh T, Alm P, Westman J (1997) Topical application of insulin like growth factor-1 reduces edema and upregulation of neuronal nitric oxide synthase following trauma to the rat spinal cord. Acta Neurochir Suppl 70, 130–133
31. Sharma HS, Sjöquist P-O (2002) A new antioxidant compound H-290/51 modulates glutamate and GABA immunoreactivity in the rat spinal cord following trauma. Amino Acids 23, 261–272
32. Allen AR (1911) Surgery of experimental lesion of spinal cord equivalent to crush injury of fracture dislocation of spinal column. Preliminary report. JAMA 57, 878–880
33. Bresnahan CJ, Beattie MS, Todd FD, Noyes DH (1987) A behavioural and anatomical analysis of spinal cord injury produced by a feedback-controlled impaction device. Exp Neurol 95, 548–570
34. Sharma HS, Dey PK (1982) Correlation of spinal cord tissue 5-HT with edema development following surgical spinal cord trauma in rats. Indian J Physiol Pharmacol 26, Supplement I, p. 8–10

Index

Stephen D. Skaper (ed.), *Neurotrophic Factors: Methods and Protocols*, Methods in Molecular Biology, vol. 846, DOI 10.1007/978-1-61779-536-7, © Springer Science+Business Media, LLC 2012

I

K

L

M

N

O

MIX
Papier aus verantwortungsvollen Quellen
Paper from responsible sources
FSC® C105338

If you have any concerns about our products,
you can contact us on
ProductSafety@springernature.com

In case Publisher is established outside the EU,
the EU authorized representative is:
Springer Nature Customer Service Center GmbH
Europaplatz 3, 69115 Heidelberg, Germany

Printed by Libri Plureos GmbH
in Hamburg, Germany